Metallurgy

CARL G. JOHNSON
WILLIAM R. WEEKS

5th Edition

Revised By JOHN G. ANDERSON

AMERICAN TECHNICAL PUBLISHERS, INC.
HOMEWOOD, ILLINOIS 60430

COPYRIGHT ©1938, 1942, 1946, 1956, 1977

BY AMERICAN TECHNICAL PUBLISHERS, INC.

Library of Congress Catalog Number: 76-24520
ISBN: 0-8269-3482-X

No portion of this publication may be reproduced by *any* process such as photocopying, recording, storage in a retrieval system or transmitted by any means without permission of the publisher

56789-77-151413121110

PRINTED IN THE UNITED STATES OF AMERICA

PREFACE

Knowledge of metallurgy is essential to modern industry. Nearly every type of manufacturing today depends upon what we know about the behavior of metals and their alloys, and upon our skill in using them.

This book presents the basic information about metallurgy—how ferrous and non-ferrous metals behave and why. It defines terms, explains guiding principles, and outlines manufacturing processes. The theories discussed are the organized results of the best shop practices; therefore, not only the practical man in industry but also the student in the classroom will find it an accurate source of important information.

For the beginner who wants to develop his skill and judgement, the systematic and logical organization of the text material will prove particularly helpful. The concise writing and selected illustrations should give a clear and simple introduction to metallurgy. For anyone who wants to learn, this book provides a working knowledge of the behavior of metals and alloys in modern manufacturing.

Because of its recognized practical value to everyone concerned with the fabrications of metal products, Physical Metallurgy is stressed rather than Chemical Metallurgy.

To help the reader systematically study metallurgy, this text presents the subject in four sections. The first three sections relate to ferrous metallurgy and the fourth section relates to non-ferrous metallurgy, including bearing metals.

Section I, Chapters 1 through 4, covers the early history of iron and steel, the raw materials and equipment used in producing basic iron, the refining processes used to produce steel and the various shaping and forming processes used to supply the required steel shapes used in modern industry.

Section II, Chapters 5 through 10, covers physical metallurgy, the metallic state of pure metals, the mechanical properties of ferrous metals, the methods, theory and equipment related to testing metals, the theory of alloys and equilibrium diagrams, heat treatment for steel and surface treatments of ferrous metals.

Section III, Chapters 11 through 15, includes alloys and special steels, the SAE and AISI classifications of steels, cast irons and foundry practice, welding metallurgy and related processes and powder metallurgy.

Section IV, Chapters 16 through 20, covers the non-ferrous metals. This section includes the production of non-ferrous metals, light metals and alloys, copper and its alloys, bearing metals and lesser known metals, pure or used as alloys of the major metals.

At the end of each chapter questions are available for the student to use in the review of the material in the chapter. All weights, measures and temperatures in the text are given in both the common English system and the equivalent SI (metric) system in parentheses. At the end of the text there is a Glossary of Terms and an Index.

The Publisher

TABLE OF CONTENTS

CHAPTER	CONTENTS	PAGE
1.	**Early History of Iron and Steel** Direct processes: beginnings of production furnaces ■ Early steel-making processes: steeling, cementation, blister steel, shear steel, Wootz steel, Huntsman's process ■	1
2.	**Production of Iron and Steel** The smelting process ■ Raw materials: iron ores, coke, limestone, scrap ■ The blast furnace: construction and operation ■ Direct reduction processes: kilns, retorts, fluidized bed techniques ■	13
3.	**Refining into Steel** Basic oxygen process, Bessemer, electric arc ■ Induction furnaces, open hearth, specialty melting and refining ■ Ingot production: types of ingot steel; continuous casting ■	27
4.	**Shaping and Forming Metals** Hot rolling, forming structural shapes ■ Hot forging and forging process; extrusion ■ Effects of hot working, cold working, cold rolling, drawing, tubular products ■ Metal forming: bending, flat drawing, roll forming, effects of cold working ■	45
5.	**Physical Metallurgy** Pure metals, crystalline or solid state ■ Crystallization, dendritic growth, grain structure ■ Deformation, work hardening, recrystallization, germination and grain growth ■ Cold crystallization ■	74
6.	**Mechanical Properties of Metals** Carbon steels: minor constituents in steel; gaseous impurities ■ Selection of carbon steels by type; mechanical properties of carbon steels ■	87
7.	**Testing of Materials** Destructive tests: tensile, impact, shock, fracture tests, fatigue, hardness, corrosion tests, salt spray, macro-etching, microscope ■ Non-Destructive tests: magnetic inspection, ultrasonic, radiographic ■ Temperature measuring ■	105
8.	**Theory of Alloys** Equilibrium diagrams, thermal curves, time-temperature curves ■ Freezing of alloys, solid solution alloys, eutectic alloys, hetero-crystal alloys, immiscible alloys, peritectic alloys, changes in the solid state, compound-forming alloys ■ Properties of alloys: iron-carbon diagram, solidification, allotropy ■ The eutectic, decomposition of cementite, grain size changes, critical temperature changes ■	131
9.	**Heat Treatment of Steel** Normalizing, annealing, spheroidizing, stress-relief treatment ■ Hardening of carbon steel: martensite, retained austenite, cryogenic treatment ■ Quenching media ■ Effect of carbon: shallow hardening, effect of grain size, effect of alloys ■ Heat-treat furnaces, tempering, furnace atmospheres, liquid treating baths ■	165

10.	**Surface Treatments**	205
	Surface hardening, carburizing, Chapmanizing, nitriding ■ Other diffusion processes: flame hardening, induction hardening ■ Coatings: painting, metal spray, hard-facing, vapor deposition, oxide coatings ■	
11.	**Alloy or Special Steels**	231
	Behavior and influence of special elements ■ Alloy steels: nickel, chromium, manganese, tungsten, molybdenum ■ High-speed steels: vanadium, silicon ■ Magnetic steels ■	
12.	**Classification of Steels**	253
	SAE and AISI systems ■ Tool steel classification ■ Stainless and heat-resisting steels ■ Classification by hardenability ■	
13.	**Cast Irons**	269
	Gray cast irons, white and chilled cast irons ■ Malleable, ductile, wrought iron, alloy cast irons ■ Foundary practice: melting practices, innoculants ■	
14.	**Welding Metallurgy**	303
	Gas welding: oxy-acetylene, arc and gas shielded ■ Solid state welding; ultrasonic ■ Resistance: butt, seam, projection, flash ■ Fusion: plasma arc, electron beam, laser ■ Brazing and soldering ■ Effects of welding on metal structures ■	
15.	**Powder Metallurgy**	329
	Metal powders and their production ■ Characteristics of metal powders ■ Molding, welding and pressing; sintering furnaces and atmospheres ■ Applications of powdered metallurgy products ■	
16	**Producing Non-Ferrous Metals**	347
	Copper, lead, zinc, cadmium, nickel, aluminum, magnesium, tin, manganese, chromium, tungsten, cobalt, antimony, bismuth, ferro-alloys, precious metals ■	
17.	**Light Metals and Alloys**	374
	Aluminum and its alloys, properties ■ Machining light metals: casting alloys, wrought aluminum alloys ■ Die-casting methods and equipment ■ Magnesium and its alloys ■ Beryllium metal ■ Titanium ■	
18.	**Copper and Its Alloys**	403
	Commercial grades, impurities, brasses, Muntz metal, Admiralty metal, copper-tin bronze, leaded bronze, aluminum bronze, beryllium-copper alloys, copper-manganese-nickel, Monel metal, copper nickel, nickel-silver, nickel-iron ■	
19.	**Bearing Metals**	427
	Properties ■ Lead base, tin base, lead-calcium alloys, cadmium alloys, copper-lead alloy ■ Composite bearing, zinc base alloys, self-lubricating bearings ■ Silver-lead-indium bearings ■	
20.	**Zirconium, Indium and Vanadium**	437
	Ores, refining processes, mechanical processes, shaping and forming ■ Application of the metals to industry ■	
	Glossary	460
	Index	469

CHAPTER 1
EARLY HISTORY OF IRON AND STEEL

Iron has been known and used for over five thousand years. An iron object, apparently a dagger, was found at the site of the ancient city, Ur of the Chaldees. It is believed to date from about 3,700 B.C. References to iron in the early literature of India and China indicate that iron was known in these countries at least 2,000 years before the Christian era.

The first iron that was used is believed to be of meteoric origin. Meteors, the shooting stars that flash across the night skies, are of two types: stone and metallic. Ancient iron articles have been chemically analyzed and found to contain a high percentage of nickel (7 to 15 percent) in the iron base; this is very close to the nickel content of metallic meteors, as in Fig. 1-1.

The common use of iron could not begin until a method of removing iron from its

Fig. 1-1. Structure of iron-nickel meteorite showing crystalline structure. (Field Museum of Natural History)

ore was learned. Iron ore is a chemical compound of iron and oxygen, technically referred to as iron oxide. This oxide of

Metallurgy

iron is the rust that develops on steel and iron products when they are exposed to moist air. A method of removing the oxygen from the compound is called a *reduction* process.

Early hunters probably learned the reduction technique by accident when a camp fire was built over an out-cropping of iron ore, Fig. 1-2. The glowing charcoal in the fire would unite chemically with oxygen in the ore to form a common gas, carbon dioxide (CO_2), and would leave a spongy mass of metallic iron mixed with ash and other matter. The hunters found that the mass could be shaped by pounding it while it was still white hot. The pounding not only shaped the mass but also squeezed out most of the ash and impurities. The metal that resulted was a crude form of *wrought iron*. This was not iron as we define it today because, while there was enough heat to remove the oxygen from the iron ore, the fire was not hot enough to melt the iron.

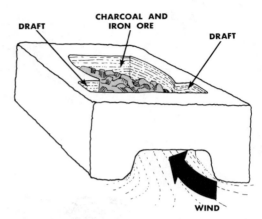

Fig. 1-3. Earliest furnace design took advantage of wind power.

Fig. 1-2. Prehistoric man first smelted iron accidentally by building a fire over an ore outcropping.

Later on, sufficient heat to melt the iron became available with the building of small, rough furnaces. These were stone and clay boxes into which were put iron ore and charcoal, Fig. 1-3. These box furnaces were placed in areas noted for high winds and the openings faced the usual wind direction. The wind-driven air increased the intensity of the fire and melted the iron. Later furnaces utilized bellows operated by hand as in Fig. 1-4, or by power from water wheels or windmills.

Over the centuries, improvements were made upon this basic furnace. The shape of the furnace was made more efficient in retaining heat and stacks were contructed to add a draft effect to the incoming air from the bellows. The charcoal used was the source of the carbon which not only was the fuel for the furnace but also, as it became very hot, supplied the chemical carbon for the reduction process. As the metallic iron formed it melted, ran down

Early History of Iron and Steel

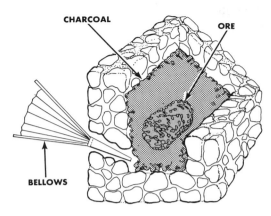

Fig. 1-4. When winds fail, the bellows creates its own supply of air under man power.

Fig. 1-5. Damascus swords were banded layers of tough wrought iron and hard steel edges.

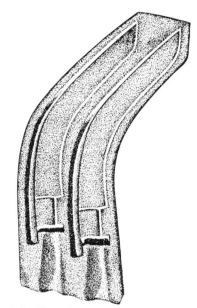

Fig. 1-6. Fourth century iron molds for casting sickle blades from China.

through the red-hot coals and out of the bottom to form an early form of *pig iron*.

The first iron was relatively soft, of uneven quality and hard to work. However, iron ores are plentiful and widely distributed throughout the world. Some historians believe that the Hittites brought the knowledge of ironmaking with them when, about 1500 B.C. they migrated west from the area now known as Afghanistan. Ancient Egyptian writings indicate that the highly sophisticated civilization of the Pharoahs knew nothing of iron and negotiated with the Hittite kings to learn the techniques of ironmaking. Iron is mentioned in the earliest books of the Old Testament and the reverberatory furnaces of Ezion-Geber at the head of the Red Sea have drawn the attention of archeologists.

The ancient ironmakers occasionally produced steel, possibly by accident. By definition, steel is an alloy of iron with a small amount of carbon (usually less than 1 percent). Certain percentages of other elements were also found in steel, some valuable and beneficial, others harmful. Thus, it is not hard to imagine an occasional heat of iron containing carbon and other elements that would produce the much stronger metal, steel.

However, it is probable that most of the steel produced in ancient times resulted

Metallurgy

from heating the iron in a forge fire in contact with red hot carbon (Fig. 1-4). The surface layer of the iron would absorb some carbon on the outer surface. This is now referred to as a *case*. The red hot iron bar with its case of high-carbon iron could then be hammered. After many repetitions of heating and hammering, a fine steel would be produced, such as the famous swords of Damascus and Toledo, Fig. 1-5. There is evidence of steel made by this process as early as 1000 B.C. Recently found Chinese iron sickle molds have been dated before 400 B.C., Fig. 1-6.

DIRECT PROCESSES FOR MAKING WROUGHT IRON

From about 1350 B.C. until 1300 A.D., practically all iron that was worked into tools and weapons in Europe and the Near East was produced directly from iron ore. This method of production is called the *direct process* to distinguish it from the process now used, in which iron is first smelted, then purified, and then finally cast or wrought. This is an *indirect process*. The product of the direct process is called *sponge iron*. Today *wrought iron* is produced by the indirect method which will be described later in the book.

In the early direct process, alternate layers of charcoal, and mixtures of charcoal and iron ore were placed in the furnace until several layers of core built up. Then the whole mass was covered with fine charcoal. The charcoal had three functions: (1) to supply heat during combustion, (2) to serve as a reducing agent on the ore, and (3) to shield the hot metal from the oxidizing influences of the air.

Air for combustion was induced through a pipe or nozzle directly into the bed of coals first built up before the furnace was loaded (charged) with charcoal and ore.

Very early in the history of iron making, combustion was aided by an artificial draft supplied by bellows. The air (20 percent oxygen) in the blast combined with carbon in the incandescent charcoal to produce carbon monoxide (CO), a reducing gas. As this gas ascended through the mixture of charcoal and iron ore it combined with the oxygen in the ore (Fe_2O_3) to produce carbon dioxide (CO_2) which passed out of the forge into the atmosphere.

As the process continued, tiny crystals of iron would form and drip down into the bottom of the forge. The temperatures were only sufficient to achieve a partial reduction of the iron from its ore. The rest of the iron content (between 50 and 60 percent) was lost in the slag (refuse). Reduction of iron ore starts at about 2200°F (1200°C) and the melting of iron at 2800°F (1540°C). Since this latter temperature could not be maintained in the primitive furnace (Fig. 1-4) the iron was not liquefied. Besides iron oxide, the slag would contain fused impurities from the ore and the fuel. As the iron droplets settled to the bottom of the hearth, the slag, being light-

Early History of Iron and Steel

er, would float on top of the iron and protect it from oxidation.

After several hours, a spongy mass of iron would have accumulated at the bottom of the furnace. This was then pried out or withdrawn with tongs. White hot, it was taken to the anvil and hammered vigorously and repeatedly to squeeze out as much slag as possible and to weld the iron crystals into a solid mass. Between the forging periods, the iron would be reheated to soften or liquefy the remaining slag rendering the iron more malleable.

The iron mass produced was called a bloom. It was very small, some weighing only one or two pounds. This weight was increased to more than ten pounds after 1000 A.D. Furnaces produced only one bloom per day as late as 1350 A.D. although more than one furnace might be operated at an iron works.

In the later Middle Ages, the process was improved by adding a fluxing material to the charge. The was usually limestone, but slags from old furnaces were reused. The added fluxing material markedly cut down the waste of iron and conserved charcoal. In reacting chemically with the gangue (impurities in the ores), the flux also prevented some of the impurities from dissolving in the fluid iron and this improved the quality of the iron. This addition of a flux to the charge represented a major step forward in iron working.

BEGINNING OF PRODUCTION FURNACES

In the early furnaces, recharging or continuous charging was not frequently practiced. This practice limited the size of the bloom produced and the amount of iron that could be produced in one day. Old records indicate that as late as 1350 A.D., English forges were producing an average of less than three tons per year, and the average bloom weighed less than 33 pounds.

About 1300 A.D., the famous Catalan forge (furnace) was developed in northern Spain, Fig. 1-7. This furnace employed the principle of repeated charging and also utilized water power to force the draft. This furnace is regarded as the ancestor

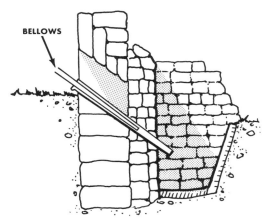

Fig. 1-7. Catalan furnace provided principle of repeated charging without shutting down.

Metallurgy

of the blast furnace. The hearth apparently ranged from 20 to 40 inches (or 0.8 x 1.6 m) square and from 16 to 30 inches (0.6 x 1.2 m) in depth. The furnace was enclosed within a building but had no chimney.

The operation was started by covering the hearth with several inches of fine charcoal. Then a vertical column of charcoal was placed against the back of the furnace. In front of the charcoal was placed a column of lump ore. Air was first blown gently through the ignited charcoal and then gradually increased in force and volume. In passing through the glowing charcoal, the oxygen in the air would react with carbon to form hot carbon monoxide (CO) gas. This gas would pass through the pile of ore where it reacted with the oxygen in the iron ore to produce carbon dioxide gas (CO_2) which then dispersed into the atmosphere.

Crystals of iron produced in the process gradually melted and settled to the base of the hearth and from time to time charcoal and additional ore would be added to their respective columns on the hearth. After five or six hours, a soft, white mass of iron containing a mixture of slag was pried out of the hearth. Usually the blooms weighed between 200 and 350 pounds. The blooms were then forced on an anvil by a powered hammer operatd by water power, Fig. 1-8.

The influence of the Catalan forge was felt throughout Europe. Between 1300 and 1400, a revolution in forge construction and operation took place. Furnaces be-

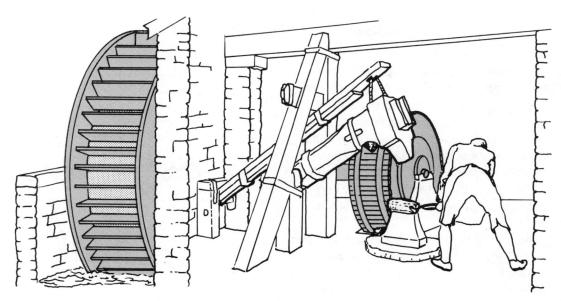

Fig. 1-8. Water power raised a massive forging hammer by winching it up, then dropping it on red hot bloom.

came larger, continuous charging more common and the use of water power quite general. As a result, the production per furnace increased many fold. In England the new methods and equipment came into use somewhat later. In 1408 an English forge produced 278 blooms in a year, each bloom weighing 195 pounds, for an annual production of over 27 tons. Compared to a national average of less than three tons fifty years earlier, this was a meaningful increase.

Also evolved from the Catalan furnace was the *stuckofen* or *old high bloomery*, Fig. 1-9. This furnace was popular in northern Europe in the fourteenth century. The development of the stuckofen furnace consisted in gradually increasing the height of the furnace, providing a stronger blast, introducing the charge through an opening near the top of the stack and providing a drawing hole at the bottom of the shaft for extracting the blooms. Fuel and ore were added at regular intervals. The draw hole was closed by a brick or a stone wall which was easily torn down, then rebuilt.

As these furnaces were built higher and the air blast increased they began to generate higher heat and melted the iron for easier extraction. Variations in charge and operation permitted the production of either low-carbon wrought iron or molten high-carbon pig iron. A similar furnace called a *flussofen* was developed to produce molten, high-carbon steel. These furnaces were referred to as *shaft furnaces* and were prototypes of the modern blast furnace and the cupola now used world wide.

Several centuries later American bloomeries represented the highest development of the Catalan or simple hearth-type funace. These bloomeries were open at the front like a domestic fireplace. The hearth was rectangular shaped, about three feet wide and two feet deep, front to back (91 x 61 cm). The blast was preheated to save fuel by passing it through pipes around which the hot gases rising from the hearth passed. A *tuyere* or blowing hole was placed either on one side or at the back of the hearth which was protected by a water-cooled bottom plate.

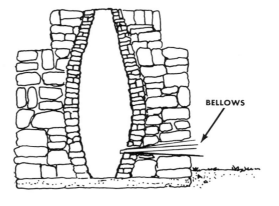

Fig. 1-9. Old high stuckofen was beginning of modern blast furnace, providing greater blast, and separate charging and tapping of the molten metal.

Metallurgy

Fig. 1-10. The Saugus Iron Works, America's oldest colonial bloomery.

After a bed of charcoal had been ignited and was burning well, alternate layers of ore and charcoal were charged and smelting continued until a sufficient amount of metal had collected on the hearth bottom. This mass of spongy metal, in which considerable slag was entrapped, was then taken to a separate forge to be hammered into a bloom. A typical bloomery would produce about a ton of wrought iron in 24 hours. Fig. 1-10 represents the historic colonial Saugus bloomery. The last American bloomery ceased operations in 1901.

BEGINNING STEELMAKING PROCESSES

Iron produced by the direct process was relatively pure, containing about 99.5 percent iron (ferrite) and less than 0.1 percent carbon. Since carbon is the principal substance that gives iron its hardenability, the iron as originally reduced could not be hardened by heat treatment or any amount of forging. This iron was not suit-

Early History of Iron and Steel

able for making cutting tools or weapons.

Early smiths achieved some success in hardening iron when they discovered that if iron was heated for a sufficient length of time in a deep bed of charcoal embers the iron could be hardened by plunging the red-hot metal into cold water. Why this happened was a mystery to the smiths although they probably didn't give it a second thought. The process, however has great metallurgical meaning.

The glowing bed of charcoal not only supplies the heat to raise the iron to a redhot temperature but it also protects the surface from the oxidizing effect of the atmosphere. In addition, the glowing charcoal permits the transfer of carbon atoms into the surface of the iron which raises the carbon content of the surface iron and penetrates to a significant depth. This increase in carbon transforms the soft iron into steel, at least on the surface.

This early carburizing process permitted weapons and cutting tools to be hardened and sharpened. However, the rapid quenching induced brittleness which caused the tools and swords to break under stress. The Romans, who needed reliable swords, are credited with inventing a process to reduce the brittleness while retaining most of the hardness. This process is also a heat treatment called *tempering*. With this invention, hardened steel became the wonder metal of the Western world.

The ancient Vikings understood the process of partial carburization and are believed to have introduced this process into England during their invasions. The metal of a Danish battle-axe of about 895 A.D. found in England has a carbon content of only 0.05 percent, typically wrought iron. The cutting edge of the blade, however, had been locally carburized and then hardened by quenching in cold water.

An interesting cost reduction process dating from the first century A.D. is *steeling*. This consisted in welding steel strips onto an iron base. Hammers have been found in England which had thin plates of steel welded to the face of the hammer head, as in Fig. 1-11. Steeling for the conservation of an expensive metal was a common practice until this century.

The sword makers of the Near East developed a rather complicated method of increasing the carbon content of wrought iron. This process consisted of piling high-carbon steel bars together with pieces of

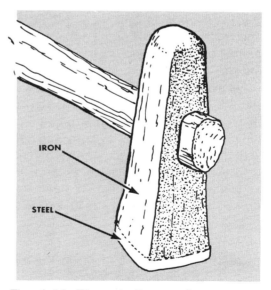

Fig. 1-11. Through the steeling process, an old English hammer was faced with steel for more lasting service.

Metallurgy

low-carbon wrought iron, heating the pile to a red-hot welding temperature and then forging the pile into a single solid piece. This process would be repeated until the smith judged that the correct amount of carbon, referred to as qualities, had been imparted to the metal. The diffusion of carbon from one layer of metal to another was incomplete and the steel exhibited a peculiar wavy appearance called *watering*. This was a visible characteristic of the famous Damascus blades.

Any number of methods of increasing the carbon content of wrought iron came to be called the *cementation* process. The process is basically the heating of wrought iron in close contact with any carbonaceous material, such as charcoal, up to a red heat (1550 to 1750°F) (850 to 960°C). Depending on the length of time the part is held at the red heat a carbon content up to 1.7 percent can be obtained. This process was the principal means of producing steel up to the middle of the 18th century. In most of the world no other process was known.

The cementation process practiced in the 17th and 18th centuries consisted of placing alternate layers of wrought iron and charcoal in sealed clay pots. The pots were then placed in a type of reverberatory furnace where the flames would have no contact with the pots. The temperatures varied from 1470° to 2010°F (800°C to 1100°C) and the total time in the furnace ranged from 7 to 12 days.

The bars of carburized wrought iron this process produced would be covered with small blisters and the steel was known as *blister steel*. Steel carburized by this special cementation method was also referred to as *cement steel*. Blister or cement steel was uneven in composition and more carburized on the surface than in the center.

To obtain a more homogeneous product the English steel makers, early in the eighteenth century developed a process called *shearing*. In this process, cement steel bars would be broken in two, the two halves laid together in a bundle, covered with clay or sand to prevent oxidation, reheated to a welding temperature and then hammered into a solid bar. This was called *single shear* steel. For still higher homogeneity of the metal the process would be repeated to produce *double shear* steel.

Shear steel proved to be superior to the best steel manufactured elsewhere in Europe. It was used in the production of cutlery, shears, razors, engraving tools or wherever the highest quality steel was required. Techniques used in producing shear steel were similar in some ways to the ancient process of producing Damascus steel.

The only other early method of producing steel was developed in India many centuries ago. The famous *wootz* steel was produced in fire clay crucibles in a single heating process and produced a homogeneous high-carbon steel. It is assumed that high purity iron ore was ground together with charcoal and sealed in the crucible. During the first phase of the heating period, the carbon in the charcoal would reduce the iron oxide to pure iron. Then the iron would absorb carbon. As the iron absorbed greater amounts of carbon, its melting point was reduced. Whereas pure iron melts at 2800°F

(1540°C), a highly carburized iron will have a melting point several hundred degrees lower.

As the carburization continued and the melting point lowered, the metal would begin to melt and the added carbon would be equally distributed within the melted steel. When the crucibles cooled, they were broken open and a small ingot of high-carbon steel was removed. The crude ingots could then be reheated to the welding temperature and forged into bars. This ancient method, which does not appear to have been known to the Western world, was lost to the world before modern times. One of the remarkable things about this process is the fact that the iron workers of ancient India, unlike those in Europe and the Near East, were able to consistently attain temperatures high enough to melt a highly carbonized iron.

No steel made by the cementation process could be totally homogeneous and products made from this steel would be unpredictable in hardness and hardenability. This bad characteristic of cement steel led Huntsman, an English clockmaker to seek a way of obtaining a more uniform steel suitable for delicate clock springs. Huntsman conceived the idea of melting steel in a closed container so that the metal would lose none of its carbon or other elements and become completely homogeneous throughout. Huntsman's process, introduced in 1742, was similar to the wootz process of ancient India.

Huntsman started with cement steel with carbon averages that would give him the desired carbon percentage in the final product. The bars were broken into small pieces and placed in sealed crucibles. The crucibles were placed in a coke fired furnace and heated until the bars were thoroughly melted and fused. The crucibles were removed from the furnace and opened. The small amount of impurities (slag) floating on top of the steel was first skimmed off and the steel then poured into cast iron ingot molds. The ingots were then forged to desired shape.

The steel produced by Huntsman's method was not only homogeneous but free from slag and dirt. In fact, the steel was admirably suited to Huntsman's purpose and others soon recognized it as the best steel available, Fig. 1-12. Where high quality steels are required, the crucible method has held a position of supremacy for over two centuries. Today, little crucible steel is made in crucibles of fire clay or graphite fired in a open flame furnace.

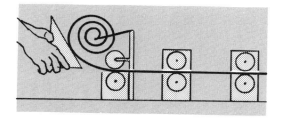

Fig. 1-12. Huntsman's pure steel made better clock springs and found widespread use.

Metallurgy

New types of furnaces, electric fired and with controlled atmosphere, turn out specialty steels that Huntsman couldn't dream of matching with his crucible process.

REVIEW

1. About how many centuries ago was iron known about and used?
2. Metallic meteors contain nickel in what percentage range?
3. What is common iron ore, a compound of iron and oxygen, chemically named? What is the common name for this compound?
4. What was used long ago to remove oxygen from the iron ore?
5. Why is a blast of air necessary to melt iron?
6. What is the definition of steel?
7. Name two types of famous swords.
8. What percentage of oxygen is in the air we breathe?
9. Define slag.
10. Define a bloom.
11. When and where was the Catalan forge developed?
12. What is the principal substance that gives iron its hardenability?
13. Describe the *wootz* or Huntsman's process of producing steel.

CHAPTER 2  PRODUCTION OF IRON AND STEEL

The United States has been the world leader in steel production since the beginning of the twentieth century. Annual production of ingot steel in the U.S.A. has exceeded 100 million net tons several times in the last three decades. The nation's steelmaking capacity increased at an unprecedented rate during and after World War II. An average of $1.3 billion a year was spent in those years to improve and expand on iron and steel-making facilities.

Increased requirements for steel created a need for larger supplies of raw materials and for blast furnaces, coke ovens, rolling mills and other facilities. The nation's iron and steelmaking facilities today number over 150 blast furnaces for producing basic pig iron and more than 1200 refining furnaces to convert pig iron into steel.

RAW MATERIALS

Huge quantities of raw material are needed and consumed annually by the steel industry. A ton of steel produced in any one of the more than 150 steel mills may use iron ore mined in Minnesota, limestone from Michigan, coal from Montana, manganese from India, fuel oil from Oklahoma, scrap steel from many industrial centers, magnesite from Canada, dolomite from Oregon, ferrosilicon from New York and iron ore and raw materials from other countries and other continents.

Metallurgy

Iron ores. Iron ore is the term applied to an iron-bearing material in which the content of iron is sufficient for commercial use. About five percent of the earth's crust is composed of iron. Unlike other metals, such as copper and gold, iron is rarely found as a metal. Iron is usually found as a chemical compound with oxygen, sulfur and other elements. In addition, iron compounds are usually found mixed with other materials such as clay, sand, gravel, or rock. This material is worthless and is called *gangue*.

The most common iron minerals occurring are listed in Table 2-1.

furnaces. Stony ores such as *taconite*, which contain 22 to 40 percent iron, were not used in earlier years as long as better ores were plentiful. Today taconite ores are beneficiated by crushing and grinding to a flour-like consistency. The powdered ore is then passed over a series of magnetic separators where the magnetic iron ore particles are separated from the non-magnetic impurities.

The concentrated taconite, which contains 60 to 66 percent iron, is partially dried and formed into round pellets after adding a suitable binder material, Fig. 2-2. The pellets are then baked into hard,

TABLE 2-1. COMMON IRON MINERALS IN NATURE.

NAME	FORMULA	IRON CONTENT (PERCENT)
Hematite	Fe_2O_3	70
Magnetite	Fe_3O_4	72.4
Limonite	$Fe_2O_3 + H_2O$	52–66
Siderite	$FeCO_3$	48
Taconite		22–40
Jasper	(a wide variety of minerals)	

The high grade iron ore from the Mesabi mines, Fig. 2-1, in northern Minnesota is relatively free of worthless gangue (non-ore rock) and is dug up from open pits and shipped to the steel mills. Lower grade ores require a beneficiation process to remove the worthless gangue before shipping.

Ores are up-graded by crushing, screening and washing. Complete removal of very fine particles is necessary as such fines create problems in the blast

dense, blast furnace charge material. The pellets are grey in color and sized from 3/8 to 5/8 inches in diameter for processing.

A second method of forming ore fines into usable blast furnace charge material is sintering. Sinter is a clinker-like product obtained by fusion on a traveling grate machine passing over a vacuum induced by large fans. Modern sintermaking practice generally incorporates sufficient fluxes to render the sinter self-fluxing. This

Production of Iron and Steel

Fig. 2-1. Iron ore being removed from open pit mine located near Lake Superior. (American Iron & Steel Institute)

Fig 2.2 Rich taconite ore powder, with non-iron bearing material removed, is rolled with coal in huge drums to make small balls in Minnesota iron beneficiation operation (Allis-Chalmers)

material, when charged into a blast furnace, is pre-calcined and reduces the furnace's thermal load. This thermal reduction results in an overall lowering of coke consumption in the blast furnace.

Coke. Coke is not only the fuel and

Metallurgy

source of heat for a blast furnace but it also actively enters into the complex chemical reactions that occur in the operation of the blast furnace. Coke is produced from coal called *metallurgical coal* which has a low sulfur content. Coke is the skeleton of coal and is the solid that remains from a complex chemical process called *coking*.

The coking process, in addition to making coke, produces gaseous products which are piped to a chemical plant where they yield industrial fuel gas, tar, ammonia, oils, coal tar dyes and other valuable by-products. The coke ovens are air-tight ovens, rectangular in shape, 30' to 40' (9 to 12m) long and 6' to 14' (1.8 to 4.3 m) deep. They are only 11" to 22" (28 to 56 cm) wide and are built in rows containing up to 100 ovens. Metallurgical (low sulfur) coal is loaded in rotation and the ovens are sealed. Heat is applied to release the volatile gases in the coal, Fig. 2-3.

A modern coke oven can receive a charge of 16 to 20 tons of coal. Tempera-

Fig. 2-3. Incandescent coke from a long, thin oven, falls into quenching car. (American Iron & Steel Institute)

Production of Iron and Steel

tures of 1600 to 2100°F (870 to 1140°C) are maintained up to 17 hours to complete the coking. The oven doors are opened and a pusher ram shoves the entire charge of coke (about 12 tons) into a waiting quench car. The glowing coke is taken to a quenching station where it is cooled by a deluge of cold water.

Limestone. The third major raw material in iron and steelmaking is limestone. Most states have limestone deposits but much of the 30-plus million tons used annually as fluxing material comes from Michigan, Pennsylvania and Ohio. Limestone is blasted from rock formations at the quarry, crushed and then screened and sorted to matching sizes, Fig. 2-4.

Limestone is a natural compound, calcium carbonate ($CaCO_3$) which when heated in the blast furnace decomposes into lime, CaO, and carbon dioxide (CO_2). Limestone is often referred to as *flux stone*.

Any metallurgical operation in which metal is separated by fusion (melting) from the impurities with which it may be chemically combined or physically mixed (as in ores) is called *smelting*. Both conditions exist in iron smelting. The impurities are difficult to melt. The primary function of limestone as a flux is to render these impurities more easily fusible.

Some elements in the iron ore, being reduced with the iron, dissolve in or even combine chemically with it. Some other compounds, already combined with the

Fig. 2-4. Limestone is sized for fluxing use. (American Iron & Steel Institute)

Metallurgy

metal in the raw materials, refuse to be separated from it unless there is present a substance for which they have a greater chemical affinity. To furnish a substance with which these elements or compounds may combine in preference to the metal is the secondary function of a flux such as limestone.

Scrap. Scrap iron and steel are not strictly raw materials but are used like raw materials. Over 60 million tons of scrap ferrous metal is used annually. In modern terminology this old steel industry practice of reuse of scrap is called recycling.

Basically there are three different kinds of iron and steel scrap; (a) *home*, that which is the iron and steel scrap generated within a steel plant, (b) *prompt industrial*, scrap steel returned to the steelmaker by a customer and, (c) *dormant*, scrap steel in the market from junked autos, building demolition and environmental recycling activities, Fig. 2-5.

Steel mills are their own best suppliers of steel scrap inasmuch as about 30 percent scrap is generated in steelmaking operations. On an average, 60 percent of the scrap required by the steel industry is generated within steel plants and the remaining 40 percent is purchased from outside sources. In mills which utilize scrap only, as in the case of electric furnace steel, about 75 percent of the scrap is purchased.

The quality and analysis of scrap steel is an important factor in the use of scrap. It must be sorted and graded as to particle size, cleanliness and rust as well as to chemical composition. Certain nonferrous metals such as zinc and tin are harmful to furnace refractories. Copper in excess of 0.30 percent is detrimental to

Fig. 2-5. Scrap iron for charging into open hearth furnace. (American Iron & Steel Institute)

Production of Iron and Steel

the hot-forming properties of flat rolled products. Scrap containing alloying elements such as nickel, chromium and molybdenum in known quantities is more valuable than carbon steel scrap in the manufacture of many alloy steels because its use recycles and helps conserve expensive and scarce alloy elements.

SMELTING—THE BLAST FURNACE

Iron is extracted from its ore by means of the blast furnace. Taller than a twelve story building, the ironmaking blast furnace takes its name from the blast of hot air and gases that are forced up through layers of iron ore, coke and limestone. Fig. 2-6 shows a typical view of a blast furnace including the auxiliary charging equipment. Fig. 2-7 shows a cross-section of the furnace proper with the alternating layers of iron ore, coke and limestone reacting with a blast of hot air which generates temperatures necessary to liquefy the iron as well as promoting the complex chemical processes. The burning of the coke furnishes the necessary heat and the carbon monoxide (CO) gas generated by incomplete combustion of the coke, together with the red-hot, glowing coke, reduces iron ore to the element, iron.

The passage of the hot carbon monoxide gas up through the furnace charge causes the following actions:

1. Drives off moisture in the materials charged in at the top.
2. Chemically changes (reduces) the oxides of iron (the ore).
3. Calcines the flux (limestone): $CaCO_3 = CaO + CO_2$.
4. Melts the slag and iron.
5. Reduces the oxides of manganese, silicon and phosphorus.
6. Removes sulfur from the molten iron.

Blast furnaces require many auxiliary facilities to support their operation. In simplest terms the furnace itself is a huge steel shell almost cylindrical in shape and lined with heat-resistant fire brick. Once

Fig. 2-6. Exterior view of a blast furnace. (Inland Steel Corporation)

Metallurgy

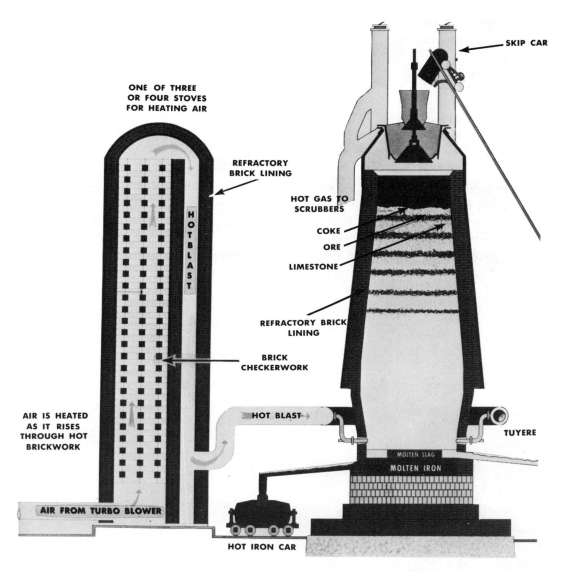

Fig. 2-7. Diagram of a blast furnace. (Bethlehem Steel Corp.)

started, or blown in, the furnace operates continuously until the refractory lining needs repairs or until the demand for iron drops to the point where the operation is closed down. The duration of furnace operations from start to finish is referred to as a *campaign* and may last several years.

Outline of operations. Iron ore, coke

and limestone are charged alternately into the furnace thru a double-bell hopper at the top and work their way down against the up-rising hot gases, becoming hotter as they descend. In the top half of the furnace, the carbon monoxide gas removes a great deal of the oxygen from the iron ore. About halfway down, the limestone begins to react with the impurities in the ore and coke to form a slag. The ash from the burnt coke is absorbed into the slag.

Silica (SiO_2 or sand) in the ore is also reduced by the gas to form silicon which dissolves in the iron as does some of the carbon from the coke. At the furnace bottom the temperatures rise well over 3,000°F (1643°C). The melted iron and slag trickle down into the furnace hearth where lighter slag floats on the heavy iron. This separation makes it possible to drain the slag off through a slag notch into ladles for disposal. The molten iron is likewise released from the hearth through a tap hole into a large ladle or a thermos-bottle-like hot metal railroad car for transportation to the refining furnaces.

The hot air blown into the hearth through tuyeres is heated to 1400-2100°F (760-1140°C) by passing it through stoves. Stoves are built in pairs, each serving in two alternating functions; (1) being heated by burning exhaust carbon monoxide from the blast furnace; (2) heating ambient air blown through the heated brick checkerboard in the stoves. The stoves may contain up to 250,000 ft^2 (2,322,500 m^2) of heating surface. As much as four and a half tons of air may be needed to make one ton of pig iron.

Material requirements. To produce one ton of pig iron requires about 1.6 net tons of iron ore, 0.65 ton of coke, 0.2 ton of

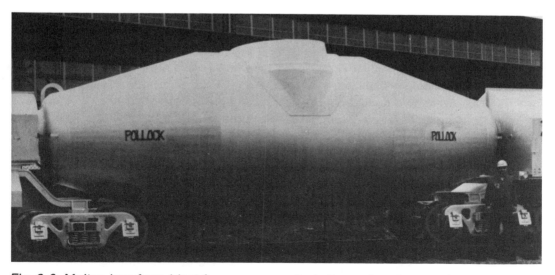

Fig. 2-8. Molten iron from blast furnace moves by ladle car to other steel making department. (William B. Pollock Co., Div. GATX Co.)

Metallurgy

Fig. 2-9. Casting machine delivers molten iron to molds for pig iron production, later to be made into pipe. (Deere & Co.)

limestone, 0.05 ton miscellaneous iron and steel scrap. In addition, the water cooling system for the furnace shaft may use 10 to 12 million gallons of water per day. At an average production rate of 3000 tons per day for one blast furnace, this means that up to 12,000 gallons of water are used to produce one ton of pig iron.

The term *pig iron* may seem obscure to modern city dwellers but its origin was perfectly clear in earlier farming communities with a small blast furnace. In those days the molten iron from the tap hole ran through channels in the sand floor with small slots at their sides. To the farmers the set-up resembled newborn pigs at the sow's side, suckling. Today most of the iron from the blast furnace is kept in a liquid state and taken to the refining process, or may be pigged for cast iron production, Fig. 2-8. Commercial customers for cold pig iron are foundrymen who cast pipe, automobile engine parts and the many cast iron products in industry, Fig. 2-9.

DIRECT REDUCTION PROCESSES

The term *direct reduction* as currently used in the industry refers to any process, other than the conventional blast furnace, for extracting iron from its ores. The iron in use commercially up to the latter half of the 19th century was wrought iron which

Production of Iron and Steel

Fig. 2-10. Sponge iron is obtained by direct reduction of iron minerals without melting the metal. (American Iron & Steel Institute)

originally was the product of primitive forges unable to produce temperatures high enough to melt the iron.

While the earliest blast furnaces came into use around 1350 A.D., direct reduction of iron ore in charcoal hearths had been used for at least 1500 years to produce *sponge iron*, Fig. 2-10. This metal, containing various impurities such as slag (silicates), could be hammered while hot to produce useful implements. Direct reduction was the original ironmaking process; however, no direct process has yet attained mass production capabilities that could compete with the speed and large tonnage capacity of the blast furnace.

The comparative ease with which iron ores can be reduced to metallic iron makes a direct reduction process desirable if it can eliminate the large investment, maintenance and operating costs of a blast furnace. There are over fifty direct reduction processes being developed, tested and scaled up for high production. The industry is looking at these efforts to choose those which are commercially viable.

The various processes for direct reduction can be divided into three general types:

1. Those that utilize a kiln for reduction of the ore.
2. Those that utilize retorts in a batch process.
3. Those that use the fluidized bed techniques.

Kiln reduction. One of the most advanced processes of this type to be tested on a large scale has produced up to 100 tons per day in a pilot plant built to demonstrate its commercial potential.

The process derives its fuel energy and reducing potential from coal and natural gas (or other suitable hydrocarbons). A variety of bituminous, sub-bituminous and anthracite coals have been tested, and if proper charging techniques are utilized, almost any coal can be used successfully.

The raw material is high grade ore pellets containing 67 percent iron, dolomite (limestone) and coal. The pellets and the stone are screened to remove over-sized and under-sized particles. These raw materials are fed continuously in desired proportion into a common conveyor leading to the kiln, Fig. 2-11.

The kiln is 115 ft (35m) long with an internal diameter of about 7 1/2 ft. (2.3 m) and can be rotated at constant speed over a wide range of speeds. The kiln is very much like those used in the Portland ce-

Metallurgy

Fig. 2-11. Conveyor-fed grate kiln iron ore pelletizing facility. (Allis-Chalmers)

ment industry. It is lined with fire brick about 8 1/2 inches (22 cm) thick. It is fired by natural gas.

The materials enter at the slightly elevated head end of the kiln and the gas flame moves up from the lower, discharge end. The rotation moves the material through the kiln in about five hours. The counter-flow of the raw materials to the ascending heat permits sequential drying, chemical reduction of the ore, carbonizing of the coal and sintering to take place.

At discharge the material is about 2000°F (1095°C). It is passed through a secondary cooler-kiln which is about 65 ft (20m) long, 5 ft (1.5 m) O.D. and rotated at a constant speed. The first third of the cooler-kiln is lined with fire brick, and water sprays are directed on the outer surface of the unlined two-thirds. The cooled material is then magnetically separated to segregate the metallic iron.

Retort reduction. This process involves the batch reduction of high grade iron ore in sealed retorts. A 500 ton per day plant has been in production since 1960. The process makes use of reformed natural gas containing 73 percent hydrogen, 16 percent carbon monoxide, 7 percent carbon dioxide and 4 percent unconverted methane.

In the 500 ton per day plant, the ore

charge is 100 tons of ore in each of four retorts and the full cycle time is 12 hours, equally divided into the four stages of the process. These are: (1) charging, (2) secondary reduction, (3) primary reduction and (4) cooling. In the process, the purified reformed gas is preheated to 1600° to 1800°F (870-980°C). The reduction retorts are arranged in parallel, and the preheated gas enters one of the retorts in the primary stage of ore reduction (final reduction) and passes downward through the fixed bed of ore where a portion of the hydrogen is converted into water. At this point, the partially spent reducing gas is used to preheat and partially reduce the ore in a second retort. The final degree of reduction averages about 90 percent but varies from 96 percent at the top to 73 percent at the bottom of the ore bed.

Fluidized bed process. There are three major fluidized bed processes in use. One process is a simple multi-stage, continuous fluidized bed process operating at essentially atmospheric pressure. This method utilizes a long, moving grate type furnace where the ore enters at one end. It moves through a flame resulting from the partial combustion of natural gas with air, generating a reducing gas of 21 percent carbon monoxide, 41 percent hydrogen and 38 percent nitrogen. The air is preheated to 1500° to 1600°F (815°-870°C), and the natural gas to 600 to 700°F (310-370°C). The partial combustion of the natural gas provides the heat required to accelerate the reduction reactions.

A second process uses the fluidized bed technique for reducing powdered ores of a minus 20 mesh fineness. It differs from other processes in that it is a semi-batch method and operates at a lower temperature of 900°F (480°C) and a higher pressure of 500 psi (3447.5 kPa).

The third method uses the fluidized bed technique for reducing minus 10-mesh ore by hydrogen. This is done in a two stage process at 1200° to 1300°F (650 to 700°C) and at a pressure of 30 psi (207 kPa). Hydrogen for the process may be manufactured from natural gas, fuel oil, powdered coal or other materials by several common commercial methods.

In the process, iron ore is preheated to about 1700°F (930°C) and enters the primary-stage reactor where the hematite ore (Fe_2O_3) is reduced to the oxide FeO at 1300°F (700°C) by the partly spent off-gas from the secondary-stage reactor. The original reducing gas, containing not less than 85 percent hydrogen, enters the secondary-stage reactor at about 1550°F (840°C) and 30 psi (207 kPa) where it reduces the FeO from the primary stage to metallic iron at 1200° to 1300°F (650° to 700°C).

Product quality. Sponge iron contains two or three percent oxygen combined with the iron. While not desirable, this small amount of oxygen is not necessarily harmful as the refining process that follows is an oxidizing process. A major advantage of sponge iron is the absence of trace alloying elements such as nickel, chromium, copper, tin, zinc, lead and sulfur which are usually present in scrap purchased on the open market.

A distinct disadvantage in sponge iron is that the gangue initially present in the ore is retained in the final product. This is undesirable in refining operations as it retards melting, increases fuel consumption,

Metallurgy

is generally acid in nature and must be fluxed by added lime, thus increasing the slag volume.

From the physical property standpoint, low temperature direct reduction of fine ores produces a pyrophoric material that can be neither stored, shipped nor exposed to moisture without the possibility of extremely rapid oxidation even to the point of self-ignition. This pyrophoric character of fine iron powders can be minimized or eliminated by reduction at temperatures above 1300°F (700°C) or by hot-briquetting the freshly reduced iron powder under a neutral or protective atmosphere.

REVIEW

1. Define gangue.
2. Name the most common iron ores and their percentage of iron content.
3. Describe the taconite beneficiation process.
4. Define sinter.
5. Define coke and describe its manufacturing process.
6. What is the primary function of limestone in smelting?
7. Name the three principal sources of iron and steel scrap.
8. How did the blast furnace get its name?
9. What three materials are charged into a blast furnace?
10. Which substance is the fuel of the blast furnace?
11. What action changes the iron ore to iron?
12. How hot does the bottom portion of a blast furnace get?
13. Iron and slag melt in a blast furnace. Which substance drops to the bottom and what happens to the other?
14. How is the molten slag and iron released from the bottom of the blast furnace?
15. How is molten iron transported from the blast furnace to the refining operation in a modern integrated steel mill?
16. What is pig iron? How did it get its name?
17. What process produces sponge iron?
18. Discuss the advantages and the disadvantages of sponge iron as a product.

CHAPTER 3

REFINING INTO STEEL

MAJOR REFINING PROCESSES

Over 150 million tons of steel per year is currently being produced in the United States and Canada. Steel is produced in a metallurgical process called *refining* and is basically a method of reducing the high carbon content of pig iron to the lower requirements of steel.

Pig iron contains up to 5 percent carbon and variable amounts of silicon, sulfur, phosphorus, manganese and other trace elements. The amounts of silicon and sulfur are governed to some extent by the operation of the blast furnace, but the phosphorus content depends solely upon the raw materials used.

Sulfur and phosphorus are the two most undesirable impurities in steel. Much of the sulfur comes from the coke, particularly if the coke is made from coal with a high or even moderate sulfur content. Sulfur in steel causes brittleness at high temperatures, known as *hot-short* steel. Such steel causes difficulties in hot working operations such as forging. Phosphorus increases the tendency toward a coarse-grained steel. This weakens the steel at any temperature and is known as *cold short* steel.

Steel was first produced in high-tonnage in the United States in 1864 at Wyandotte, Michigan using an air blowing process invented by William Kelly. Kelly began his development of the process in 1847 unaware that Henry Bessemer in England was developing a similar process. Both men were granted patents in their respective countries but Kelly went bankrupt in the Panic of 1857 and sold his patents to the company that built a plant and produced the first ingots of what was later to be known as Bessemer steel in 1864. America's first steel railroad tracks were made in 1865 at the North Chicago Rolling Mill from steel made by Kelly's process. Steel was also being made by another company using Bessemer's patents. In

Metallurgy

1866, the Kelly and Bessemer interests were merged and the process became known as the *Bessemer process*. Tonnage increased sharply as 3000 tons (2721 tonnes) of steel were produced in 1867 and one million tons (907,000 tonnes) a year by 1880, only 13 years later.

Close on the heels of the Bessemer process came the *open hearth process,* based on development work done by the Siemens brothers in England and the Martin brothers in France in the 1860's. The open hearth furnace made its debut in the United States in 1868. Large tonnages of steel have been produced by both methods but both the open hearth and the Bessemer processes where challenged a century later (1950's) by the *basic oxygen process* developed in Austria after World War II. The basic oxygen process gives the steel industry several major advantages.

The Basic Oxygen Process

The main advantage of the basic oxygen process is a shortened heat time. Forty-five minute tap-to-tap heat cycles out-pace the average open hearth's ten hour cycle, with the quantity and quality of each heat being approximately equal. While the Bessemer converter is able to produce steel about as fast as the basic oxygen furnace (BOF), selected raw materials are required and the properties of Bessemer steel strictly limit its use in industry.

Another economic advantage is the lower total investment required for a BOF. Beginning in the 1960's, many steel makers opted to install new BOF facilities to replace aging facilities of other kinds. New federal requirements to retro-fit old facilities with air pollution equipment, coupled with increasing maintenance costs on the old equipment, led to the increase in BOF installations.

The basic oxygen furnace has a capacity up to 300 tons of molten metal. It is a steel shell, lined with refractory materials. The body of the furnace is cylindrical; the bottom slightly cupped and the top is shaped like a truncated cone set on top of the cylinder, Fig. 3-1. The narrow top of the truncated cone is open for charging and the insertion of an oxygen lance. As the *burn* proceeds, the waste gases exit the furnace through the opening and are treated in extensive air pollution control equipment to remove particulate matter.

In operation, the furnace is tilted on trunnions so that the top opening can be positioned for charging. Cold scrap and hot metal (liquid pig iron) is loaded into the furnace. The percentage of cold scrap steel varies with different steel producers. From a low limit of 30 percent scrap, its use has climbed to 50 percent scrap under certain operating and economic conditions.

After charging, the furnace is raised to a vertical position and the oxygen lance is lowered. The lance is 16 to 20 inches (40 to 50 cm) in diameter, 50 to 60 ft (15-18 m) long and has a replaceable bronze tip. The tip and the lance tube are water cooled. As the lance tip nears the surface of the liquid iron, the oxygen is turned on to a flow of 5000 to 6000 cfm (approximately 23,600 to 28,300 m^3/s) at a pressure of between 140 and 160 psi (965 to 1100 kPa). When the lance has reached the de-

Refining into Steel

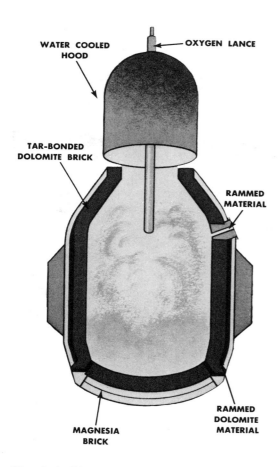

Fig. 3-1. Diagram of a basic oxygen furnace. (Bethlehem Steel Corp.)

sired distance from the liquid surface, it is clamped in position. This distance varies from 60" to 100" (150-250 cm) depending on the chemical composition (blowing quality) of the scrap and pig iron used.

Shortly after the lance has been positioned, a *light* will be obtained as the oxygen starts burning the excess carbon. This is the start of the *blowing* reaction. About one minute after the light, lime, fluorspar and mill scale (iron oxide) is added. These are the fluxes required. The end-point of the blow is determined by a visual drop in the flame, accompanied by a definite change in the sound being emitted from the furnace. In newer, computer controlled furnaces, devices to sense the reduction in the weight of the charge due to the oxidation of solids are used to determine the end-point.

When the end-point is reached, the oxygen is shut off, the clamps are released and the lance is retracted through the hood. The furnace is then rotated toward the charging floor until the slag, floating on top of the steel, is even with the lip. The slag quantity, amounting to about 250 pounds per ton (114 kg/tonne) of steel, makes it unnecessary to pour off at this point. The temperature of the liquid is then determined with an immersion thermocouple. The desirable temperature range for low-carbon steel is 2900°F (about 1600°C). With steel over 0.15 percent carbon, a test sample is taken at this time and if necessary the metal is recarburized to the required percentage.

When both temperature and carbon content are found to be correct, the furnace is rotated until the slag approaches the pouring lip and the metal flows under the slag and into the ladle without any admixture of slag. Alloys are dropped into the ladle and it is transferred to the pouring *(teeming)* aisle.

The Bessemer Process

First introduced in the U.S. in 1864, the Bessemer refining process was the first

29

Metallurgy

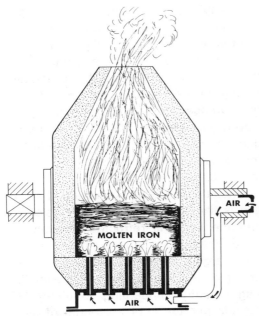

Fig. 3-2. Section of bessemer converter with detachable bottom.

The oxygen in the air burns out the excess carbon, and the silicon and manganese present in the pig iron. The silicon and manganese are the first to oxidize and the products of reaction, with some oxide of iron produced at the same time, combine to form a slag.

Bessemer steel is likely to be highly oxidized and dirty. Although it is usually considered inferior to steel produced by other methods, it is used for products calling for a low-carbon steel with not very high requirements for strength, ductility and toughness. Bessemer steel is now used in low-grade sheets, fence wire and screw stock. Due to the economics and high quality of basic oxygen steel, relatively small amounts of Bessemer steel are made today, less than 3 percent of total U.S. production, but that is still millions of tons.

Electric Arc Furnace

high production method of producing steel although it is now a minor process. Its air blowing methods are related to the basic oxygen furnace with its oxygen.

The Bessemer converter is a large, pear-shaped vessel made of steel plate, lined with refractory materials and mounted on trunnions to permit its tilting to load and discharge. As shown in Fig. 3-2, the converter has a double bottom through which air is blown into and through molten pig-iron poured into the converter top.

Converters range in capacity from 5 to 25 tons and the air is blown in at a pressure of 22 to 28 psi (150 to 200 kPa) and at the rate of 25,000 to 35,000 ft^3/min (700-1000 m^3/min).

Using electricity solely for the production of heat, the electric arc furnace has a reputation for producing tool and die steel, roller and ball bearing steel, stainless and heat resisting steel and other specialty alloy steels. The first electric steel making furnace was built in 1906 and the process has grown in capacity from small units for specialty steel to large furnaces with capacities of 150 to 200 tons (135 to 180 tonnes) of steel per heat at a rate of up to 800 tons (725 tonnes) per 24 hours. Today the process accounts for one quarter of all U.S. production.

The electric furnace has proved to be an ideal melting and refining unit. Its advantages are the non-oxidizing condition of

Refining into Steel

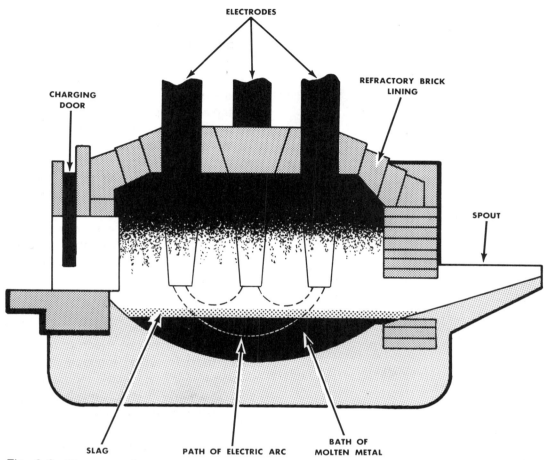

Fig. 3-3. Diagram of an electric furnace. Temperatures can be accurately controlled. (Bethlehem Steel Corp.)

the carbon arc, which is pure heat, the construction possibility of a tightly closed furnace for maintaining special atmosphere, the high temperature attainable, close limits of quality control and extremely high efficiency in its operation.

The body of this furnace is a circular steel shell shaped like a kettle and mounted on rollers so the molten steel and slag can be poured out, Fig. 3-3. The side walls and bottom are lined with refractory materials as is the removable roof, shaped like a flat dome. Three large cylindrical electrodes of carbon or graphite enter vertically through holes in the roof. The electrodes are up to 24 inches (60 cm) in diameter and carry the electric current to the steel charge in the furnace.

Metallurgy

Fig. 3-4. Electric steelmaking furnace. (Inland Steel Company)

Electric furnaces are usually charged by removing the roof and charged with 100 percent steel scrap by overhead crane buckets. The roof is swung back into position, the electrodes are lowered into close proximity with the scrap and the alternating current turned on. The heat resulting from the arcing between electrodes and the metal begins to melt the solid steel scrap, Fig. 3-4.

At a given point, iron ore is added to reduce the carbon content. This causes a violent boiling action and is the period where phosphorus, silicon, manganese and carbon are oxidized. Limestone and other fluxes are then added to form a molten slag that floats on top of the steel and absorbs the oxides and other impurities. When the heat is completed, the electrodes are raised to their maximum height after the electric power is shut off. The tap hole is opened and the furnace tilted so that the steel is drained from under the slag into a ladle. Special alloys, in the form of ferro-alloys, are added to the melt in the ladle at this time prior to pouring the ingots.

Electric furnace use and tonnage are rising because of their adaptability to continuous pouring and continuous slabbing operations.

Induction Furnace

A different type of electric furnace, usually used for small tonnage melts, is the high-frequency, electric induction furnace. This furnace is essentially a transformer with the metal charge acting as the core or secondary windings. The furnace consists of a refractory crucible around which a coil of copper tubing is wound. The copper tubing serves as a water-cooled primary winding, Fig. 3-5.

A high-frequency AC current, usually 960 Hertz (Hz), is applied to the primary windings. This induces a much heavier secondary current in the scrap steel placed in the crucible. The resistance of the scrap metal causes it to melt and thoroughly mix by violent convection action. This mixing action is a natural phenomenon which takes place as soon as the molten state is reached.

Because melting is so rapid, there is on-

Refining into Steel

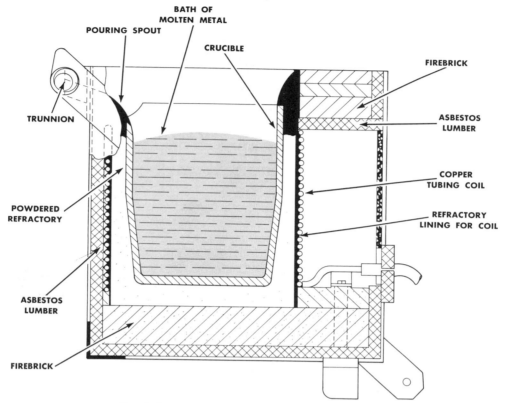

Fig. 3-5. Cross-section of induction furnace.

ly a slight loss of alloying elements. With a power input of 300 kw, it is possible to melt down a 1000 pound (450 kg) charge in 45 minutes. After melting is complete, the charge is further heated for about 15 minutes to reach the tapping temperature. During this time, required alloys and deoxidizers may be added as the mixing action continues. The furnace is usually mounted on trunnions so that the charge may be poured into a ladle for transport to the pouring floor.

Open Hearth Steelmaking

The open hearth furnace was developed in England in 1864 and after a decade or two overtook the Bessemer process in annual tonnage produced. The Siemens-Martin or *open hearth* furnace is so named because the hearth or floor of the furnace is shaped like an elongated saucer. Limestone, scrap steel and molten pig iron are charged into the hearth and are exposed to the open sweep of flames from

Metallurgy

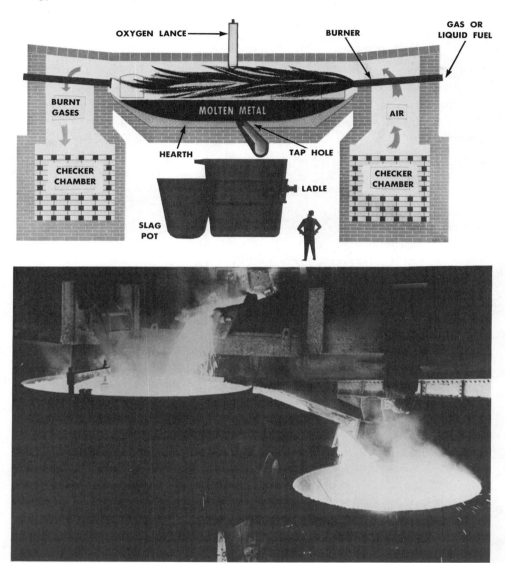

Fig. 3-6. Diagram of open hearth furnace. Flames fired first from one side and then the other, sweep over the top of the metal. (Bethlehem Steel Corp.) Tapping of 550-ton open hearth furnace. (Weirton Steel Co.)

gas or oil burners which emanate alternately from opposite ends of the furnace, as in Fig. 3-6.

The open hearth furnace is an extremely large and costly piece of equipment. A furnace capable of producing 350 tons of

Refining into Steel

steel (320 tonnes) in five to eight hours is likely to measure 90 ft (27 meters) long by 30 ft (9 meters) wide. The auxilliary equipment, the air heating *checker board* regenerators, the air flues, air regulating dampers, together with the hearth, represents a large capital investment with corresponding high maintenance expense.

Open hearth furnaces may be charged in three ways; (1) steel scrap and molten pig iron (about a 50-50 mix), (2) steel scrap and solid pig iron, and (3) 100 percent steel scrap. About 25 percent of ingot weight produced is recycled to supply the steel scrap of the refining operation. Methods 2 and 3 also utilize a limestone flux. While sufficient oxygen is usually generated from the oxides on the scrap metal or by the introduction of iron ore into the hearth, most modern open hearth furnaces employ oxygen lances to provide gaseous oxygen during the working period after melting.

The manufacture of steel in open hearth furnaces takes much more time per ton than by the basic oxygen process. Basic hearth is capable of producing very large batches—up to 600 tons (545 tonnes) of steel with close control of analysis. Alloy additions are made in the furnace or in the ladle if they are oxidizable. Nickle or copper is added into the furnace while manganese and chromium, as ferroalloys, are added to the steel in the ladle.

The open hearth furnace (and Bessemers) today account for about twenty-five percent of the steel industry tonnage. Various economic factors, such as fuel availability, keep it in use, especially after the addition of the oxygen lance capability, which shortens its cycle.

The Crucible Process

High-grade tool steels and some alloy steels are still made by the crucible process, although the electric furnace is capable of making steel equal in quality to the earlier, famed Swedish crucible steel. In the crucible process, wrought iron or good, clean scrap, together with a small amount of high purity pig iron, ferromanganese, the necessary alloying metals and a flux are placed in a clay or clay-graphite pot with a tight cover and melted in an oven.

The earlier process of melting utilized a floor furnace, essentially a firebrick-lined trench into which the crucibles were placed and a stream of hot gases from an oil or gas flame were passed down the trench and around the crucibles. Single crucibles holding fifty to a hundred pounds of steel can also be melted down in induction type furnaces or conventional above-the-floor furnaces. The convection currents, which are always produced when a liquid is heated, completely mix the alloys during the melt.

After the charge has melted, with sufficient time for the mixing to take place and the gases and impurities to rise to the surface of the crucible, the crucible is withdrawn from the furnace and the slag floating on the top of the steel is removed. The steel is then poured into a small ingot of fifty to a hundred pounds. The ingot is then ready for forging or rolling to desired shape.

The crucible process differs from other steel-making processes in that very little refining is done. The purity of the steel is directly related to the purity of the materi-

Metallurgy

als placed into the crucible at the start of the process. The chief advantage of the process is that it removes practically all of the impurities including oxygen, dissolved gases and entangled particles. Crucible steel is used in the production of fine cutlery, surgical instruments, and fine cutting tools for the machinist.

Specialty Melting and Refining Processes

When high quality steel such as that used in tool-and-die, cutlery, or ball and roller bearings is required, special type furnaces and melting techniques are employed. These facilities can be grouped in two general categories. First are those which receive molten metal from the conventional steel making furnaces and remove the impurities. Among these facilities are: vacuum stream degassing, vacuum ladle degassing, argon-oxygen decarburization, Fig. 3-7. The second group includes: vacuum induction melting, consumable electrode melting and electroslag melting, Fig. 3-8.

Two other processes that can be used with either molten or solid metal as a charge are: electron beam melting, Fig. 3-9, and vacuum induction melting. Both of these processes are carried out in a vacuum as are the processes in the first group. Basically, the process of vacuum degassing is used to draw off the undesirable gases before the steel solidifies. The exception in group one is the argon-oxygen decarburization which is not operated in a vacuum but rather in a controlled atmosphere which minimizes the oxidation of chromium alloy.

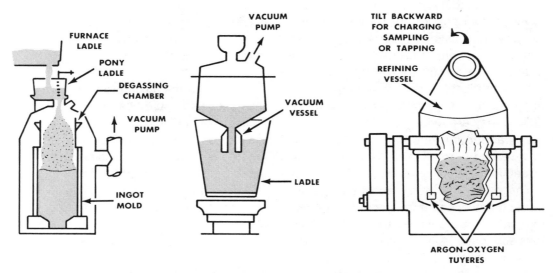

Fig. 3-7. Vacuum stream degassing; vacuum ladle degassing, and argon-oxygen decarburization schematics. (AISI)

Refining into Steel

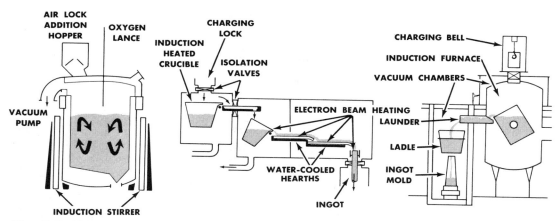

Fig. 3-8. Vacuum oxygen decarburization; electron beam melting; vacuum induction melting schematics. (AISI)

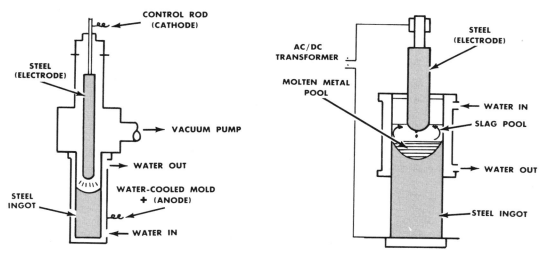

Fig. 3-9. Consumable electrode melting; electroslag melting schematics. (AISI)

INGOT PRODUCTION

Pouring the Ingot

The pouring operation is critical, first in controlling the molten steel so as not to endanger personnel or damage equipment; second, because poor practice in the assembly, adjustment or manipulation of the pouring equipment may seriously

Metallurgy

and adversely affect the quality of the end-product.

In the teeming or pouring operation, the ladle of molten steel is carefully poured into a refractory lined storage and distribution trough called a *tundish*. The tundish is positioned over the ingot molds and the metal is gated out of the tundish to fill the molds. The ingot molds are massive structures of cast iron. They range from eight feet (2.5 m) high by 3 feet (0.8 m) in diameter for high tonnage usage, down to three feet (0.8 m) high by six inches (15 cm) in diameter for specialty steels used in the manufacture of tools, Fig. 3-10.

The cross-section of most ingots is either square or rectangular with round corners and corrugated sides. All ingot molds are tapered to facilitate removal of the solid steel ingot. Molds usually are tapered from one end to the other so that the top of the cast ingot is either smaller or larger than the bottom. In the case of the big-end-up type, the taper helps in the solidification process by permitting freezing from the sides and bottom in such a manner that the last part of the ingot remaining liquid is under the hot top, thus promoting soundness.

The hot top is a separate fire clay mold shaped like an upside-down, large bottomless flower pot. The metal in contact with the hot top does not freeze as fast as the metal in the ingot mold and permits the still molten metal to feed down into the contracting ingot and at the same time permits the unwanted oxides and impurities to rise up into the hot top. The solid hot-top section with the accumulated impurities (dirt) is later sheared from the ingot prior to rolling.

Fig. 3-10. Molten steel is transferred by pouring or teeming from the ladle into ingot molds. The ingot eventually will be processed through a rolling mill into one of many useful steel products as, for example, plate, sheet, strip or structural shapes. (Bethlehem Steel Co.)

Stripping The Ingot

The inside surface of the ingot mold is carefully coated with a releasing agent to prevent droplets, created during teeming, from splashing up on the chilled cast iron sides and sticking to them, thus spoiling the surface quality of the ingot. The coating also facilitates extracting the solidified ingot from the mold.

The stripping operation of a big-end-up ingot requires a special hydraulic ram table with clamping devices to hold the mold upright while the ram pushes the ingot up. Tongs from an overhead bridge crane at-

Refining into Steel

Fig. 3-11. Removing ingot molds. The operation is called stripping (American Iron and Steel Inst.)

tach to the red-hot ingot and lift it free of the mold and carry it to the soaking pit. In the case of a big-end-down ingot, a plunger descends from the tongs. The tongs grab the ingot mold and, while the plunger pushes against the top of the ingot, the tongs lift the mold up, Fig. 3-11. The ingot remains upright on its base plate while the mold is laid to one side. Then the tongs return to carry the red-hot ingot to the soaking pit.

The making of good ingots is one of the most difficult and important steps in the fabrication of steel, for often defects in the ingot cannot be eliminated by subsequent hot working operations such as rolling or forging. The following items are some of the major defects:
1. Pipe
2. Dendritic structure
3. Blowholes
4. Segregation
5. Slag inclusions
6. Checking, cracks and scabs
7. Internal stresses

Pipe. After the molten metal has been poured into the mold, it begins to solidify from the outside due to the chilling effect of the cast iron mold. Because of the shrinkage that occurs when metals solidify, the solid metal occupies less space than the molten metal. This shrinkage results in a cavity or hollow in the cast ingot. This cavity is called the *pipe* of the ingot and is usually found in the section that is last to freeze, the central upper portion of the ingot. This is shown in the second, third and fourth ingots in Fig. 3-12. The only way to eliminate this pipe section is to shear it off the ingot just prior to rolling.

Metallurgy

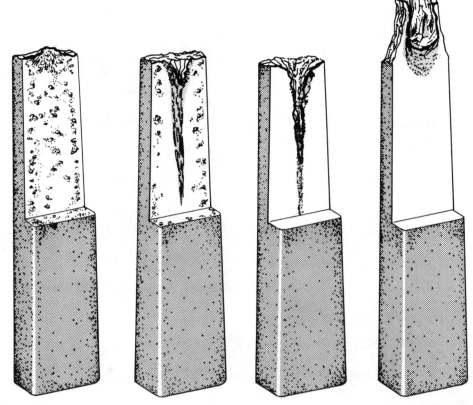

Fig. 3-12. Formation of pipe in central upper portion of ingot. Left to right: *rimmed steel (many blowholes); capped steel (some blowholes and some piping); semikilled steel (practically no blowholes but marked piping); killed steel (highly uniform composition).* (American Institute of Mining Engineers)

Dendritic structure. The chilling effect of the cold iron mold also causes crystallization of the molten metal to occur in an abnormal manner. The crystals form first on the surfaces of the mold and grow inwardly to form a *dendritic* (from the Greek word *dendron,* meaning tree) grain structure. This leaves the steel inherently weak because of the obviously weak planes so formed. Fortunately, in most cases the chilling does not extend to the center of the ingot. The central section of the ingot has a characteristic random crystal orientation and resultant tough structure.

Blowholes. Blowholes cause defective steel. These blowholes are formed by the liberation of gases during the solidification of the ingot. Gases from the molten steel may rise as bubbles and escape from the ingot or may be trapped in the solidifying metal. Blowholes near the center of the ingot, known as deep-seated, are less harmful than blowholes on or near the surface of the ingot. Deep-seated holes will normally weld shut during rolling or forging operations but surface blowholes will elongate during hot rolling, producing seams in the finished steel.

Segregation. During solidification, liquid solutions separate out or *segregate,* the first portion to freeze being relatively pure metal (compared to the last to freeze). This selective freezing results in uneven concentration of the elements found in the alloy and is called *segregation.* In ingot manufacture, the segregation is aggravated by the chilling action of the mold. The first portions in contact with the mold freeze relatively pure. The central portions of the ingot, which are the last to freeze, are richer in the lower freezing point constituents. This is another good reason for cropping the upper portion of the ingot prior to rolling. Cutting to remove *pipe* also removes the section of the ingot where segregation is worst. The degree of segregation is influenced by the type of steel, pouring temperature and ingot size. It will vary with position in the ingot and according to the tendency of the individual element to segregate.

Types of Ingot Steel

In most steel-making processes the primary reaction involved is the combination of carbon and oxygen to form a gas. If the oxygen available for this reaction is not removed before or during casting (by the addition of ferrosilicon, aluminum or some other deoxidizer) the gaseous products continue to evolve during solidification. Proper control of the amount of gas evolved during solidification determines the type of steel. If no gas is evolved, the steel is termed *killed,* because it lies quietly in the molds. Increasing degrees of gas-evolution results in *semikilled, capped* or *rimmed* steel.

Killed steel. These steels are strongly deoxidized and are characterized by a relatively high degree of uniformity in composition and properties. The metal shrinks during solidification, thereby forming a cavity or *pipe* in the extreme upper portion of the ingot. Generally, these grades are poured in big-end-up molds. Use of a *hot-top* will usually result in the pipe being in the hot-top section which is removed prior to rolling. Severe segregation will also be in the hot-top area.

The uniformity of killed steel renders it most suitable for applications involving such operations as forging, piercing, carburizing and heat treatment.

Semikilled steel. These steels are intermediate in deoxidation between the killed and rimmed grades. Sufficient oxygen is retained so that its evolution counteracts the shrinkage on solidification. The composition is fairly uniform but there is a greater possibility of segregation than in killed steels. Semikilled steels are used where the greater uniformity of killed steel is not required. This steel is used in the manufacture of structural steels such as plate, rail steel and forgings.

Rimmed Steel. These steels are only slightly deoxidized, so that a brisk effervescence or evolution of gas occurs as the metal begins to solidify. The gas is a product of a reaction between the carbon and oxygen in the molten steel which occurs at

Metallurgy

the boundary between the solidified metal and the remaining molten metal. As a result, the outer rim of the ingot is practically free of carbon. The center portion of the ingot has a composition somewhat above that of the original molten metal as a result of segregation.

The low-carbon surface layer of rimmed steel is very ductile. Proper control of the rimming action will result in a very sound surface in subsequent rolling. Consequently, rimmed grades are particularly adaptable to applications involving cold forming and where surface is of prime importance.

Capped Steel. These steels are much the same as rimmed steels except that the duration of the rimming action is curtailed. A deoxidizer is usually added during the pouring of the ingot with the result that a sufficient amount of gas is entrapped in the solidifying steel to cause the metal to rise in the mold.

Capped steels have a thin, low-carbon rim which imparts the surface and cold-forming characteristics of rimmed steel. The remainder of the ingot approaches the degree of uniformity typical of semi-killed steels. This combination of properties has resulted in a great increase in the use of capped steels in recent years.

Continuous Casting or Continuous Slabbing

The continuous casting process is in use in the steel and non-ferrous metal producing industry. The number of installations in North America is growing. There is a distinct economic advantage in the system in that production slabs are produced that are ready for rolling; and costly and labor consuming ingot pouring equipment, blooming mills and slabbing mills are eliminated.

The process is illustrated in the schematic drawing, Fig. 3-13. The process is of European origin based on the Junghans-Rossi designs and the machine depicted is capable of casting 5″-8½″ (13-22 cm)

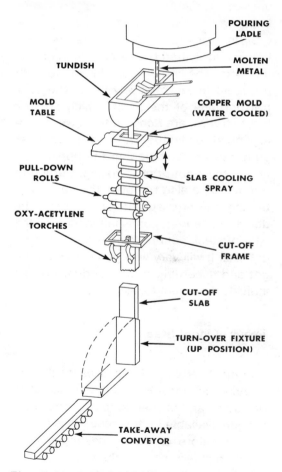

Fig. 3-13. Billets emerge from 4-strand continuous caster. (Inland Steel Company)

squares up to 6⅝" x 24" (17-60 cm) slabs. The unit is vertical, utilizing gravity for the flow of the metal. The pouring area is approximately 50' (15.5 m) above the run-out conveyor.

Molten steel in a ladle is hoisted to the refractory lined tundish, a horizontal storage trough. The liquid metal is shielded from oxidation by a cover over the tundish and an inert gas atmosphere. At the start of the operation, a dummy bar with same cross-section as the casting and about 20 feet (6m) long is inserted into the water-cooled copper mold. This starting bar is pulled down the mold by the pinch rolls as the molten steel flows from the tundish into the mold. The water-cooled copper surface chills the molten steel to form a shell that is ¼" to ½" (6-12 mm) thick. The mold is oscillated vertically through ¾" (19 mm), lowering at the same speed as the billet is dropping and then rising at three times this speed.

The billet emerges from the mold with a solid shell and a liquid metal core. It is then sprayed with cold water to hasten total solidification. At the point of cut-off, the billet is solid metal and oxyacetylene torches cut it to a desired length up to 16' (5 m). The billet is lowered into a discharge basket which rotates 90° and places the exposed portion of the billet on the run-out conveyor. The billets, cooled to about 1500°F (815°C) may now be removed to the rolling mill. Casting speed runs to 175" (445 cm) per minute and up to 60 tons (54.4 metric tons) per hour is cast.

Fig. 3-14. Billets emerge from 4-strand continuous caster. (Inland Steel Company)

A variation of continuous or strand casting is shown in Fig. 3-14. Here the descending billet of steel is gradually curved while in the plastic condition to a horizontal position for the cut-off operation.

REVIEW

1. Pig iron is not used as an industrial metal because of its high carbon content and impurities. What is its average carbon content percentage and what are the major impurities in pig iron?

2. What is *hot short* steel and what element causes this particular condition?

3. When was large tonnage steel first produced in the United States and what type of process was used? What was the process called?

4. What was the second type of large

Metallurgy

tonnage steel refining process called?

5. Describe the design and operation of a basic oxygen furnace.

6. Name some of the advantages of the BOF over its earlier predecessors named in questions 4 and 5 above.

7. What is charged into the BOF furnace at the start of the operation?

8. What happens when oxygen flows through the lance positioned near the surface of the charge in a BOF?

9. At what point are alloying materials added to steel produced in the various refining processes? Why?

10. Define the Bessemer converter. How is it charged and operated?

11. What are the advantages of electric arc furnaces?

12. What is the size and from what material are the electrodes in an arc furnace made?

13. What is the charge put into an arc furnace?

14. Describe an electric induction furnace.

15. What is the charge put into an induction furnace?

16. When are the alloys added in the induction process?

17. What is the charge put into an open hearth furnace?

18. What are the fuels used in an open hearth furnace?

19. What is used to add supplementary oxygen during the working period of an open hearth furnace?

20. What is the maximum batch size of steel produced in an open hearth furnace?

21. What type of steel is made in the crucible furnace?

22. What is the charge placed into the crucible in the crucible process?

23. What is the batch weight of an average crucible load?

24. How does the crucible process differ from other steel-making processes?

25. Name the types of specialty melting and refining processes.

26. Define an ingot casting.

27. Of what material is an ingot mold made?

28. Why is a *hot-top* used in teeming an ingot?

29. How is an ingot casting removed from its mold?

30. What is the nature of defects called pipes and how may they be avoided?

31. What effect does the cold iron mold have upon the grain structure of a cast ingot?

32. What becomes of deep-seated blowholes when an ingot is rolled?

33. What is the cause of segregation in an ingot?

34. Name the four types of ingot steel.

35. What is the economic advantage of continuous casting?

CHAPTER 4

SHAPING AND FORMING METALS

With the exception of powdered metal products, all the available shapes and sizes obtainable in metals have their beginning in the form of a casting. That is, the metals start from a liquid or molten state. The molten metal is poured into a suitable mold and allowed to solidify into a mass of desired shape.

Cast iron is an example of obtaining a desired shape which in many cases is of unbelievable complexity. Cast steel products such as the massive coupler of a railroad car also get their intricate shapes from the casting process. Casting techniques are further discussed in Chapter 11.

All steel starts out as a casting, the steel ingot described in Chapter 3. As cast, the steel ingot is a large (up to 5 ton) mass of steel with impurities at its top and a non-uniform, segregated crystalline structure. In modern, integrated steel mills, ingots are not allowed to cool after pouring but are placed in soaking pits to conserve heat energy. The reheating of cold ingots or rough slabs as done by some older mills is an expensive process and is avoided if at all possible.

The ingots in the soaking pits are kept at a uniform temperature of 2200°F (about 1200°C) to insure an optimum plastic condition for rolling or forging processes. The ingot is removed from the soaking pit by overhead crane, Fig. 4-1,

Fig. 4-1. Soaking pit reheats ingots to 2200°F (1205°C) for rolling mill. (Bethlehem Steel Corp.)

Metallurgy

and placed on a powered roller conveyor in the horizontal position. This conveyor moves the ingot to the shear where a skilled operator crops off the hot-cap or upper section which contains unwanted *pipe* with its oxides and other dirt. The cropped ingot is then conveyed to preparatory shaping in the cogging mill where the irregular ingot is roughly squared as it passes through a series of rollers which have a rough or cogged surface.

HOT WORKING

Hot Rolling

Rolling and hammer forging serve two purposes. First the processes serve the purely mechanical purpose of getting the steel into the desired shape. Secondly, the process improves the mechanical properties of the steel by breaking up the *as-cast structure*. This refining the grain is important because it makes the steel stronger and more ductile and gives it greater shock resistance.

Where simple shapes are to be made in large quantities, rolling is the most economical process, Fig. 4-2. It is used to produce plates, sheets, rails and structural shapes. It is also used to produce intermediate bar shapes to be used in the production of rods, forging billets and wire. The process of rolling consists of passing the red-hot ductile steel between two large, steel rolls revolving in opposite directions but at the same peripheral (outer edge) speed. The rolls are spaced so that the distance between them is somewhat less than the thickness of the steel entering them. The rolls grip the metal and reduce the thickness as it passes between them. As the thickness is reduced, the steel will increase in length in proportion to the reduction. The amount of reduction and the width of the piece will govern the amount of lateral spreading, Fig. 4-2.

The cogged ingot is then conveyed to the blooming mills. These mills are usually two-high electrically driven rolls that keep reversing their direction of rotation. The ingot passes back and forth as its size is being reduced. An ingot 25" by 27" (62 to 65 cm) in cross-section can be rolled into a *bloom* 9" (23 cm) square, in about 16 passes and in less than five minutes. Fingers, called manipulators, turn the steel bloom so that its thickness and width may be controlled. Controls for the manipulators and the screw-down which regulates the thickness are located in a control pulpit where the mill operator is situated. This control pulpit is a platform with a glass enclosed, air-conditioned office located above the mill where a clear view of the operation is possible, Fig. 4-3.

Three-high mills use three rollers, one above the other, Fig. 4-4. By the use of el-

Shaping and Forming Metals

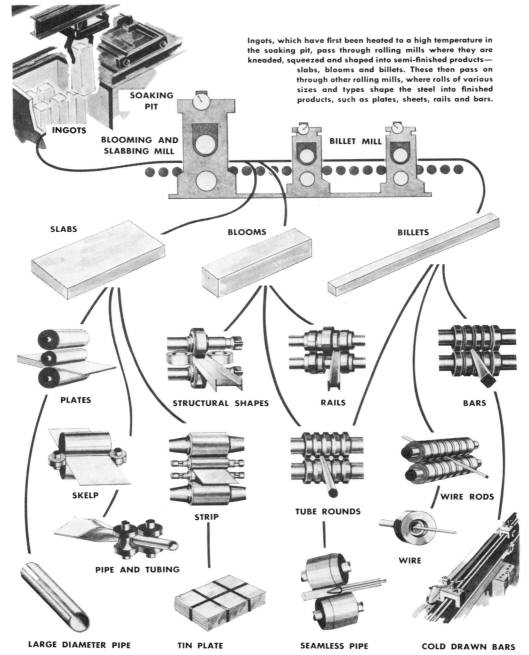

Fig. 4-2. Shaping steel ingots into primary metal products. (American Iron and Steel Institute)

Metallurgy

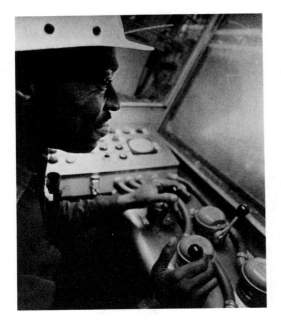

Fig. 4-3. Control pulpit above rolling mill floor. (Inland Steel Company)

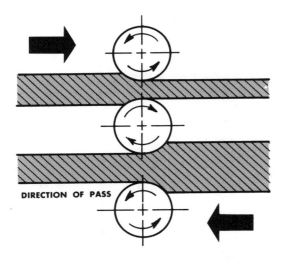

Fig. 4-4. Principle of a three-high rolling mill.

evating sections of the powered roller conveyors, the steel is lifted from the lower opening to the upper. Thus no reversing of the rolls is necessary. The product of the blooming mill, while still red-hot, must be sheared into shorter lengths. The

Shaping and Forming Metals

shears are an important part of every blooming mill. The lower knife blade, made of very tough steel, is fixed, while the upper blade is actuated. A blooming mill can produce either blooms, slabs or billets. The width of a blooming mill is somewhat limited so that special wider mills roll billets and are called billet mills while slabbing mills are used to produce the thick, wide slabs. Slabs are used to feed the hot mill which is used to produce sheets and strip.

Hot strip mill. Hot rolled sheet and strip are thin, flat products less than 3/16'' (0.48 cm) thick. Strip is customarily much narrower in width than is sheet and is often produced to more closely controlled thicknesses. At steel warehouses wide sheet is often slit into strip to customer specifications. A continuous hot-strip mill with its auxiliaries may be many city blocks long and capable of reducing a heated slab about 7'' (18 cm) thick and weighing more than 12 tons to a coil of thin sheet in a matter of minutes. The delivery speed of a hot-rolled sheet may be 3500' (1066.8 m) per minute.

The hot-strip mill consists of numerous stands of rolls, set in tandem. The first stand of rolls is known as the *roughing train* and consists of four big stands, each housing four rolls placed one above the other, and known as a *four-high*. The four-highs make the first reduction in thickness of the metal. The two middle or working rolls squeeze the heated slab down to about one fourth or more of its original thickness. The two center rolls are relatively small in diameter. The two back-up rolls, placed above and below the work rolls, Fig. 4-5, exert the pressure necessary to keep the work rolls from bending, which aids in the production of sheet of uniform gage.

The slab, now considerably longer and thinner, next enters the *finishing train* of rolls. Each of the finishing stands reduces the thickness of the metal, resulting in a thinner and much longer piece. The finishing stands look much like the roughing stands but are only a few feet apart. It is not uncommon to see seven of these stands in the operation. The rolls in each stand move a little faster than in the stand preceding it in order to take up the extra length of steel as it is rolled. The exiting temperature of the steel sheet is about 1500°F (815°C). The hot-rolled sheet is delivered to the cooling table and thence to pickling.

The removal of the heavy oxide, *mill scale*, is done by *pickling* which is the process of passing the steel through a bath of dilute hydrochloric or sulfuric acid. A modern pickler may be several thousand feet long and is operated continuously. When the steel emerges from the tank it is free of hard scale. It is then thoroughly washed and, when required by the customer, is oiled to inhibit rusting in storage. This final preparation prior to coiling, complies with a basic industry specification *Hot Rolled, Pickled and Oiled* abbreviated *H.R.P.O.*

Forming structural shapes. The principal shapes of structural steel are angles, beams, columns, channels, tees, zees, sheet piling and rebar. The formation of a structural shape takes place in three hot rolling stages; roughing, blooming and shaping. The first two have been described earlier in this chapter.

Metallurgy

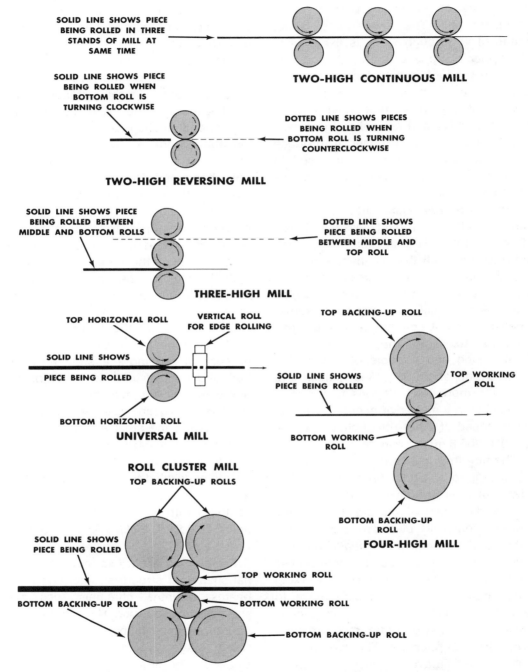

Fig. 4-5. Diagram of roll arrangements in a rolling mill.

Shaping and Forming Metals

Fig. 4-6. Rolls of angle iron.

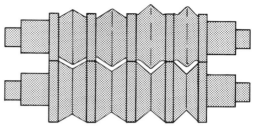

ROLLS FOR ANGLE IRON

Fig. 4-7. Rolls for H-beams.

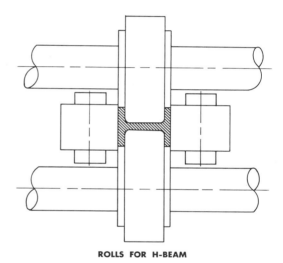

ROLLS FOR H-BEAM

The rolls for producing structural shapes differ from those that produce sheet in that the roll surface is machined with shaping grooves. The rolls for producing angle iron, for example, are shown in Fig. 4-6. The red-hot, rough billet enters the rolls at the left and is reduced in the first pass by about 20 percent. This rough shape is then passed back through the next groove in the rolls by reversing the direction of rotation of the rolls. Another 20 percent reduction takes place and the section begins to resemble an angle iron. Further reduction of area and continued shaping in the third and fourth passes produces the angle iron finished to commercial tolerances.

The production of wide flange W-beams, shaped like the letter *H* as differing from the *I* shape of conventional S-beams, has been made possible by the development of the *universal mill*, Fig. 4-7. In addition to having the conventional grooved horizontal rolls, such a mill has

Metallurgy

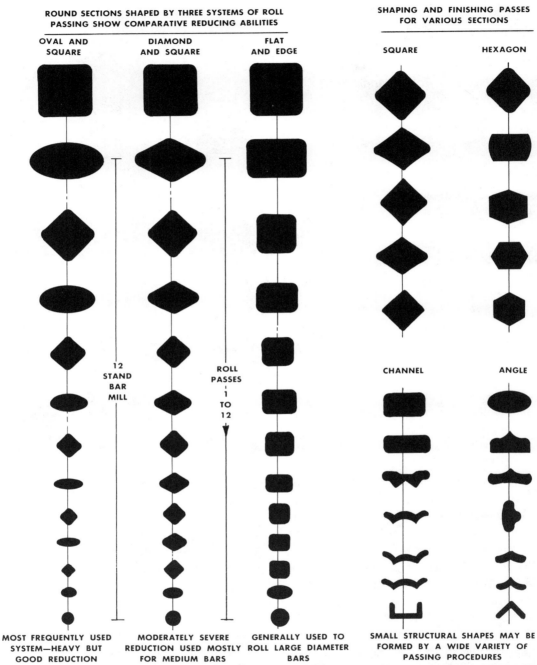

Fig. 4-8. Bar mill roll passes. Each vertical line of roll passes indicates steps in rolling the bar or section shown at bottom.

vertical rolls to finish the outer surface of the H.

The steel cribs of machine shops contain a multitude of hot-rolled steel shapes. These are all produced on mills with special grooved forming rolls. Fig. 4-8 shows some of the common commercial shapes and how they are progressively rolled. As each family of shapes requires a special set of forming rolls, it is easily seen that a modern steel mill must have on hand hundreds of these roll sets. Production runs must be of long duration to reduce the costly labor of changing roll sets to produce other shapes. This operational procedure of the steel mills makes it necessary to have warehouses, strategically located across the country to serve the many small shops and fabricators quickly.

Hot Forging

Forging or hammering is the oldest and most simple way of reducing metal to the desired shape. Hot-working is defined as the mechanical deformation of metal carried out above the temperature of recrystallization. Forging is *deformation by compressive force*.

Two basic dies are used in forging: (1) the flat or *smith dies* which are similar to the old blacksmith anvil and hammer with flat surfaces and, (2) cavity or *closed dies*. Forging in closed dies forms the metal to the shape and designs that are on the inner surface of the die cavity. When smith dies are used there is little or no restriction to lateral flow and the final shape depends on the hammer operator.

When metal is forged, strength and duc-

Fig. 4-9. Axle forging, sectioned and etched to show flow of metal fibers.

tility increase significantly along the lines of flow or grain. A deep-etched transverse section of a steel forging is illustrated in Fig. 4-9 and shows the flow lines. Good forging design controls the lines of metal flow to put the greatest strength where it is needed. Steel forgings are used in mechanism where extreme shock loading is anticipated, as in the steering linkage and the spindles of an automobile.

Forging machines. There are four basic types of forging machines in use. These are illustrated in Fig. 4-10. Drop hammers depend on the weight of the hammer and gravity for the compressive force. The board drop hammer has a long history in metal working and consists of a heavy wooden board which is raised vertically by pincher rolls controlled by the operator. When the rolls are loosened with the board in the *UP* position, the board and the hammer drops down to the anvil with a force of up to 10,000 pounds (44.5 kN).

The steam drop hammer is essentially a steam engine whose connecting rod is fastened to the hammer. Steam pressure at the bottom of the cylinder raises the hammer and when the valve is reversed,

Metallurgy

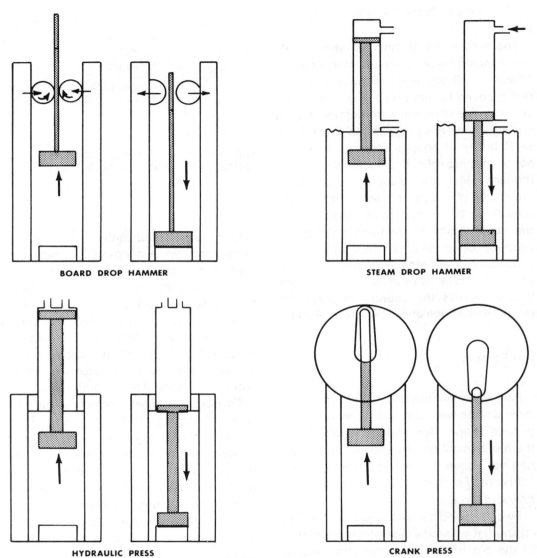

Fig. 4-10. Power hammer arrangements.

the expanding steam forces the hammer down with great compressive force. These hammers exert forces of up to 50,000 pounds (222 kilonewtons).

The hydraulic press exerts a continuous pressure on the dies, forcing the hot metal into the die cavities. Hydraulic presses exert forces up to 50,000 tons (222 mega

Shaping and Forming Metals

newtons). The mechanical or toggle press combines the impact action of a drop hammer together with the finishing squeeze of the hydraulic press. Either smith or cavity dies can be used and the press capacity ranges in size from 200 to 8,000 tons (1,779 to 71,171 kilonewtons) at speeds of 45 to 125 strokes per minute.

Extrusion

Extrusion covers the conversion of a solid length of metal, a billet, into lengths of desired cross-sections: tubes, bars, rounds, flats, ovals, tees, wire and special shapes. This is accomplished by forcing the plastic metal through the orifice of a die shaped to the desired cross-section configuration. Ferrous metals must be heated to red-hot forging temperatures to obtain the necessary plasticity. Non-ferrous metals, being more ductile, are usually extruded cold.

The operation consists of placing the heated billet into the chamber of the extrusion machine shown in Fig. 4-11. The advancing ram, under high hydraulic pressure, pushes the metal through the forming die. The emerging piece takes the shape of the die opening and is conveyed along a take-away conveyor while cooling. As a large billet will produce great lengths of some sections, the product is sheared into standard stock lengths after cooling.

Extrusion dies for steel extrusions are generally made of hot work steels, tungsten carbide, refractory alloys or high tungsten alloys, as the process produces a die temperature during the run in excess

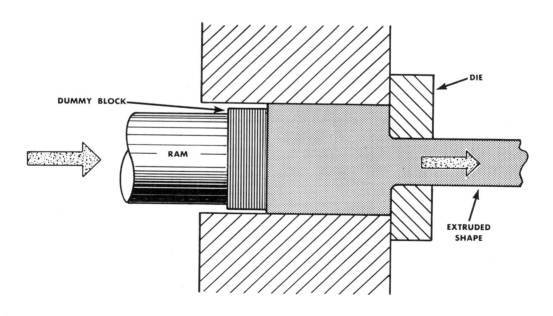

Fig. 4-11. Direct extrusion press.

Metallurgy

of 1000°F (538°C). The life of a die depends on the quality of the die material. The more expensive, high-quality die steels will produce up to 100 billet loads of product before replacing.

EFFECTS OF HOT WORKING

Although the primary object of hot working of steel is to shape it into a useful article, the working of metals by hot, plastic deformation deeply affects the structure and therefore, the physical properties of the metals, as illustrated in Fig. 4-12 to 4-15. Some of the resulting changes of internal structure are:

1. Refinement of grain
2. Better distribution of the constituent
3. Development of fiber
4. Improvement of soundness

Grain refinement. If it is possible to cause the grain structure of a metal to form a new set of crystals, and at the same time prevent any marked growth of the newly formed grains, the result is a materially refined grain or crystal structure of the metal. Grain size is of utmost importance in metallurgy. A rule-of-thumb is: *"the smaller the grain, the stronger the metal."*

The formation of new crystals from old is referred to as *recrystallization,* and the most practical way to bring about recrystallization and the refinement of grain structure is by means of *plastic deformation.* Plastic deformation resulting from forging, rolling and other mechanical operations will cause spontaneous recrystallization if the temperature of the metal is above the temperature of recrystallization.

The temperature of recrystallization is affected by the chemistry of the metal and by the amount of plastic deformation. But if hot working is carried out at relatively low temperatures and at such a rate of deformation that little time is allowed for recrystallization, the resulting structure will resemble, more or less, that of cold-worked metal and the metal will become work-hardened. However, at relatively high temperatures, recrystallization is very rapid and sudden quenching of the deformed metal from the forging temperature will not prevent recrystallization and some grain growth. Large grain size produced by a high finishing temperature and very slow cooling results in soft, weak metal. Factors which influence the grain size obtained from hot deformation are:

1. Initial grain size
2. Amount of deformation
3. Finishing temperature
4. Rate of cooling

Improvement of uniformity. The lack of uniformity, referred to as *segregation*, which is found in the ingot is little altered by hot working, owing to the great distance that the segregated particles would have to diffuse to improve this condition. However the dendritic, or tree-like crystal

Shaping and Forming Metals

Fig. 4-12. Microstructure of steel containing 0.30 to 0.35 percent carbon, in the as-cast condition. White ferrite, dark pearlitic structure, and magnified 100 times. Coarse, weak grain structure.

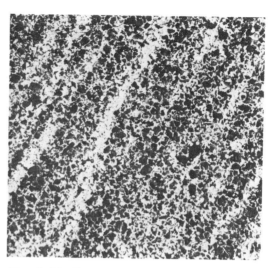

Fig. 4-14. Same steel as Fig. 4-12, magnified 100 times, aircraft drop forging. Unsatisfactory microstructure due to banded grain structure.

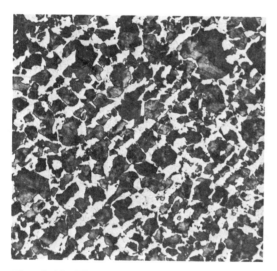

Fig. 4-13. The same steel as in Fig. 4-12, magnified 100 times. Section of drop-forged aircraft forging. Unsatisfactory microstructure due to coarse or banded grain structure.

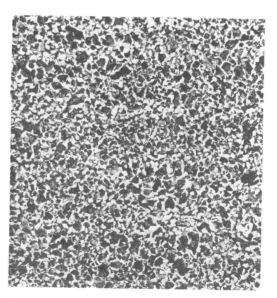

Fig. 4-15. Same steel as Fig. 4-12, magnified 100 times, aircraft drop forging. Desirable microstructure.

Metallurgy

segregation produced at the cooling of the ingot, is improved by hot working. The plastic deformation breaks up the segregated shell of impurities surrounding the grains and acts as an aid to the diffusion of the different elements. This diffusion is never completed and the different segregated areas become elongated into thin fibers called a *banded structure*. This banded structure is called the *fiber* of the forging and produces directional properties in the product. The slag particles and other insoluble materials are rolled out into the fibers, forming weak planes where splitting may occur. This fibrous type of structure is found in all plastically shaped steel.

Hot working, if properly carried out, will further improve metals by increasing the density, forcing the structural particles into more intimate contact and by closing up blowholes and other cavities; welding may occur if the local conditions are favorable.

COLD WORKING

The shaping of common metals by working at ordinary temperatures was practiced soon after the discovery of methods to smelt the metals from their ores. Early people made use of cold-working in the shaping and hardening of their metals. The process of hammering (forging) was used first; rolling and drawing processes were developed much later.

Cold rolling

Used largely to produce a fine finish for hot-rolled metals, cold rolling imparts a bright, smooth surface, insures accuracy in thickness dimension (gage) and appreciably increases the tensile strength of the finished product. Cold rolling is used in the production of sheets, strip steel, and bar-stock such as shafting, flat wire, etc.

Much of the sheet and strip that has been hot-rolled and pickled, as described earlier in this chapter, goes through a continuous cold reducing mill, Fig. 4-16. The coils of pickled hot-rolled sheets are started through special mills to make a product that is thinner, smoother and with a higher strength-to-weight ratio than can be made on a hot mill.

A modern five-stand tandem cold reducing mill may receive hot rolled sheet about the thickness of a half dollar and three-quarters of a mile long. Two minutes later that sheet will have been squeezed to the thickness of 0.2000" to 0.0060" (5.0 to 0.15 mm) and will be more than two miles long, Fig. 4-17.

Cold rolling, like hot rolling, may be carried out on a two-high or a four-high rolling mill, usually in a tandem, continuous set-up. Cold rolling is continued until the sheet reaches its desired thickness or un-

Shaping and Forming Metals

Fig. 4-16. Cold strip mill. (Inland Steel Company)

Fig. 4-17. Cold strip end-product. (Inland Steel Company)

Metallurgy

til it becomes too hard to continue the process. Steel hardened by cold rolling must be annealed to soften it to a workable condition. If annealing is carried out in open furnaces, oxide scale will form on the surface. Repickling will then be necessary to remove the oxide before continuing the rolling.

The alternative to open furnace annealing is continuous annealing in long, controlled-atmosphere furnaces that complete the softening without the formation of surface scale, or even discoloration. Continuous annealing furnaces are imposing structures, often longer than a city block and several stories high. The coils of hard cold-rolled steel are uncoiled and led up and down through towers in the annealer. Subjected to the heat within the towers, the steel is softened in preparation for further processing.

Temper mill. If the end-product desired is cold-rolled sheet and there is no intention of coating the product with zinc or tin, the annealed cold-rolled sheet goes to the *temper mills*, Fig. 4-18. A typical temper mill contains two stands of four-high mills and imparts the desired flatness and surface quality for sheet steel used for automobile bodies, refrigerators, washers and dryers and many other consumer goods, Fig. 4-19.

Fig. 4-18. Sheet steel enters temper mill through a single strand to restore stiffness lost in annealing process. (Bethlehem Steel Corp.)

Shaping and Forming Metals

Fig. 4-19. Inspection of cold-rolled sheet at tin mill. (Inland Steel Company)

Metallic Coating

Tinning. The food-processing industry uses millions of *tin cans* as containers for their products. The metal of a tin can is a coated steel product. The metal, tin, has been known for centuries as one of the few metals that will not react chemically with the acids of foods. The first true tin cans were used by Napoleon's quartermaster to provide edible meat for the army on maneuvers.

Tin plate is thin cold-rolled sheet steel, coated with a very thin layer of tin. Tin plate starts with a product called *black plate* which is actually a form of cold-rolled sheet which has been annealed, Fig. 4-20. Early annealing processes coated the steel with a black oxide, hence the name. Present methods produce an oxide-free product, although the term black plate is retained.

The annealed sheet steel goes to a temper mill for further reduction in thickness. The coating process originally used was a hot-dip process where sheets were fed through a bath of molten tin. Today most of the tin plate is produced by the electrolytic method.

There are three continuous *electrolytic tinning* processes in general use today involving the electrodeposition of tin on sheet steel. These are the halogen, alkaline and acid processes. The three processes are basically the same, differing only in the type of electrolyte used. The halogen process uses an electrolyte of tin halide salt. The sheet steel enters the pla-

61

Metallurgy

Fig. 4-20. Delivery end of annealing line. (American Iron and Steel Institute)

ter and is washed, scrubbed and cleaned in a dilute acid bath before it enters the plating section. In the plating section, the positively charged tin anodes release the element tin and it is deposited on the negatively charged sheet steel. The thickness of the coating may be controlled by the speed of the metal through the bath and the electrical current used. The coating may be as thin as 0.000015" (0.000038 mm) or as thick as required. The coating is uniformly applied to both sides.

Once the sheet steel is coated with tin, it passes through high frequency induction heaters which melt the tin to form an impervious and lustrous coating. This tin coating is quenched in water, given an electro-chemical treatment and then rinsed in water. Tin plate is shipped in coils and in boxes of cut lengths.

Terne plate is an important coated metal used in the automobile industry, Terne is a sheet steel, thicker than the steel used in tin-plate, which is coated with a lead-tin alloy. The coating is applied in a similar fashion to tin plate. Terne plate is extremely corrosion-resistant and is used in the production of automobile gasoline tanks and in a special type of laminated rolled tubing used for gasoline lines and hydraulic brake lines. The alloy resists the internal corrosive action of the fluids contained, and also the exterior corrosion of roadway contaminants.

Shaping and Forming Metals

Galvanizing. Certain end-uses of sheet steel require that the steel be coated with a corrosion-resistant coating. Coating with zinc, galvanizing, is the most common process in use. Large quantities of hot-rolled sheet and strip are galvanized. While there are two methods of applying the zinc coating, continuous hot-dip galvanizing is by far the most prevalent method used. Several million tons of galvanized sheet steel is produced yearly for applications where corrosion resistance is required.

The continuous hot-dip method consists of uncoiling, cleaning and moving the metal through a bath of molten zinc. Control of the speed of the sheet and the temperature of the zinc gives close control of the coating thickness and improved adherence of the coating. In the fabrication of home and farm utensils such as buckets, pails, etc., the product is fabricated and then hot-dipped as an assembly into the molten zinc bath. Zinc-coated auto body steel has also been used in many underbody parts that are subjected to extreme corrosion. The coated steel is treated to permit the forming of the part in the press without rupturing the continuity of the protective zinc coating. The zinc has a self-healing characteristic that makes galvanizing such an important process. Minor scratches into the surface of the coated steel heal by the formation of oxides of zinc which cover the scratch and continue the protection of the under metal.

Cold Drawing

The rolling of bars into rounds, squares, rectangles and other special shapes is a common mill procedure. These products of the hot-rolling mill would require considerable machining and grinding by users to get bars with required finish and dimensional accuracy. It was early discovered that if a hot-rolled bar with the usual surface oxides and varying dimensional tolerances were pulled through a die of smaller dimension, several beneficial actions occured. These were:

1. Bars of various cross-sections may be elongated and reduced in section to an extent not readily attainable by other means.

2. The surface of the bar becomes uniformly smoother and more highly polished than a hot-rolled bar.

3. The mechanical properties of the metal are changed. Hardness, stiffness, tensile strength and elastic limit will be increased, while ductility will be correspondingly decreased.

Metal preparation. Hot-rolled bars from the mill are generally coated with a heavy oxide scale and are relatively hard. The hardness must be reduced by a heat treatment such as annealing. In this operation, the steel is heated to a red hot temperature and allowed to cool slowly. Following annealing, the heavy oxide scale is removed by passing the bar through a hot, dilute sulfuric acid bath. The acid reacts with the unwanted oxide and dirt on the steel surface and the steel emerges from the bath with a clean surface.

The next step in the preparation is to coat the surface of the bars with lime to neutralize any trace of the acid from the bath. The lime adheres to the surface and functions as a lubricant for the drawing operation. The lime-coated bars are then

Metallurgy

baked to dry the coating and to remove any hydrogen absorbed by the steel during the acid bath (pickling) operation. The absorption of hydrogen into the steel can cause brittleness resulting in failure during the drawing operation.

Drawing operation. Cold-drawn round bars are made in one of two ways. If the dia is 3/4" (19 mm) or less, the bars are handled in coils. The coils are annealed, cleaned and recoiled after drawing. Steel over 3/4" (19 mm) is handled in straight lengths.

The die, as used in the drawing operation, is a very efficient tool. It has no moving parts and does not waste metal as in a cutting operation. Dies are made of cemented tungsten carbide, a material most able to withstand the abrasive action of the steel. For certain reductions in diameter there is a die angle where the drawing force is at a minimum. Fig. 4-21 shows a typical die. Die angles usually range up to 15 degrees, depending on the material of the die and the stock being drawn. There is a maximum reduction (approximately 45 percent) after which the steel will fail under the drawing force.

The end of the coil or bar must first be reduced in size to permit entrance into the die. This is done by a hammering operation called *swaging* or by grinding. The entering coil end is fitted to a capstan or power reel and the small diameter bar is cold drawn off one reel and recoiled on the other. Individual bar lengths are drawn through a succession of dies on a *draw bench*. The dies are arranged in a line with progressively smaller die openings. Large pieces are drawn through a single die at a time.

Shapes varying in size from the finest wire to those having a cross-sectional area of several square inches (mm^2) are

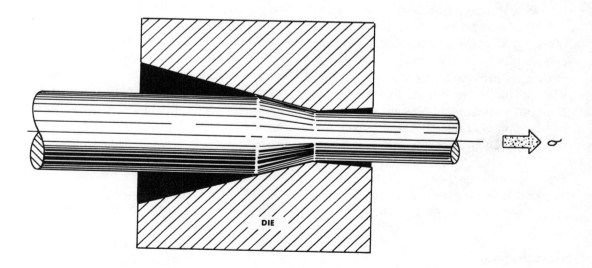

Fig. 4-21. Schematic of die drawing for reducing rod and wire diameter.

Shaping and Forming Metals

commonly drawn. Due to the difficulty of making dies and the small demand for any other form, the finer sizes are drawn only to a round cross-section. Larger sizes of round, square or irregular cross-section are drawn on a drawbench.

With wire, multiple die machines are used. In these machines the wire passes through one die, around a capstan, through a second die to a second capstan, etc. As many as twelve stations may be employed. The speed in multiple die machines reaches 10,000 feet (3048 m) per minute on fine wire.

Tubular Products

Steel pipe and tubing are among the most versatile and widely used products of the steel industry. In contrast to some other large tonnage steel products which may be die-drawn, shaped, machined or otherwise fabricated by the user, pipe and tubing are generally furnished in finished form for use. However, they may be reworked, reshaped or machined in a variety of ways.

Sizes of tubular products range from those used for hypodermic needles to those used for pipe lines. Two methods are used in the manufacture of tubular products. These are the *welded* and the *seamless processes*. Seamless processes comprise three general methods for producing tubular products, all without welded seams. These methods are:

Hot piercing. The blooms, billets or rounds are heated to a red hot temperature and an initial hole is then pierced through the metal in an axial direction. The rough tube is then rolled through a piercing mill over a mandrel to form the outer and inner diameters.

Hot extrusion. Pre-pierced, red-hot billets are forced through an orifice and simultaneously over an internal shaping mandrel. The outer surface of the tube acquires the size and shape of the die while the inner surface conforms to the size and shape of the mandrel.

Cupping and drawing. By this process, steel plates are hot-cupped and hot-drawn to the desired finished size of tubing.

Welding processes start with the forming of a tube from flat strip or *skelp* by passing the metal through a succession of shaped rollers. The open joint is welded shut by basically four methods: *Furnace* welding where the red hot edges of the joint are forge welded by the mechanical pressure of the rolls; *Electric Resistance* welding where the joint is welded by passing an electric current through a set of round pass welding rolls; *Electric Flash* welding where the joint surfaces are flashed to welding temperature by a flow of electric current, followed by application of mechanical upsetting pressure by a set of final rolls. The fourth method is a continuous, automatic fusion process where the joint is welded in a conventional electric arc or gas torch method.

Tubing formed by hot-working as well as electric welded tubing may be given a cold-working to produce a finer surface finish and more accurate dimensions. The cold drawing process for tubing is much the same as that used for solid stock. When there is no mandrel inside the tubing during the draw, the outer diameter is reduced but the wall thickness is un-

Metallurgy

changed and unfinished. When a mandrel is used during the drawing, both outside diameter and inside diameter can be accurately maintained.

Metal Forming

Steel in the form of cold-rolled or hot-rolled sheets or strip (narrow sheets) may be formed into many intricate shapes for consumer products by a number of processes. Some of the more common processes are described below.

Bending. Press-break forming is used to produce parts as shown in Fig. 4-22. These parts have simple curves in a parallel, linear configuration. Two important factors affecting bending are the minimum bend radius and springback. The minimum bend radius determines the limit of forming, after which the metal will crack. Springback is the elastic movement of a bent metal at the moment of release from the bending die. It results in a decrease of the bend angle. Compensation is accomplished by designing the dies to over-bend the part slightly so it will spring back to the desired angle on release.

Flat drawing. This process, referred to in the automobile industry as *stamping*, is used in the production of metal vessels and compound curved sheet metal parts. Flat sheet metal is placed between the two halves of a forming die. The lower half usually is the cavity, forming the outer surface of the part. The upper half is the punch which, when pressed into the cavity, causes the flat metal to flow into the die cavity and assume the required shape. This process is used in the production of such auto body parts as fenders and

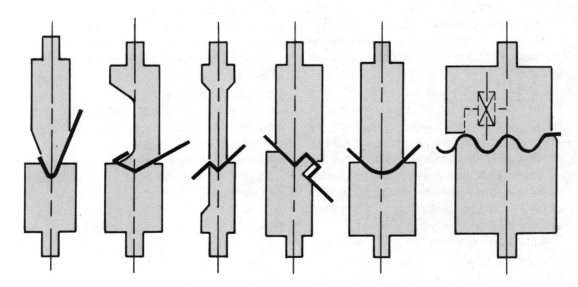

Fig. 4-22. Press break forming dies and parts: acute angle; obtuse goose-neck die; V-bend die; channel die; radius die; self-gaging corrugating die.

Shaping and Forming Metals

hoods, plumbing fixtures such as bathtubs, and many other metallic shapes.

To draw properly, metal must be properly conditioned. Annealing provides steel with the necessary ductility to prevent cracking. Flex-rolling is often used in an automobile stamping plant to condition coil stock and press blanks (preforms) for drawing. Lubrication is also required and is applied to the surface of the metal to reduce friction during drawing.

Flanging. Essentially a bending operation, this process is used to provide flanges on drawn parts such as automobile doors, and in the production of various brackets and reinforcing members in metal assemblies.

Upsetting. This is a hot or cold process used to form a head or to increase the cross-section of a round bar. Cold upsetting is used to form the heads on bolts, rivets and nails in high production automatic presses called upsetters.

Punching and blanking. These related processes consist mainly of a shearing operation using special shaped punches and dies. When round, square or irregular shaped holes are required in a sheet metal part, punches are designed into the die set to shear the holes. Blanking is a preliminary operation to press work when a specially shaped flat sheet is cut out of a coil of sheet metal. The blank is usually irregular at its outer edges to provide optimum gripping action by the grippers of the forming die. The blanking die set is es-

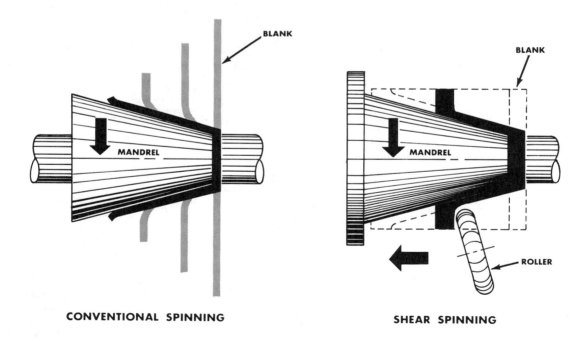

Fig. 4-23. Conventional metal spinning process (left) and shear spinning process (right).

67

Metallurgy

sentially a very large punch, shaped to the required contour of the blank. It shears the blank through the die.

Spinning. This process involves the forming of parts having rotational symmetry over a shaped mandrel by means of a forming tool or roller. Two spinning processes are shown in Fig. 4-23; conventional spinning, and shear spinning. Some of the major applications of this process are in the production of aircraft and aerospace components and missile cones.

Stretch forming. The stretch-forming process shown in Fig. 4-24 gives shallow shapes to large areas. Drawing alone does not strain the metal sufficiently to give it the desired permanent set. Stretching overcomes this problem by means of grip-

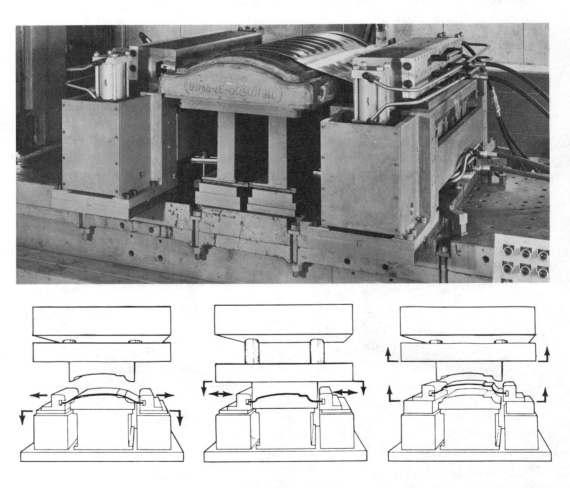

Fig. 4-24. Stretch forming. The pre-stretch blank hydraulically grips the sheet, stretches it and pulls it over the male die. The mating die presses the piece. Schematic of process below. (The Cyril Bath Company)

Shaping and Forming Metals

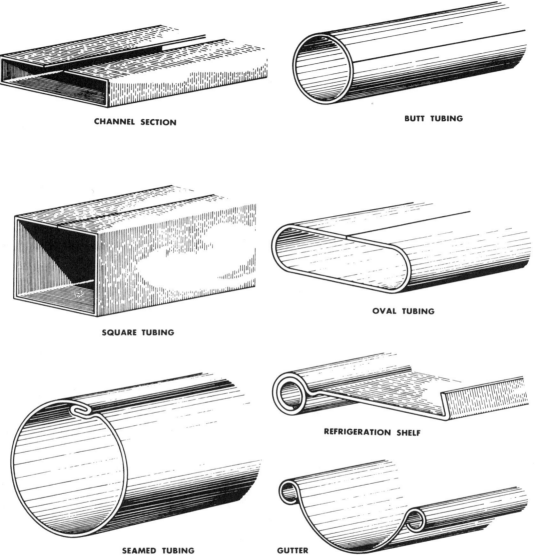

Fig. 4-25. Shapes produced by continuous roll forming.

ping the metal while at the same time forcing the forming punch into the metal. The resulting form has minimum springback.

Roll forming. Continuous roll-forming is a high-production cold-forming process used in the production of tubing, metal

Metallurgy

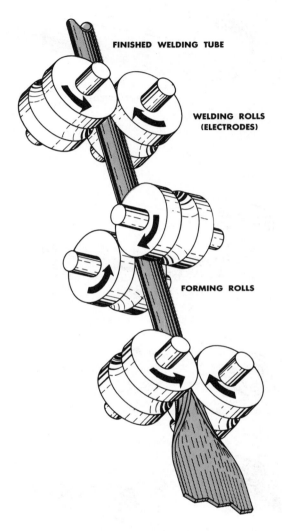

Fig. 4-26. Skelp (flat strip) being rolled into finished welding tube.

moldings, aluminum siding and roof gutters and other constant section shapes. Typical rolled sections are shown in Fig. 4-25.

The roll-forming machine consists of a series of matched rolls that progressively shape a flat strip of cold metal into the desired cross-sectional shape, Fig. 4-26. Because of the high cost of roll-forming machines and the special tooling for each section produced, roll-forming is not economical when production runs are less than 25,000 lineal feet (7500 m).

Shaping and Forming Metals

Effects of Cold Working

When metal is at room (ambient) temperature, it may be deformed permanently by the plastic flow behavior of the metal. The metallic crystals elongate by a process of shearing on weak planes, called *slip planes*. If metal is sufficiently cold-worked, all the available slip planes will be used up and further deformation will produce failure by rupture. Therefore, in shaping metals cold, plastic deformation must not be carried beyond a certain point or cracking is likely to result.

Cold-working operations may be divided into two broad classes:

1. Where the cold-working is carried out for the purpose of shaping the articles only; and where the hardening effect is not desired and must be removed at various stages of plastic shaping as well as from the finished article by annealing.

2. Where the object of cold-working is not only to obtain the required shape but to harden and strengthen the metal, and where the final annealing operation is omitted. High-strength wire for suspension cables, Fig. 4-27, is thus obtained.

In order to continue to shape metals by cold-working, they must be annealed at proper intervals. Otherwise deformation must be carried out at temperatures where annealing is simultaneous with hardening, as in hot-working of metals. The method adopted will depend on the particular metal, as well as on the shape of the end-product. See *recrystallization*, pages 82 and 83.

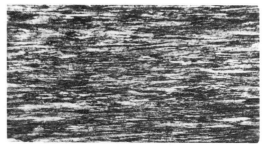

Fig. 4-27. Photograph of a 0.60 to 0.70 percent carbon steel in the form of hard drawn wire, magnified 100 times.

Overworked metal. After a metal has been subjected to severe cold-working operations, further cold deformation may cause a decrease in its strength. When this occurs, the metal is said to be *overworked*. This loss of strength is usually due to failure of the internal structure of the metal, such as shearing apart or splitting. Normally, the strength increases as the cold-working progresses, and continues as long as the continuity of the metal is maintained.

Cold work and corrosion. The corrosion rate of metals is accelerated by cold-working, particularly if such cold deformation is localized. For example, when strained metal comes in contact with unstrained metal, the strained metal is more subject to corrosion. The action of chemi-

Metallurgy

cals upon strained and unstrained metals is different.

Deformation of aggregates. Steel and many other alloys are structurally composed of mixed crystals having different chemical and physical characteristics. The weak, ductile constituents deform readily. The stronger and less plastic constituents resist deformation, but ultimately, if the deformation is severe enough, the harder constituents are also deformed, either by plastic flow or by rupture; that is, by breaking up into smaller parts and flowing with the more plastic and weaker constituents. Many alloys in the cast state are rather weak and brittle; but in hot-working, the brittle constituents surrounding the more ductile grains are broken up and are enveloped in a ductile phase, and the metal can be drastically worked.

Several authorities maintain that cold-working breaks up brittle constituents even more effectively than hot-working, and that the plastic constituents weld the internal structure together, even when working below the annealing temperature. Thus, the structure remains sound. Cold-working with annealing can be considered as a means of developing marked plasticity in metals that are structurally composed of an aggregate of plastic and brittle constituents.

REVIEW

1. In what physical form does all steel begin?

2. Why are soaking pits used in integrated steel mills?

3. What is the temperature maintained in the soaking pits? Why?

4. Why is the upper end of an ingot sheared off?

5. Why is the ingot run through a cogging mill?

6. What are the basic hot working processes?

7. What two purposes are served by hot working?

8. What type of shapes are produced by rolling?

9. What is the maximum thickness of sheet or strip?

10. How is the heavy oxide and mill scale removed from hot-rolled sheets? Describe the process.

11. What do the initials H.R.P.O. stand for?

12. Name the principal structural steel shapes.

13. Define hot working.

14. What are the two basic die shapes used in forging?

15. How does forging increase the strength of steel?

16. Name four basic types of forging machines.

17. Define the operation of extrusion.

18. Name four improvements result from the mechanical treatment of steel by hot working.

Shaping and Forming Metals

19. What are three physical results obtained by cold rolling?

20. Describe the process of making tin plate.

21. Describe terne plate and name some industrial products made from it.

22. Describe the galvanizing process and name some common articles made from galvanized iron.

23. Describe the drawing of wire.

24. Name the three seamless methods in the production of tubing.

25. Name the four methods used to produce welded tubing.

26. What products are made by flat drawing (stamping)?

27. What products are formed by the upset forging process?

28. What means is employed to restore plasticity to metal that is hardened by coldworking?

CHAPTER 5  PHYSICAL METALLURGY

Strength together with plasticity (ability to deform without rupture) is the combination of properties that makes metals of great importance in the mechanical and structural field.

Many substances, glass for instance, have high tensile strength, but lack plasticity when loaded rapidly, as in shock; these are known as brittle substances. Brittle substances may flow slowly or exhibit plasticity under tremendous pressure. If an attempt were made to use a brittle substance such as glass in the structure of a bridge, it would fail and fall into the river when it became unevenly loaded. However, various metals are used in the construction of bridges. The structurally used metals have the ability to deform without rupture, both elastically and plastically, when subjected to overloads due to uneven alignment. This deformation occurs without loss of strength; indeed, in many cases metals so deformed become stronger.

Although the materials used in machines and structures in most cases are subject to loads so low that the materials behave in a nearly elastic manner, a number of conditions may occur under which metals are subjected to loads so great that plastic action occurs. Therefore, the behavior of metals in the plastic range becomes of utmost importance. This is true during the shaping and fabricating operations performed upon metals in the creation of structures, or in finished structures that occasionally become overloaded. Materials that can deform plastically when overloaded are much safer structural materials than those having greater strength that are lacking this property.

In addition to strength and plasticity, metals exhibit many favorable characteristics, such as resistance to corrosion, electrical and heat conductivity, a good luster, etc. The characteristics of metals are due to two structural factors: first, the atoms of which the metallic state is com-

Physical Metallurgy

posed; and second, the way in which these atoms are arranged. A crude comparison can be made to a house, the character of which is in part determined by the type or quality of raw materials, such as bricks and boards (the atoms), and in part by the workmanship or assembly of the raw materials (the arrangement of the atoms.)

THE SOLID OR CRYSTALLINE STATE OF METALS

When a metal freezes (changes from the liquid to the solid state), it crystallizes. During the process of solidification, the atoms of the liquid metal arrange or group themselves into an orderly arrangement, forming a definite space pattern. That is, the atoms regiment themselves into groups similar to a group of soldiers on parade. The pattern formed by this orderly arrangement of the atoms is known as the *space lattice*, and it is made up of a series of points in space, giving birth to a geometrical structure of atom groups.

Although the atoms are too small to be seen by ordinary means, such as under the microscope, we may visualize a crystal as being made up of atoms, these atoms being arranged in space in some predetermined pattern or space lattice. The space lattice of the crystal state of metals has been determined by X-ray crystal analysis methods. By means of X-rays, the internal atomic arrangement of the crystal state may be studied, and a picture of the space lattice may be drawn as shown in Fig. 5-1.

There are a number of different patterns or space lattice groups into which the atoms of a crystal may arrange themselves. A very common space lattice is that of the *cubic* pattern which many of the common metals, such as copper, iron, aluminum,

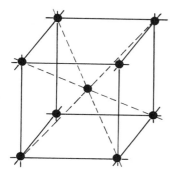
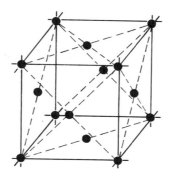
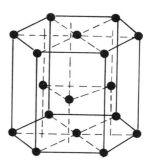

Fig. 5-1. Common crystal patterns.

Metallurgy

lead, nickel, molybdenum, and tungsten, form. Some of the metals forming the cubic pattern assume the *body-centered* cubic pattern, an atom at the center of the cube and an atom at each corner of the cube, Fig. 5-1 left. Other metals crystallize into the *face-centered* pattern having an atom in the center of the cube face and an atom at each corner, similar to Fig. 5-1 center. Also, some metals have the capacity of crystallizing in one form but, upon cooling, change to another form; some change from a face-centered to a body-centered pattern. Iron undergoes such a change of space lattice during heating and cooling, and within a temperature range that still maintains iron in a solid or crystal form without melting. When a metal undergoes a change from one crystal pattern to another, it is known as an *allotropic change*. When a metal undergoes an allotropic change, such as the change from a body-centered to a face-centered cubic lattice, this is accompanied by a marked change in the characteristics and properties of the metal involved in the change.

The change from a body-centered to a face-centered pattern completely rearranges the space lattice and atoms within the crystal, with the body-centered cube with 9 atoms in its pattern changing to a face-centered cube with 14 atoms in its pattern. Such a change of atomic arrangement brings about a complete change of characteristics.

While the cubic pattern is the most prominent with the common metals, there are many other space lattice arrangements. A rather common arrangement is the *hexagonal* pattern, known as the *close-packed* because of the number of atoms involved in this arrangement. Metals such as zinc, magnesium, cobalt, cadmium, antimony, and bismuth assume this arrangement upon freezing or crystallizing. This arrangement is illustrated in Fig. 5-1 right.

The properties of metals are dependent, in a large measure, upon the type of space lattice formed during solidification. In general, metals with the face-centered lattice lend themselves to a ductile, plastic, workable crystal state. Metals with the close-packed hexagonal lattice exhibit, in general, a lack of plasticity and lose their plastic nature rapidly upon shaping, such as in a cold-forming operation. Hexagonally latticed antimony and bismuth are brittle.

Cleavage and Slip Planes

Some of the most striking effects of the crystal pattern of atoms are the directional properties created by the planes formed by the orientation of the atoms. Between the rows of atoms lies the path of rupture, so the fracture or failure of a crystal occurs by the separation of the atoms along a path parallel to the atom orientation; such a path is called *cleavage plane*. Proof of crystallinity is obtained when a material, such as a metal, is ruptured, and the ruptured surfaces reveal facets or surfaces forming various planes, each plane being the path between the row of atoms formed by the crystals.

When an attempt is made to fracture metals of the plastic crystal state, instead of a brittle fracture taking place between the rows of atoms, the atom groups slide over one another along the planes formed by the pattern of atoms, causing a plastic

Physical Metallurgy

flow or *deformation* to take place without rupture. This behavior leads to the plastic nature of the more common metals and allows us to change the shape of the crystals without rupture in a forming operation, such as the bending of a piece of metal. The planes along which slips occur are known as the *slip planes*. The action of the plastic flow of the crystal state will be discussed more fully in the paragraph covering the deformation of metals.

As we have seen, the crystal state consists of a group of atoms arranged in some definite pattern or orientation. We will now see how a liquid which is cooled in a crucible or mold gives birth to the crystal state in which there are a great number of individual crystals called *grains*.

Manner of Crystallization

A crystal may be visualized as forming from a center of freezing, or nucleus, which is composed of a small group of atoms oriented into one of the common crystal patterns, Figs. 5-2 and 5-3.

In the process of solidification many of these nuclei spring up, each nucleus being a potential crystal and able to grow to form a crystal large enough to be seen with the unaided eye. As each nucleus is a growing crystal, and the atoms within it are all similarly oriented, it should be appreciated by the student that no nucleus within the freezing melt is likely to form with its planes or groups of atoms the same as those of any other nucleus. Thus, when the individual crystals have grown to the point where they have absorbed all of the liquid atoms and have come in contact with each other along their boundaries, they do not line up; i.e., their planes of atoms change direction in going from one crystal to another. This results in a solid state composed of a number of crystals of different orientation, and we have a crystal aggregate or *mixed crystal*. Each crystal is composed of a group of similarly oriented atoms, but on going from one crystal to the neighboring crystals the orientation changes.

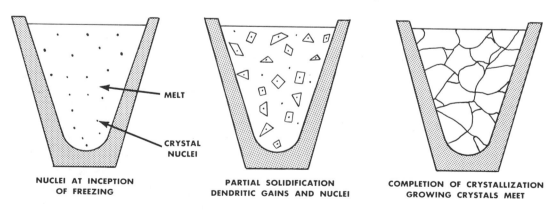

NUCLEI AT INCEPTION OF FREEZING

PARTIAL SOLIDIFICATION DENDRITIC GAINS AND NUCLEI

COMPLETION OF CRYSTALLIZATION GROWING CRYSTALS MEET

Fig. 5-2. Progressive freezing of a uniformly cooled melt.

Metallurgy

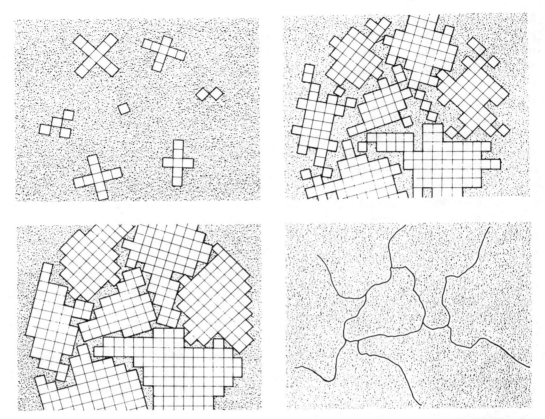

Fig. 5-3. Growth of crystalline grains. (Introduction to Physical Metallurgy by Walter Rosenhain)

The nature of the crystal border is still more or less of an unknown, but we may assume that it is an interlocking border line where the atoms of one crystal change orientation from the atoms of another crystal. It may be that there are some left-over atoms along this border line separating the differently oriented crystals; such unattached atoms act as a noncrystalline cement between the various crystals. This condition may account for the greater strength of the crystal boundaries as compared to the strength of the individual crystals, and for many of the actions that take place at the crystal boundaries.

In order to help in the visualization of the change from a molten state to the crystal aggregate state, Fig. 5-4 may be studied. A crystal nucleus forms as shown in (A) and then proceeds to send out shoots or axes of solidification as shown

Physical Metallurgy

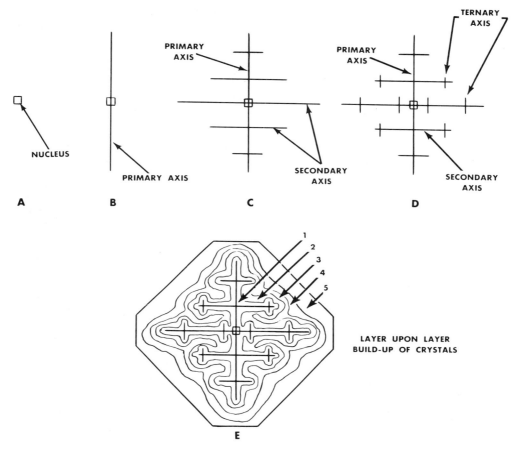

Fig. 5-4. *Steps in formation of a crystal.*

in (B), (C), and (D), forming the skeleton of a crystal in much the same way that frost patterns form. Atoms then attach themselves to the axes of the growing crystal from the melt in progressive layers, finally filling up these axes, and thus forming a completed solid, as shown in (E).

The student may observe the nature of crystal growth in the freezing of water which frequently shows a surface pattern resembling a tree (trunk, branches, etc.). Also, the crystalline nature of metals is apparent to the eye in the study of the surface of galvanized iron (zinc or iron). This treelike freezing pattern is called *dendritic*, taken from the Greek word meaning treelike, and the term is frequently used in referring to a structure that has a treelike appearance.

A crystal may have almost any external

79

Metallurgy

shape, which is controlled by the conditions leading to its formation. The most important factor affecting the external shape of any crystal is the influence of the other growing crystals which surround it. However, regardless of the external shape of the crystal, the internal arrangement of the atom is the same. The perfect external shape would be a cube for a cubic pattern, and a hexagon for a hexagonal orientation, etc. The crystals found in all commercial metals and alloys are commonly called *grains* because of this variation in their external shape. A grain is, therefore, a crystal with almost any external shape, but with an internal atomic structure based upon the space lattice with which it began.

Single vs. Polycrystalline State

In commercial metals and alloys, the control of crystal size is considered very important. The reason for this is revealed in the properties exhibited by a fine-grained crystal aggregate, as compared with the properties of a coarse or larger-grained metal. A fine-grained metal is stronger and in many applications proves more satisfactory than a coarse-grained metal.

A single crystal exhibits a strength characteristic of its resistance to slip or cleavage along any of its atomic planes, whereas a polycrystalline or many-crystal aggregate has no smooth, straight line throughout its section for slip or fracture, due to the change of orientation taking place at each grain or crystal boundary. Thus, a fracture travelling through a polycrystalline aggregate changes direction when going from one crystal to the next. The effect of this is similar to the break in the masonry line in a brick wall. Each crystal keys to its neighbor and builds up a resistance to slip or failure. On the other hand, a coarse grain size may exhibit more plasticity than a finer-grained aggregate. The coarse grains have longer slip planes, and therefore greater capacity to slip during plastic deformation than the smaller and shorter slip planes of a finer-grained state.

DEFORMATION OF METALS

Knowing that metals are composed of many crystals or grains, and that each crystal in turn is composed of a great many atoms all arranged in accordance with some pattern, how can we visualize the plastic flow that must take place when metals are deformed during a bending operation, or during the drawing out of a piece of metal to form a long wire? This deformation may be visualized as *shearing*. When a metal is subjected to a load exceeding its elastic limit, the crystals of the metal elongate by an action of slipping or shearing which takes place within the

Physical Metallurgy

crystals and between adjacent crystals. This action is similar to that which takes place when a deck of cards is given a shove, and sliding or shearing occurs between the individual cards. The mechanism of plastic deformation is illustrated in Figs. 5-5, 5-6, and 5-7.

Fig. 5-5 is a photomicrograph of pure iron in the unstrained condition, and Fig. 5-6 shows the same iron after the occurrence of slight plastic flow. The dark lines within the grains in Fig. 5-6 indicate the planes on which plastic deformation took place. If deformation of the metal is continued, the crystals become noticeably elongated, as illustrated in Fig. 5-7, a photomicrograph of a section of brass that was severely deformed during plastic shaping.

This plastic flow of the metal, resulting

Fig. 5-6. Commercially pure iron after slight plastic deformation. (Introduction to Physical Metallurgy by Walter Rosenhain)

Fig. 5-7. Cartridge brass, severely deformed, magnified 100 times.

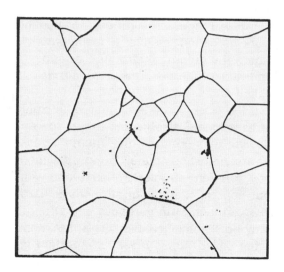

Fig. 5-5. Commercially pure iron before deformation. (Introduction to Physical Metallurgy by Walter Rosenhain)

in permanent deformation of the crystals, is accompanied by marked changes in the physical properties of the metal. The ten-

Metallurgy

sile strength, yield point, and hardness are increased, but not the scratch hardness, or difficulty of cutting, as in machine operations in a lathe. The stiffness remains about the same, though in some cases it may increase as much as 3 percent. With the increase in hardness and strength, the plasticity or formability is reduced. If deformation of the crystals is continued, the metal becomes brittle. This deformed state may retain high residual stresses and may corrode rapidly when exposed to corrosive atmospheres.

Work Hardening

Many attempts have been made to explain why metals become harder and stronger when permanently deformed. Two common theories are the *amorphous cement theory* advanced by Rosenhain, and the *slip interference theory*, by Jeffries.

The amorphous cement theory explains that work hardening is caused by the formation of a hard, noncrystalline cement (amorphous, or without form), due to the rubbing together of the weak planes within the crystals; and that this cement acts as a binder, making the weak planes stronger and therefore increasing the strength of the metal as a whole.

The slip interference theory explains that work hardening is caused by the action of local disorganization of the crystal structure, not necessarily until it is amorphous in nature, but having more crystal fragmentation. These crystal fragments, according to this theory, act as mechanical keys along the weak planes of the crystals, keying them together and thereby making them stronger. If we add to these effects the end-resistance to slipping which is offered by the differently oriented crystals, perhaps we will have some reasonable explanation of the effects of work hardening of metals.

Recrystallization

The deformed or work-hardened condition found in metals that have been stressed beyond their elastic limit is an abnormal one. If the temperature of the work-hardened metal is raised above normal, the deformation begins to disappear and the metal returns to the normal condition of structure and properties. This remarkable change occurs through the recrystallization and grain growth of structure that takes place when work-hardened metals are heated to within certain temperature ranges. The process of heating work-hardened metals to temperatures where the deformed structure disappears and the properties return to normal, is called *annealing*.

If the work-hardened metal in its highly stressed state is subjected to very low annealing temperatures, the internal stresses are almost completely removed without any apparent structural change and without appreciable loss of strength or hardness. In fact, both hardness and strength may be slightly increased. This treatment, often called *stress relief*, is important industrially.

If work-hardened metals are heated to higher temperatures, the return to the normal, soft, ductile state occurs rapidly. This change is accompanied by a marked change in structure. Such a change is one of *recrystallization;* that is, the deformed grains crystallize over again without any melting. The action is one of new crystal nuclei forming around the boundaries and throughout the old deformed crystals, in a way similar to the formation of nuclei in a melt; and these nuclei grow at the expense of the old grains until they completely obliterate them.

The mechanism of recrystallization is illustrated in several photomicrographs, as follows. Fig 5-7 is a photomicrograph of cold deformed brass showing that marked plastic flow has taken place. Fig. 5-8 shows the same specimen after heating slightly above room temperature but into the recrystallization range for this specimen, new crystals having formed in and around the old crystals. Fig. 5-9 shows the same specimen after heating to a temperature where complete recrystallization has occurred. The metal is now in excellent condition with a fine, strong, and highly plastic structure. The original grain structure of the metal may have been coarse, with envelopes of impurities around the grains and having the relative brittleness of the cast state, along with the internal stresses and distorted condition of the work-hardened state. This has all disappeared, resulting in the strong, tough, uniform, and stress-free condition which makes a forging so much superior to a casting for many industrial applications.

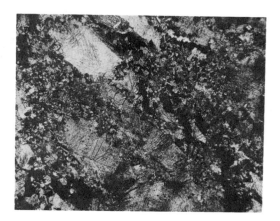

Fig. 5-8. Severely deformed cartridge brass as shown in Fig. 5-7, partly recrystallized by heating. Magnified 100 times.

Fig. 5-9. The same samples as in Fig. 5-8, completely recrystallized by heating. Magnified 100 times.

Grain Growth

If the temperature of the work-hardened metal is increased above that of the lowest temperature of recrystallization, the newborn grains grow rapidly in size by absorbing each other, as illustrated in Fig. 5-10. This action is known as *grain growth*

Metallurgy

and results in fewer but larger grains. The amount of grain growth occurring during an annealing operation depends to a great degree upon the maximum temperature and the time spent in the annealing operation. Large grains are often wanted, particularly if cold working is to be resumed after annealing, because a large crystal size does not harden as rapidly in subsequent deformation as does fine-grained metal.

The temperature at which metal recrystallizes after cold deformation depends upon the composition of the metal and the amount of *cold deformation*. The greater the amount of cold deformation, the lower the recrystallization temperature. Lead recrystallizes at room temperatures, and therefore cannot be work hardened by cold deformation.

The greater the amount of cold deformation, the finer the grain size will be after the annealing treatment. However, sometimes slight deformation (for example, 5 percent with iron) results in extremely large grain after an annealing operation.

Germination

The heating of cold-worked metal to its recrystallization temperature usually results in a very fine grain or crystal size, due to the large number of nuclei or new centers of crystal formation. The number of nuclei that form seems to govern the size of the individual grains. Also, the greater the amount of cold work or deformation before annealing, the finer and more numerous will be the grains after annealing.

However, a slight amount of cold deformation before annealing (such as 5 percent with iron) results in the forming of only a few nuclei or centers and they grow at a rapid rate, resulting in very coarse grains. This extreme growth of grain is sometimes called *germination*.

Other conditions also contribute to extreme grain or crystal growth. Coarse or large grains weaken the metal and possibly ruin it for commercial use, as for certain shaping operations such as deep drawing in power presses. Extremely large grains may cause the metal to rupture in deep-drawing operations, or may result in such a rough surface as to make the product undesirable.

Fig. 5-10. Cold worked brass, as shown in Fig. 5-7, annealed at 1475° F (about 800° C). Magnified 100 times.

Cold Crystallization

Grain growth, which is commonly termed *crystallization*, cannot take place in steel or iron while cold. The old belief that metal sections often failed in service due to cold crystallization has been discarded. This belief developed from the fact that parts breaking in service showed at the break a relatively coarse and open grain. The assumption was that this coarse grain developed in service. The fact is that the part had a relatively coarse grain, due to poor forging or heat-treating practice, *before* it was put to work in the machine. This coarse grained part, when subjected to severe stresses in service was incapable of standing up under them and failure resulted. In other words, the weak, coarse grain did not form in service, but was in the steel before it had seen any service at all.

It is true that grain growth will take place below the critical range, but only with free ferrite or iron, and this free ferrite or iron must be severely distorted and heated to a red heat before any marked grain growth will take place. No grain growth will occur at room temperature.

In chains and other machine parts that are abused and become severely strained in service, there is always danger of increasing the brittleness of the material, but this brittleness is due to the cold working effect produced by severe abuse of these parts. If a coarse grain is present in these parts, failure due to extreme brittleness is certain to result. Where the chain links and other parts have not yet failed because of severe service, the brittleness can be relieved by annealing. This annealing relieves the strains and stresses, putting the parts back in shape for further service. However, if the parts have been so strained as to cause a slight failure, such as *surface cracking*, annealing should be prohibited, as it will not strengthen nor weld broken parts.

SUMMARY

Grain size control during cold working is one of the major problems encountered in industrial uses of commercially pure metals or their alloys. The only way in which a commercially pure metal may have its properties altered is by means of cold deformation and control of the annealing operation that may follow the cold working. Therefore, a careful study should be made of all the characteristics of commercially pure metals, particularly those involved in the cold working of these metals. This knowledge can be applied also to the behavior of the more complicated alloys.

From an industrial point of view, commercially pure metals find their greatest use in the electrical industry as conductors, and in the fields where corrosion-re-

Metallurgy

sistant materials are needed. With the exception of some special stainless alloys, such as stainless steel and Monel metal, pure metals are much more resistant to atmospheric corrosion than their alloys. Such metals as copper, aluminum, nickel, chromium, and cadmium are used either alone, or as a surface coating to impede corrosion of a base metal.

REVIEW

1. What combination of properties makes metals of great importance in the mechanical and structural field?

2. The characteristics of metals are due to what two structural factors?

3. What is a space lattice?

4. How has the space lattice of the crystal state of metals been determined?

5. What is an allotropic change?

6. How does the type of space lattice formed during solidification affect the properties of metals?

7. What is meant by a cleavage plane?

8. Describe briefly the manner of crystallization of a metal.

9. What is meant by the term dendritic?

10. Why does a fine-grained metal, in many applications, prove more satisfactory than a coarse-grained metal?

11. Give a brief description of what happens when metals are deformed during a bending operation.

12. Why do metals become harder and stronger when they are permanently deformed?

13. What happens when work-hardened metals are heated to within certain temperature ranges?

14. Upon what does the temperature at which a metal recrystallizes after cold deformation depend?

15. What is germination?

Chapter 6

Mechanical Properties of Metals

A thorough knowledge of the physical characteristics of metals is required by people in the skilled trades to enable them to do their jobs efficiently and effectively. The designer, toolmaker, die maker, machinist, welder and mechanic must know the specific properties of the metals and alloys used in industry.

The knowledge of metals and their use goes back in history thousands of years, yet the use of iron and steel in large quantities is relatively recent. The American Declaration of Independence not only signaled the beginning of a new nation but was almost coincident to the start of the massive use of iron and steel. This industrial era is called the *"Iron Age"* by contemporary historians.

The reason for the increased use of metals and particularly of iron and steel products is found in their special physical properties. The most important of these properties is strength, or the ability to support weight without bending or breaking. Combined with toughness, the ability to bend rather than break under a sudden blow, these characteristics made steel the prime construction material, Fig. 6-1.

Other properties that led to increased industrial use were the resistance to atmospheric deterioration, plasticity, and the ability to be formed into desired shapes. In addition, many metals have the ability to conduct heat and electric current and to be magnetized.

Metals can be cast into varied and intricate shapes weighing from a few ounces to many tons. Steel has a special property of being weldable, which has given rise to the skill of welding, both for fabrication and repair, Fig. 6-2.

Another important metallic property in this age of impending power and material shortages is the reclamation or recycling of scrap metal. Obsolete and worn-out automobiles, ships, bridges and other

Metallurgy

Fig. 6-1. Modern skyscraper built with steel framework. (American Iron & Steel Institute)

Fig. 6-2. Using a coated steel rod to arc weld a special flange on the end of a pipe. Note the head shield, gloves and apron worn by the worker. (Fischer Controls Co.)

Mechanical Properties of Metals

Fig. 6-3. Steel can be used over and over. This scrap is on its way to the open hearth furnaces to be remelted to new steel. Special rail cars haul the loaded cars to furnace doors. It is not unusual for half of the charge in an open hearth to be in the form of scrap. (Inland Steel Company)

steel structures can be cut up, and remelted and finally worked into new industrial products, Fig. 6-3.

A knowledge of the properties of specific metals or alloys enables us to determine whether or not such materials are suitable for certain definite uses. See Table 6-1. This knowledge will also permit use of special heat and mechanical treatments of these materials in order to modify them for special uses such as forging and hardening a cutting tool.

Metallurgy

TABLE 6-1. CARBON CONTENT OF CARBON STEELS FOR DIFFERENT USES.

CARBON RANGE PERCENT	USES OF CARBON STEEL
0.05–0.12	Chain, stampings, rivets, nails, wire, pipe, welding stock, where very soft, plastic steel is needed
0.10–0.20	Very soft, tough steel. Structural steels, machine parts. For case-hardened machine parts, screws
0.20–0.30	Better grade of machine and structural steel. Gears, shafting, bars, bases, levers, etc.
0.30–0.40	Responds to heat treatment. Connecting rods, shafting, crane hooks, machine parts, axles
0.40–0.50	Crankshafts, gears, axles, shafts, and heat-treated machine parts
0.60–0.70	Low-carbon tool steel, used where a keen edge is not necessary, but where shock strength is wanted. Drop hammer dies, set screws, locomotive tires, screw drivers
0.70–0.80	Tough and hard steel. Anvil faces, band saws, hammers, wrenches, cable wire, etc.
0.80–0.90	Punches for metal, rock drills, shear blades, cold chisels, rivet sets, and many hand tools
0.90–1.00	Used for hardness and high tensile strength, springs, high tensile wire, knives, axes, dies for all purposes
1.00–1.10	Drills, taps, milling cutters, knives, etc.
1.10–1.20	Used for all tools where hardness is a prime consideration; for example, ball bearings, cold-cutting dies, drills, wood-working tools, lathe tools, etc.
1.20–1.30	Files, reamers, knives, tools for cutting brass and wood
1.25–1.40	Used where a keen cutting edge is necessary; razors, saws, instruments and machine parts where maximum resistance to wear is needed. Boring and finishing tools

CARBON STEELS

Although all steels contain carbon, the terms *carbon steel* and *plain carbon steel* are used to distinguish a steel to which no special alloying element, such as nickel, tungsten or chromium, has been added in appreciable amounts. In carbon steels, as well as in special or *alloy steel*, the constituent *carbon* more than any other constituent determines the properties and uses of steel, Table 6-1.

The carbon content of steel may vary from a few hundredths of one percent (0.008 percent) up to nearly two percent carbon. The lower percentages are in the

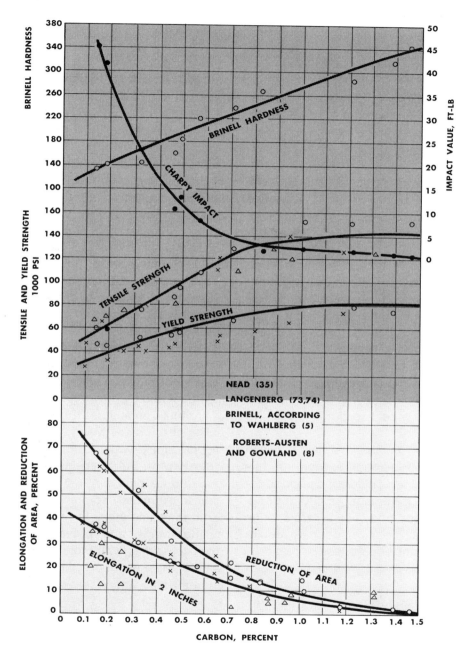

Fig. 6-4. Effects of carbon on the properties of hot-worked steel. (The Alloys of Iron and Carbon, Vol. II, by Cisco, The Engineering Foundation)

Metallurgy

range of commercially pure iron. These lower percentages of carbon steel, called *low-carbon steel* exhibit properties of nearly pure iron, that is, high ductility and inability to be hardened.

The upper limit of carbon in steel is shown in some charts as 1.7 percent. At this high percentage the steel is extremely brittle. The practical upper limit of carbon content as determined by experience in making and using cutting tools is 1.4 percent, Fig. 6-4. The object of increasing the carbon content of steel is to enhance its normal (low-carbon) qualities of strength and hardness.

Steel was early catagorized into three classes relating to its carbon content. While not an accurate definition by today's standards, steels having a carbon content of below 0.30 percent were classified as *low carbon*, Table 6-1. Low-carbon steel forms the bulk of steel tonnage produced as it has moderate strength, ductility and ease of machining. Machine steel and structural steel are prime usages of low-carbon steel.

Medium-carbon steel ranges between 0.30 and 0.70 percent carbon. In this range the steel has sufficient carbon to respond to heat treatment. As shown in Table 6-1, articles made of steel in this range are mainly machine parts that may require a limited amount of hardness or a special structure obtainable by other heat treatments.

High-carbon steels are those with more than 0.70 percent carbon content, Table 6-1. These steels constitute a small percentage of the steel tonnage produced because the articles made from high-carbon steels are mainly tools used in shaping and cutting low-carbon steels and other metals.

Minor Constituents in Steel

Carbon steel is an alloy metal consisting of iron and carbon. Carbon has a unique effect on the structure of iron, changing it from a simple, soft metal to a very complex alloy and producing an increase in hardness and strength while reducing its plasticity. However, there is no pure iron-carbon alloy as all steel contains at least four other alloys. Two of these are silicon and manganese which are added to the steel to counteract the effects of the other two undesirable elements—sulfur and phosphorus. These and other impurities are discussed below.

Manganese. This element is essential in a good steel. It promotes soundness of steel ingot castings through its deoxidizing effect and by preventing the formation of harmful iron sulfides it promotes forgeability of the steel. Manganese content in regular steels varies from about 0.30 to 0.80 percent, but in special alloy steels it may run as high as 25.0 percent. Manganese chemically combines with any sulfur in the liquid steel, forming manganese sulfide (MnS), and prevents the sulfur from combining with the iron to form iron sulfide. Steel with iron sulfide present becomes brittle and weak when hot and is known as *hot-short* steel. Any excess manganese (over the amount used to combine with the sulfur) combines with the carbon available in the steel to form manganese carbide (Mn_3C). This carbide associates with the iron carbide (Fe_3C) in

Mechanical Properties of Metals

cementite, increasing the hardness and strength of the alloy, but lowering the plasticity.

Silicon. This element is added to steel as it is poured into the ladle after refining to improve the soundness of the steel by opposing the formation of blowholes as it combines with dissolved oxygen to form silicon dioxide (common sand). This rises to the top of the ingot and is removed in the cropping operation.

The silicon content of carbon steels varies from 0.05 to 0.30 percent while special alloy steels may contain over 2.25 percent silicon. Silicon, in excess of amounts needed to deoxidize the steel, dissolves in the iron and under 0.30 percent has little influence on either the structural characteristics or the mechanical properties of the steel.

Sulfur Any sulfur content in steel is undesirable. It may vary from a trace to 0.30 percent. In the usual types of steel, the sulfur content is held below 0.06 percent. Sulfur should always be combined with manganese, for which it has a great affinity, to form manganese sulfide (MnS). Manganese sulfide is distributed through the steel as small inclusions which make the steel brittle but easily machined. Such steel is often used to make parts that do not require good mechanical properties and can be made on automatic screw machines. Hence, this steel is referred to as *screw machine steel*. This steel also has poor weldability.

Phosphorus. This element is also undesirable in steel. The phosphorus content in commercial steels varies from a trace to 0.05 percent. Phosphorus weakens steel by its tendency to segregate in the last portion of the ingot to solidify, producing a heterogeneous, dendritic structure. This segregation may cause *banding* of the steel in subsequent hot-working operations. Steel with excess amounts of phosphorus in solution is known as *cold-short* steel. While it has free-machining properties for easy forming in screw machines, the products made from cold-short steel are coarse grained and weak.

Copper. In small amounts copper is sometimes added to steel to improve resistance to atmospheric corrosion. The presence of copper in steel is detrimental to forge welding but not to arc or acetylene welding.

Lead. This metal greatly improves the machinability of steels. The lead is insoluble in steel and is randomly distributed throughout the steel in small globules which smear over the tool face when steel is being machined. This action reduces cutting friction and the steel is used mainly in screw-machine operations.

Oxides. Any oxides of iron that may become trapped in the solid steel are formed during steel-making by the additions of manganese (MnO), silicon (SiO_2), aluminum (Al_2O_3) and other metals. Iron oxide (FeO) is also present. These impurities are referred to as *dirt* in the steel and usually rise in the cooling ingot and are removed by the cropping operation. Oxides that remain in the steel are harmful because they are mixed with the steel in a mechanical manner, being entangled throughout the structure of the steel. These oxide particles break up the continuity of the structure and impart directional properties in forged steel. They also form points of weakness and may start a fracture.

Metallurgy

Gaseous impurities. Such gases as nitrogen, hydrogen, carbon monoxide and oxygen are always present in varying amounts in steel. These gases may occur as bubbles in the steel, or may be dissolved, or combined. The effects of gases, in the main, is to cause a decrease in plasticity; and in the case of hydrogen, to embrittle the steel. The more free the steel is from any gaseous impurities, the higher the quality and the more likely it is to be satisfactory.

Selection of Carbon Steels

Although the quality of steel is not determined by carbon content solely, the carbon content of the steel largely dictates the use to which the steel is to be placed. The quality of any carbon steel is largely determined by the exacting controls used by the steel maker. These are the metallurgical controls used during the steel making operations and the testing and inspection given to the semi-finished and finished steels. Close control of the amount of carbon and the control of impurities, oxides and slag, all constitute factors that contribute to good steel. In the selection of a carbon steel for any specific use, both the carbon content and the amount of minor constituents will influence the choice of a steel. In order that the reader may have some practical suggestions to follow in the selection of a steel, Table 6-1 should be studied. It will serve as an indicator of the influence of the percentage of carbon has in the type of product produced. This should help one in the choice of a steel from which to make similar products.

Ingot iron. This is a very low-carbon iron alloy. This alloy differs from wrought iron in that it does not contain an appreciable amount of slag. It is cast into ingots, similar to steel, and then rolled into plates, sheets, shapes or bars. It is used where a ductile, rust-resistant metal is required, especially for tanks, boilers, enameled ware and for galvanized sheets. It is very ductile and can be formed cold by drawing or spinning.

A very low-carbon iron, *Plastiron*, is used for molds and dies which are produced by the hobbing method. This method forces a hardened hob or pattern of the part to be produced into the soft steel to form a close dimensioned cavity. The iron is very plastic under the hob and can then be carburized and hardened for use.

Tin plate. The material from which *tin cans* are made is tin plate. The ordinary tin can is made from thin sheet steel which has been coated on both sides with a very thin layer of tin. The process of making tin plate is described in Chapter 4.

Wire, rivets and nails. These products are made from steel containing 0.10 percent or less carbon. Most of this steel is made in the Bessemer process. The chief requirements of this steel is that it be soft enough to form easily and sufficiently ductile to withstand the deformation of the cold heading process without cracking.

Structural Steel. This steel comes fabricated in such shapes as the S-beams, channels, angles, etc. used in buildings and industrial equipment. It is a common steel containing less than 0.15 percent carbon. It is produced and used in a hot-rolled condition and has a rather low strength when compared to heat-treated steels of higher carbon content. In general, the requirements for this type of steel are strength, ductility and ease of machining. Under this classification of steel may be included plates and various shapes referred to as machine steel.

Forging steel. Steel for forging, Fig. 6-4, may contain from 0.25 to 0.65 percent carbon. This steel is usually rolled into billets or rods of convenient size and then cut and forged into desired shapes. Square and rectangular forging billets may be distinguished from regular square or rectangular machine steel by the radius at the corners. This radius prevents folding and accidental forming of seams during the forging process.

Screw machine steel. This is the material from which such parts are made as screws and spacers, parts which do not require special mechanical properties. This steel has, in fact, poor mechanical properties, but has the property of *free cutting* and can be used profitably on automatic screw machines, hence its name. These steels have a low-carbon content, less than 0.20 percent, Table 6-1, and contain manganese sulfide inclusions which make the steel brittle but easily machined as the chips break into small particles. This steel is usually made in a Bessemer furnace. It is also compounded by adding sulfur to a low-carbon steel. This steel is refered to as *resulfured steel.* A second type of screw machine steel is *leaded steel* described earlier in this chapter.

Tool steels. Practically all carbon tool steel is made in electric furnaces and great care is exercised to keep impurities, such as phosphorus and sulfur, and non-metallic inclusions (dirt) at a minimum. These steels may contain 0.65 to 1.40 percent carbon, Table 6-1, depending upon the use to which they are put. Tools such as hammers and stamping dies are made from steels of the lower carbon range as they cannot be brittle nor need they be exceptionally hard. Tools such as razors, tool bits and drills are made from the higher carbon ranges. Most tool steels, both carbon and alloy steels, are sold under manufacturers' trade names.

MECHANICAL PROPERTIES

Generally we are very much concerned with the mechanical properties of metals. The mechanical properties such as hardness, tensile strength, compressive strength, ductility, etc., are those measured by mechanical methods discussed in the next chapter.

A knowledge of the properties of both ferrous amd nonferrous metals enables one not only to determine whether or not such materials are suitable for certain definite uses, but also to modify the thermal and mechanical treatments of materials in order to obtain a conditioned metal in the desired form.

Strength with plasticity is the most im-

Metallurgy

portant combination of properties a metal can possess. *Strength* is "the ability of a material to resist deformation". *Plasticity* is "the ability to be deformed without breaking". Metals, such as steel, have this combination of properties and are used in critical parts of structures, machine tools and motor vehicles that may be momentarily overloaded in service. For example, should a member of a bridge structure become overloaded, plasticity allows the overloaded member to flow, so that the load becomes redistributed to other parts of the bridge structure.

A number of strength values for a metal must be known to fully understand the strength characteristics of the metal. Among these strength values are: *tensile strength, compressive strength, fatigue strength* and *yield strength*.

Tensile strength is defined as the "maximum load in tension which a material will withstand prior to fracture". This is the value most commonly given for the strength of a material and is given in pounds per square inch (psi) or kiloNewtons per square meter (kN/m^2).

Compressive strength is defined as the "maximum load in compression which a material will withstand prior to a predetermined amount of deformation." The compressive strengths of both cast iron and concrete are greater than their tensile strengths. Due to their brittle nature, an overload will result in shattering. Steel and most other metals have tensile strengths in excess of compressive strengths.

Fatigue strength is the "maximum load a material can withstand without failure during a large number of reversals of load". For example: a rotating shaft which supports a weight, such as an automobile axle, has tensile forces along the lower portion of the shaft, due to bending, and compressive forces along the upper portion resisting the bending. As the shaft is rotated, there is a repeated cyclic change in tensile and compressive stresses around the axle. Non-rotating structures such as aircraft wings and other structures subject to rapidly fluctuating loads must have fatigue strength values considered during their design.

Yield strength is the "maximum load at which a material exhibits a specific deformation". Most engineering calculations for structures are based on yield strength values rather than on tensile strength values. Yield strength values vary from about 50 percent of the tensile strength for copper to about 85 percent for cold-drawn mild steel.

Elasticity. Since loading a material causes a change in form, *elasticity* is the "ability of a material to return to its original form after removal of the load". Theoretically, the elastic limit of a material is the limit to which a material can be loaded yet recover its original form after removal of the load. In reality, metals are not entirely elastic even under light loads; therefore, an arbitrary method of determining the commercial elastic limit must be used. If loading is increased beyond the elastic limit, the material will be permanently deformed.

Ductility. The plasticity exhibited by a material under tension loading is known as *ductility* and is measured by the amount it can be permanently elongated. Ductility is the property of many metals that permit them to be drawn into wire.

Mechanical Properties of Metals

Malleability. This is the "ability of a metal to deform permanently under compression without rupture". It is this property that permits hammering (forging) and rolling of metals into sheets and various commercial shapes (angle iron, S-beams, and forgings).

Toughness. This is a description of a metal, such as steel, which has high tensile strength and the ability to deform permanently without rupture. There is no direct and accurate method of measuring the toughness of metals. *Toughness* involves both ductility and strength and may be defined as the "ability of a metal to absorb energy without failure;" often the impact resistance or shock resistance of a material is taken as an indication of its inherent toughness.

Brittleness. This is the physical property which is the opposite of plasticity. A brittle metal is one that cannot be visibly deformed permanently as it lacks plasticity. Cast iron is brittle and fully hardened tool steel may exhibit very little plasticity and could be classified as being brittle. However, hardness is *not* a measure of plasticity. A brittle metal usually develops little strength upon tensile loading, but may be safely used in compression. Brittle metals show very little shock or impact strength.

There are no values for brittleness and a brittle metal will fail without any warning of impending failure. On the other hand, a ductile metal may fail without any visible deformation if the load becomes concentrated due to a notched effect as shown in Fig. 6-5. The test specimen A in Fig. 6-5 has sharp reduction in cross-section; when pulled apart as in a tensile test, it

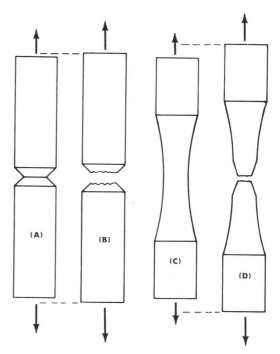

Fig. 6-5. Notching and its effect on plasticity.

ruptures with a brittle fracture (See specimen B). Specimen C, made from the same metal but with a gradual reduction in cross-section, behaves as a plastic material, elongating noticeably before rupture as in specimen D.

The *notch effect* is very important in all designs of metal parts. Sharp corners, notching or sudden changes in cross-section must be avoided. In castings, generous fillets (rounded internal corners) must be used where surfaces meet at right or sharp angles.

Hardness is not a fundamental property of a material but is related to its elastic

Metallurgy

and plastic properties. Generally, *hardness* is defined as the "resistance to indentation". The greater the hardness, the greater the resistance to penetration. The hardness test is widely used because of its simplicity and because *it can be closely correlated with the tensile and yield strength of metals*. In general, it can be assumed that a strong steel is also a hard one and is resistant to wear. Scratch or abrasion tests are sometimes used for special applications, as are elastic or rebound hardness tests. These are described in the following chapter.

OTHER PROPERTIES OF METALS

Susceptibility to Corrosion. Metals vary greatly in their resistance to atmospheric and chemical corrosion. The rusting of iron (oxidation) is a common and familiar example. A list of metallic elements in the order of their susceptibility to corrosion would start with potassium which actually catches fire when in contact with water. The list would end up with platinum and gold, which are impervious from attack by almost all chemical reagents. Such a list, known as the *Electrochemical Series*, is shown in Fig. 6-6, and further detailed in Table 6-2.

The electrochemical series is an arrangement of the metals in the diminishing order of their tendency to oxidize or corrode. Hydrogen is not a metal but is included in the series to show its activity in relation to the metals.

Oxidation or corrosion of the metals at the top of the list takes place at ordinary temperatures. Such activity decreases from top to bottom. Copper and those be-

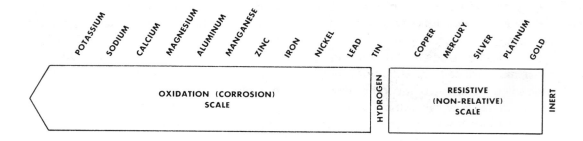

Fig. 6-6. Electrochemical series begins with the most reactive metals and ends with the most inert.

Mechanical Properties of Metals

TABLE 6-2. SOME PROPERTIES OF METALLIC ELEMENTS.
(Arranged in order of decreasing tendency to go to oxidized state)

ELEMENT SYMBOL	SYMBOL	MELTING POINT (°F)	SPECIFIC GRAVITY	ELECTRICAL CONDUCTIVITY (as percent Ag)[1]	LINEAR COEFFICIENT OF THERMAL EXPANSION[2]
Potassium	K	150	0.859	24.9	46.0
Calcium	Ca	1490	1.55	38.5	14.0
Sodium	Na	208	0.972	38.5	39.0
Magnesium	Mg	1204	1.70	35.5	14.3
Beryllium	Be	2345	1.85	9.5	6.8
Aluminum	Al	1215	2.7	53.0	13.3
Manganese	Mn	2268	7.2	35.9	12.8
Zinc	Zn	787	7.2	26.0	18.0[3]
Chromium	Cr	2750	6.92	67.6	4.6
Iron	Fe	2795	7.85	17.6	6.5
Cadmium	Cd	610	8.7	21.2	16.6
Cobalt	Co	2714	8.9	16.3	6.8
Nickel	Ni	2646	8.9	11.8	7.1
Tin	Sn	449	7.3	11.3	11.0
Lead	Pb	621	11.3	7.6	10.2
Hydrogen	H	-485	0.0695	—	—
Copper	Cu	1981	8.9	94.0	9.2
Mercury	Hg	-40	13.60	36.8	—
Silver	Ag	1761	10.5	100.0	10.5
Platinum	Pt	3224	21.45	16.4	4.9
Gold	Au	1945	19.3	66.0	7.9

[1] Silver taken as 100 percent conductivity
[2] Thermal expansion per degree F @ room temperature. Value given in millions, or $\times 10^{-6}$
[3] 18.0 for hot-rolled zinc with grain; 12.8 across grain

low it do not oxidize at ordinary temperatures when exposed to pure dry air.

Metals above hydrogen are difficult to obtain free in nature as they readily unite with other elements, hence their ores are oxides, sulfides, etc. Metals below hydrogen are usually found free in nature because they combine with other elements with difficulty. Hence, they do not corrode and are not easily oxidized.

An interesting fact about the placement of the metals in the chart above is the reaction taken when two different metals are placed in contact and exposed to salt water or acidic solutions. The metal that is nearest the beginning of the chart in respect to the second metal will corrode first. This action is known as *sacrificial action* and is used in hot water heating tanks to protect the ferrous pipes and tank from corrosion. To do this, a magnesium rod is suspended in the tank. Any

Metallurgy

corrosion that is going to take place will affect the magnesium rod. You will note that magnesium is more corrosive than iron.

Although the list indicates the varying effects or lack of effects of corrosion, it is misleading in practice. Both aluminum and zinc form protective coatings in their early stages of oxidation and are not further corroded. Iron and steel, on the other hand, if left unprotected, will rust completely away. For this reason, iron and steel, when used in exposed positions, must be protected by paint or a thin coat of zinc (galvanized sheet). A thin coating of tin, as in a tin can also protects both the outside of the can from the effect of humid air and most importantly, the inside surface of the can from the corrosive action of fruit and vegetable acids. If greater resistance to corrosion is required, copper or its alloys, brass, bronze, or nickel silver are used. Where real permanence is required, the precious metals must always be chosen.

The development of stainless steels (steels containing varying percentages of carbon, chromium, and nickel) has made it possible to use this material without protective coating against corrosion.

Electrical conductivity. Copper and aluminum are used for conducting electricity since they offer little resistance to the passage of the current. Silver offers even less resistance but is too expensive for commercial use. Copper offers less resistance than aluminum for the same size wire, but aluminum, due to its lighter weight, offers less resistance per unit of weight. Despite this fact, from a cost standpoint the advantage is still with copper. In the passage of current through a conductor, resistance results in the giving off of heat; the greater the resistance, the greater the heat for the passage of a given current. For electrical heating, metals with high electrical resistance are needed. These are found in the alloys; among the best are alloys of nickel and chromium. The relative conductivity of some metals is shown in Table 6-2. Note that copper is 94 percent of that for silver, while iron is only 17.7 percent of that for silver.

Electrical resistance. The opposition to electric current as it flows through a wire is known as the *resistance* of the wire. In order to have a standard of comparison, an arbitrary resistance unit has been agreed upon internationally. This unit is called *ohm*. The resistances of metals are expressed in terms of millionths of an ohm, or *microhms* of resistance in a cube of metal one centimeter on each side.

The resistance of an electric circuit is a property of the conductor forming the circuit and depends upon the length, cross-sectional area, temperature, and upon the material, whether copper, iron, brass, etc. Silver, as has been stated, is the best conductor known; in other words, it offers the least resistance to the passage of an electric current. Copper, gold, and aluminum are next in order of minimum resistance.

Resistance to heat. The relative electrical resistances of metals may be seen in Table 6-2. The resistance to the flow of heat is of the same order as the resistance to electricity; hence, ordinarily we use copper for boiler tubes, heating coils, and soldering irons. For cooking utensils, aluminum is used because of the combination of high-heat conduction with resist-

Mechanical Properties of Metals

TABLE 6-3. MECHANICAL PROPERTIES OF SOME METALS AND ALLOYS[1].

METAL	COMPOSITION	TENSILE STRENGTH		ELONGATION PERCENT (in inches)	BRINELL HARDNESS	MODULUS OF ELASTICITY (tension) (psi)
		Yield Point	Ultimate Strength			
Aluminum............................	Al	8,000	12,500	20	21-24	10,000,000
Brass, 70-30, annealed........	70 Cu-30 Zn	18,000	46,000	64	70	12,000,000
Brass, 70-30, cold-worked...	70 Cu-30 Zn	45,000	92,000	3	170	12,000,000
Brass, 95-5, annealed..........	95 Cu-5 Sn	13,000	46,000	67	74	11,000,000
Bronze, 95-5, cold-worked...	95 Cu-5 Sn	59,000	85,000	12	166	15,000,000
Bronze—manganese............	Cu+Zn+Mn	13,000	70,000	33	93	15,000,000
Copper, annealed................	Cu	5,000	32,000	56	47	14,000,000
Copper, cold-drawn.............	Cu	38,000	56,000	6	104	16,000,000
Chilled cast iron..................	Fe+C+Si+Mn		35-80,000		400-700	33,000,000
Duralumin, heat-treated.......	Al+Cu+Mg+Mn	48,000	56,000	18	95	10,000,000
Gray cast iron.....................	Fe+C+Si+Mn		20-60,000	1	150-225	15-20,000,000
Iron, commercial, pure........	Fe	19,000	42,000	48	69	30,000,000
Lead....................................	Pb		3,000	406		
Magnesium..........................	Mg	1,200	32,500	6	41	6,700,000
Magnesium—aluminum alloy..................................	Mg+Al	26,000	44,000	14	55	6,250,000
Malleable cast iron..............	Fe+C+Mn	37,500	57,000	22	110-145	25,000,000
Molybdenum........................	Mo		150,000	5		
Monel metal........................	Ni+Cu	50,000	90,000	40	166	25,000,000
Nickel..................................	Ni	25,000	70,000	48	85	32,000,000
Steel, structural..................	Fe+C	35,000	60,000	35	120	30,000,000
Steel, high-carbon, heat-treated................................	Fe+C	150,000	275,000	5	550	30,000,000
Steel alloy, heat-treated......	SAE 2340	174,000	282,000	8	488	30,000,000
Steel alloy, heat-treated......	SAE 4140	116,000	140,000	16	250	30,000,000
Steel alloy, heat-treated......	SAE 5150	210,000	235,000	13	455	30,000,000
Tin......................................	Sn	200	2,000	43		6,000,000
Tungsten wire.....................	W		230,000	3		60,000,000
Wrought iron.......................	Fe+Slag	30,000	50,000	35	100	27,000,000
Zinc, rolled sheet................	Zn	5,000	24,000	35		12,000,000

[1]The data in this table are based on tests made in several materials testing laboratories.

Metallurgy

ance to attack by foodstuffs. In magnetic properties, iron and its alloys stand alone.

Tables 6-2 and 6-3, listing the most important physical constants of various metals, will be helpful in selecting the proper metal for a given job.

Melting point. The melting point is the temperature at which a substance passes from a solid to a liquid condition. For water this is 32°F. Pure substances have a sharp melting point, that is, they pass from entirely solid to entirely liquid form in a very small temperature range. Alloys usually melt over a much wider temperature interval. The melting points and other temperatures in Table 6-4 are expressed in degrees Fahrenheit. According to this scale, water melts at 32° and boils at 212°. In metallurgy, temperatures are expressed frequently in degrees Centigrade. To convert degrees Centigrade to degrees Fahrenheit, use the following formula: $C = 5/9(°F-32)$

TABLE 6-4. SOME PROPERTIES OF METALLIC ELEMENTS.
(Arranged in order of melting points)

METAL	MELTING POINT (°F)	MELTING POINT (°C)	SPECIFIC GRAVITY	ELECTRICAL CONDUCTIVITY PERCENTAGE[1]	LINEAR COEF. OF THERMAL EXPANSION[2]
Tin	449	232	7.3	11.3	11.63
Bismuth	520	271	9.8	1.1	9.75
Cadmium	610	321	8.7	21.2	16.6
Lead	621	327	11.3	7.6	16.4
Zinc	787	419	7.2	26.0	18.0–12.8[3]
Magnesium	1204	651	1.7	35.5	14.3
Aluminum	1215	660	2.7	53.0	13.3
Arsenic	1497	814	5.73	4.6	2.14
Silver	1761	961	10.5	100.0	10.5
Gold	1945	1063	19.3	66.0	8.0
Copper	1981	1083	8.9	94.0	9.1
Manganese	2268	1260	7.2	....	12.8
Beryllium	2345	1350	1.85	....	6.8
Silicon	2600	1420	2.4	....	1.6–4.1
Nickel	2646	1455	8.9	11.8	7.6
Cobalt	2714	1480	8.9	16.3	6.7
Iron	2795	1535	7.87	17.7	6.6
Palladium	2831	1553	12.0	....	6.4
Platinum	3224	1774	21.45	16.4	4.3
Chromium	3275	1615	7.14	....	4.5
Molybdenum	4748	2509	10.2	32.2	3.0
Tungsten	6098	3370	19.3	28.9	2.2

[1]Silver=100 percent conductivity.

[2]Thermal expansion per degree F at room temperature. Value in millions, or $= \times 10^{-6}$.

[3]18.0 for hot-rolled zinc with grain; 12.8 for hot-rolled zinc across grain.

Mechanical Properties of Metals

Coefficient of expansion*.* With few exceptions, solids expand when they are heated and contract when cooled. They increase not only in length, but also in breadth and thickness. The number which shows the increase in unit length of a solid when it is heated one degree is called its *coefficient of linear expansion*.

Specific gravity*.* Sometimes it is an advantage to compare the density of one metal with that of another. For such a purpose, we need a standard. Water is the standard which physicists have selected with which to compare the densities of solids and liquids. Hence, the weight of a substance compared to the weight of an equal volume of water is called its *specific density* or *specific gravity*.

If we weigh one cubic foot of copper, we find that it weighs 549.12 pounds. One cubic foot of water weighs 62.4 pounds. If we divide 549.12 by 62.4, we get 8.9, which we call the specific weight of copper. The specific weight or gravity of the common metals is listed in Table 6-4. If we wish to find the weight of a cubic foot or centimeter of any of these metals, then all we have to do is to multiply 62.4 by the specific gravity as listed in the table, or with centimeters by 1.

Measuring Mechanical Properties*.* Generally we are very much concerned with the mechanical properties of metals and alloys. The mechanical properties, such as hardness, tensile strength, compressive strength, ductility, etc., are those measured by mechanical methods. These properties and their testing are discussed in Chapter 7.

REVIEW

1. Name and define the most important physical property of steel.

2. Define toughness.

3. What special property of steel has promoted its use in buildings, steel products and the repair of steel structures and products?

4. Define a carbon steel.

5. What are two properties of low-carbon steel?

6. What steel products use most of the low-carbon steel tonnage produced?

7. What is the percentage range of carbon in medium-carbon steel?

8. Name four typical products made of medium-carbon steel.

9. What is most of the high-carbon steel used for?

10. Carbon steel contains four minor constituents, two beneficial and two detrimental. Name the constituents in each area.

11. What does manganese do as an additive to improve steel?

12. What does silicon do as an additive to improve steel?

13. What is ingot iron and of what commercial use is it?

Metallurgy

14. Why do square and rectangular forging billets have rounded corners?

15. What physical property is desirable for screw machine work?

16. What are the two types of screw machine steel in use?

17. Define strength.

18. Define plasticity.

19. Define tensile strength.

20. Define elasticity.

21. Define ductility. What commercial product depends on this property?

22. Define brittleness.

23. What is the notch effect and what is done in the design of mechanical parts to avoid it?

24. Define hardness.

25. Define the electrochemical series of metals.

26. What is sacrificial action and how is it employed?

27. What two metals have high electrical conductivity and are commonly used in electrical wiring?

CHAPTER 7 — TESTING OF MATERIALS

The design of metal parts for industrial and commercial use requires precise knowledge of the mechanical and physical properties of the metal. This knowledge is obtained by physically testing representative samples of the material in various special testing machines.

Some tests involve testing the part or material sample to complete failure. This is known as *destructive testing*. Examples of destructive testing are certain types of hardness tests, tensile tests, impact tests, torsion tests and fatigue tests. Other methods of testing do not harm the sample parts nor impair them for future use. Such tests are known as *non-destructive tests*. These tests include X-ray and γ-ray inspection, ultrasonic inspection and magnetic particle (Magnaflux) inspection.

DESTRUCTIVE TESTING

Tensile Test

A tensile test is used to determine the static, mechanical properties of a material. These include ductility, tensile strength, proportional limit, elastic limit, yield point and breaking strength.

Tensile tests may be made on the machine illustrated in Fig. 7-1. This machine measures the load placed upon the speci-

Metallurgy

Fig. 7-1. Universal testing machine for determining tensile strength; equipped with electronic high magnification recorder for drawing stress-strain diagram. (Tinius Olsen Testing Machine Co.)

men, by dividing the load required to break the specimen by its area, the ultimate tensile strength of the material is obtained. Tensile strength accordingly is expressed in pounds per square inch (psi) or in Newtons per square millimeter (N/mm^2) or (kN/m^2).

If an accurate gage is clamped to the specimen and its elongation measured and plotted against the load, the point at which the elongation ceases to be proportional to the load, Fig. 7-2, is known as the *proportional limit*. The *elastic limit* is the greatest load which may be applied, after which the material will return to its unstrained condition. For practical purposes the elastic limit is the same as the proportional limit. This is an important factor in designing, as it is usually of more importance to know what load will deform a

Testing of Materials

Fig. 7-2. Stress-strain diagram.

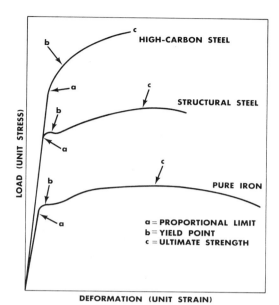

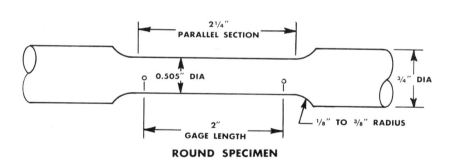

ROUND SPECIMEN

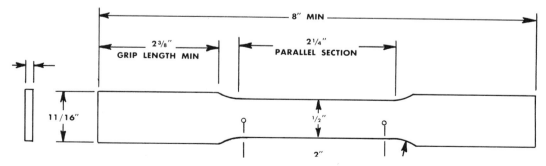

Fig. 7-3. Standard tensile test specimens.

Metallurgy

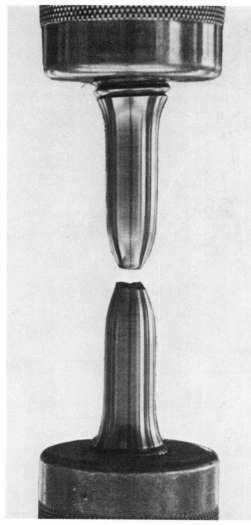

Fig. 7-4. Tensile specimen after fracture. Note necked-down portion.

structure than what load will cause rupture. Some materials, notably steel, indicate what is known as a *yield point*. This is the load at which the material will continue to elongate even though the load is not increased. It is indicated by a dropping of the weighing beam and is somewhat greater than the elastic limit.

The most common types of round and flat test specimens used in tensile strength testing are shown in Fig. 7-3. The result of a typical tensile test is shown in Fig. 7-4. Just prior to fracture, ductile materials such as low-carbon steels, usually experience a localized reduction in area known as *necking*. Necking is usually accompanied by a reduction of the load to produce subsequent straining. The reduction in area of the necked down portion of the test specimen continues until fracture occurs.

Both lever and hydraulic types of tensile testing machines are in use for pulling standard test pieces, either flat or round, bolts and medium sized bars, rods, and wire. The hydraulic type of machine is far superior and is rapidly replacing the lever type. Large hydraulic machines are needed for testing large pieces, full-sized units, and assembled machines. Lever and spring-actuated machines are more useful for light loads.

Impact Test

The behavior of metals under impact loads, or shock, may oftentimes be quite different from their behavior under loads slowly applied. The resistance of metal to shock is measured by means of a pendulum-type machine, shown in Fig. 7-5. If the pendulum is allowed to swing freely, it will swing through a known angle; if we interpose a specimen which the pendulum breaks, some energy will be used up and it

Testing of Materials

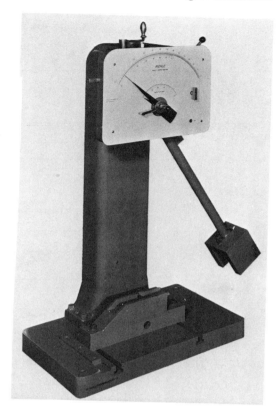

Fig. 7-5. Impact testing machine. (Riehle Testing Machines.)

will swing through a smaller angle. This machine enables us to measure the angle through which the pendulum swings, and thus to calculate the energy consumed in breaking the specimen. This is usually reported in foot-pounds. The energy consumption for a piece of cast iron (which is very fragile) will be low, and we say that the shock resistance is small; while for a piece of wrought iron, energy consumption will be high and shock resistance correspondingly large.

Standard tests exist in which notched specimens are fractured by a single impact in order to check for any tendency toward brittle behavior. Two of the most common tests are the Izod and the Charpy. In the former test, a square notched bar is gripped vertically in a vise, and a measure is made of the number of foot-pounds necessary to break the bar by a blow struck just above the notch. In the Charpy test, the specimen, usually square, is supported freely at its ends and is hit at a point just behind the notch at the middle of the span. In both cases the blow is applied by a freely swinging pendulum. See Fig. 7-6 for notched bar details.

Metallurgy

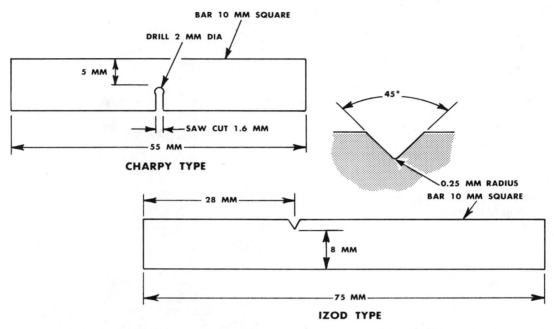

Fig. 7-6. Notched bar impact test specimen. Charpy type, keyhole notch, shown at top. Izod type, V-notch, shown at bottom.

Fatigue Test

When a sample is broken in a tensile machine, a certain definite load is required to cause that fracture. However, the same material will fail under a much smaller load if that load is applied and removed many times. In this way, an axle may break after months of use, even though its loading has not been changed. Such a failure is pictured in Fig. 7-7. The part breaks with no sign of deformation. The final part to fracture is usually quite coarse grained with adjacent sections showing signs of having rubbed together for quite some time. The fracture above shows concentric rings of sections which have rubbed together during the period of time prior to ultimate failure. The coarse grain of the final fracture has often led to the erroneous statement that the part failed due to crystallization in service. All such failures are known as *fatigue* or *progressive failures*, and in designing parts subjected to varying stresses, the fatigue limit of a material is often more important than its tensile strength or its elastic limit. The *fatigue limit* is that load, usually expressed in pounds per square inch, which may be applied during an infinite number of cycles without causing failure. In fatigue testing it is assumed, ordinarily, that a load which can be applied 100,000,000 times without failure can be applied indefinitely without failure. Fig. 7-8 shows typical diagrams for fatigue tests of metals.

Testing of Materials

Fig. 7-7. Fatigue fracture of grinding machine spindle.

A number of types of fatigue-testing machines are available. In the most common type, a round test bar is rotated horizontally with a load applied by means of a weight hung on one end, and the stresses are calculated from the applied load and the dimensions of the test bar. In such a test, the load varies from a tensile stress to a compressive stress of equal magnitude. In another type of machine, a bar is alternately compressed and elongated. For testing samples of sheet metal, a machine has been developed in which a flat specimen is clamped to one end of a calibrated spring and the other end to a shaft from an eccentric. Stress in the specimen is calculated from the movement of the calibrated spring.

Much knowledge has come to us in recent years which enables us to use metals in a safe and economical way and to meet the rigid requirements for metals in our modern machines. Of prime importance for long fatigue life is the smoothness and lack of stress-raising imperfections on metals used. Fatigue limit increases with the tensile strength and in ductile steels is approximately 50 percent of the tensile strength. Lack of surface uniformity, such as decarburization, lowers the fatigue limit. Further, roughened, notched, or threaded surfaces serve to decrease pronouncedly the fatigue limit. Damaging corrosive conditions will greatly decrease the fatigue strength of a metal at even small stresses.

In figuring allowable stresses under repeated stress, it must be remembered that stresses above the yield strength of the metal will cause elastic failure. The designer and user of steel parts must be on guard against both elastic and fatigue failure. Machinists, machine operators, and inspectors share the responsibility of keeping stress raisers out of parts and assemblies.

Metallurgy

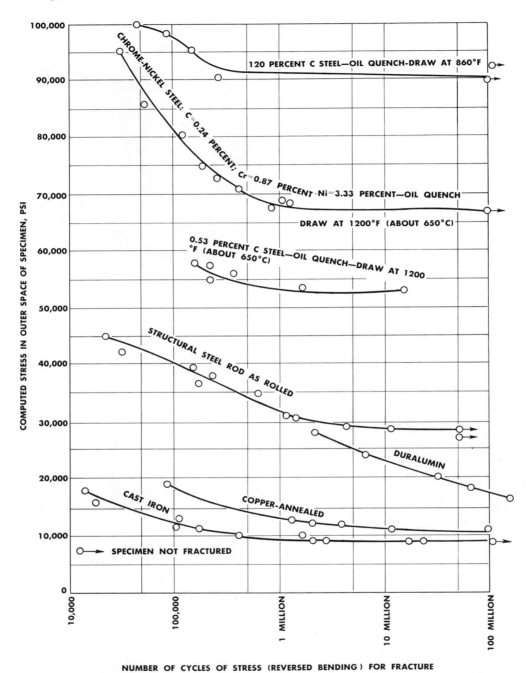

Fig. 7-8. Typical S-N (stress-cycle) diagrams for fatigue tests of various metals. (Materials of Engineering by Moore, McGraw-Hill Book Co., Inc.)

Testing of Materials

It is a known fact that offsets, slots, holes, notches, and sharp re-entrant angles build up stresses excessively when they are loaded. Use of generous fillets, rounding off the ends of keyways and slots, smoothing out corners and shoulders, and avoiding sharp tool cut marks will do much to eliminate stress raisers. Any sharp grooves or notches may so change the distribution of stresses as to alter greatly the usual mechanical properties of the material and cause the part to react to loading in a wholly unexpected way.

Hardness Test

Hardness is the property of a metal which gives it the ability to resist being permanently deformed when a load is applied. The greater the hardness, the greater the resistance to deformation. Because the definitions for ultimate strength and hardness are somewhat synonymous, in general it can be assumed that a strong metal is also hard.

One way of measuring the hardness of a metal is to determine its resistance to the penetration of a nondeformable ball or cone, or to determine the depth to which such a ball or cone will sink into the metal under a given load. The *Brinell hardness test* is made by pressing a hardened steel ball, usually 10 millimeters in diameter, into the test material by weight of a known load—500 kilograms for soft materials such as copper and brass, and 3000 kilograms for materials such as iron and steel—and measuring the diameter of the resulting impression. The hardness is reported as the load divided by the area of the impression, and tables are available from which the hardness may be read, once the diameter of the impression is obtained. A small optical viewer is used for measuring these impressions.

Fig. 7-9 shows a Brinell hardness tester. The specimen is placed upon the anvil,

Fig. 7-9. Brinell hardness tester. (Tinius Olsen Testing Machine Co.)

113

Metallurgy

which is raised or lowered by means of the screw. The anvil is then raised until the specimen is in contact with the steel ball. The load is then applied by pumping oil; the handle of the oil pump is shown directly under the dial. The load is applied for 30 seconds; then the diameter of the resulting impression is measured. The Brinell hardness of annealed copper is about 40, of annealed tool steel about 200, and of hardened tool steel about 650.

The Vickers hardness testing method is almost the same as the Brinell method. The penetrator used in the Vickers machine is a diamond pyramid, and the impression made by this penetrator is a dark square on a light background. This impression is more easily read than the circular impression of the Brinell method. There is also the advantage that the diamond point does not deform.

In making the Vickers test, a predetermined load is applied to the specimen. After removal of the load the ratio of the impressed load to the area of the resulting impression gives the hardness number. The operation of applying and removing the load is controlled automatically. Several loadings give practically identical hardness numbers on uniform material; this is much better than the arbitrary changing of scale with the other hardness testing machines. Although the Vickers tester is thoroughly precise and very adaptable for testing the softest and hardest materials under varying loads, its cost precludes its use where a Rockwell suffices and is far more rapid. The Vickers machine, Fig. 7-10, is limited definitely to size and shape of the object to be tested, and is extremely well suited to very hard

Fig. 7-10. Vickers diamond pyramid hardness tester. (Riehle Testing Machines.)

materials. The hardness is given by the relation $H=P/A$ where P is the load in kilograms and A the area in square millimeters of the surface of indentation.

Hardness as determined by the Vickers method is about the same as the Brinell hardness numbers up to a value of 500.

Testing of Materials

The Rockwell hardness tester, illustrated in Fig. 7-11 measures resistance to penetration as does the Brinell test, but the depth of impression is measured instead of the diameter, and the hardness is indicated directly on the scale attached to the machine. This scale is really a depth gage graduated in special units. In testing soft materials a 1/16" (1.6 mm) steel ball with a 100-kilogram load is used, and the hardness is read on the B scale; for testing hard materials, a diamond cone is used with a 150-kilogram load, and the hardness is read on the C scale. The advantages of the Rockwell tester are that the test can be made quickly; that only a small mark is left on the sample; and that very hard material can be tested with the diamond cone.

The Rockwell superficial hardness tester employs a light load, causing very little penetration of the diamond cone. This superficial hardness tester does not differ in principle from the standard machine. The hardness number is based on the additional depth to which a test point penetrates the material under a given load beyond the depth to which the penetrator has been driven by a small initial load. The initial load is 3 kilograms, while the major load varies and may be 15, 30, or 45 kilograms, depending on the thickness of the hard surface. The machine is very useful in testing exceptionally high surface hardness, such as case-hardened or nitrided surfaces. Thus, sections of razor blades, clock springs, etc. react favorably to this method of testing.

With the Shore scleroscope, Fig. 7-12, the hardness is measured by the height of rebound of the diamond-pointed hammer after it has been dropped on the sample. The harder the material used, the greater the rebound. The height of rebound is read on the gage. The scleroscope can be used for large sections; it is portable; and

Fig. 7-11. Rockwell hardness tester. (Wilson Mechanical Instrument Div., American Chain & Cable Co.)

Metallurgy

Fig. 7-12. Shore scleroscope model D shown with standard test block. (Shore Instruments & Mfg. Co.)

the indentation made by the test is very slight, an advantage where surface finish is important. The amount of rebound is a factor more of the elastic limit of the specimen than of its tensile strength, and therefore the machine does not measure exactly the same type of hardness as do the indentation methods. The specimen must be solidly held in a horizontal position, and the hammer must fall exactly vertically. The smallness of the impression made by the Shore scleroscope leads to variations in reading, thus necessitating a range of readings rather than any single one. More than one penetration should not be made on the same spot. Because of its portability, it can be transported to the work and tests can be made on specimens too large to be taken conveniently to the other machines.

Another useful hardness-testing machine is called the Monotron. This tester measures the indentation made by pressing a hard penetrator into the material under test. The difference between it and other testing machines is that the depth of impression is made constant instead of the pressure. Two dials at the top of the machine enable the operator to read depth of impression on one and pressure on the other. Pressure is applied until the penetrator sinks to the standard depth, 9/5000''; the amount of pressure so needed is read on the upper dial. The penetrator used is a diamond having a spherical point about 0.025'' in diameter. One advantage for this method, based on constant depth of impression, is that the same amount of cold work is performed regardless of whether the material tested is extremely hard or comparatively soft. Because of this, a wider range of materials can be tested without making any change in the penetrator or the methods of the tests.

Testing of Materials

Corrosion Tests

The ability of metals to resist atmospheric corrosion, or corrosion by liquids or gases, often is of primary importance. Common types of corrosion are the pitting or localized type, the direct chemical or solution type, and electrolytic or galvanic corrosion. Accelerated corrosion tests have been devised whereby the behavior of the materials in actual service may be deduced from conditions applied for weeks, instead of for years as in actual use. Corrosion may be measured by determining the loss of tensile strength of specimens, by loss of weight in materials which dissolve in the corroding medium, or by gain in weight when a heavy coating of rust is formed. In applying such tests, it is important to know just what the service will be, for a given material might be attacked rapidly by fruit acids and not by nitric acid, for instance; while with another material the reverse might be true.

Although no standard tests for corrosion exist, many tests now exist which have proved to be of great practical value. A few of these tests are discussed briefly in the following material. In all such tests, it is of utmost importance to see to it that corrosion test conditions nearly duplicate the conditions of service.

Salt-Spray Test. One of the most generally accepted tests is the salt-spray test. For this test, specimens that have been thoroughly cleaned are placed in a specially prepared box or chamber and exposed to a fog or spray of salt water for a designated number of hours. The concentration of the spray and the exposure period vary depending on the severity of the service required. Such a test is suitable for testing the corrosion resistance of all of the coatings used on steel. Coated steel will withstand the attack for varying periods depending on the protection the coating affords. More resistant coatings will withstand attack for some hundreds of hours.

Strauss Solution Corrosion Test

Attacking a metal with the so-called Strauss solution and microscopic examination for precipitation of carbides are two of the tests employed in looking for intergranular corrosion. In the former test, the steel is boiled in the Strauss solution for about 24 hours. The solution usually consists of 47 milliliters of concentrated sulfuric acid and 13 grams of copper sulfate crystals per liter. If intergranular corrosion is present, the attacking reagent will cause the metal to lose its metallic resonance, cause a considerable decrease in the electrical conductivity, or cause the metal to develop cracks following a bend test.

In the test employing microscopic examination for precipitation of carbides, one side of the specimen is polished and then electrolytically etched in an oxalic acid or sodium cyanide solution. Such an etch causes precipitated carbides to appear as black particles. If carbides are found, the specimen is given an embrittlement test.

Brass and bronze objects are tested by the mercurous nitrate test. The specimen is dipped in a 40 percent solution of nitric acid, and then immersed in a solution con-

Metallurgy

taining 1 percent nitric acid and 1 percent mercurous nitrate. After a few minutes, the specimen is removed, cleaned, and dried. Upon low-power microscopic examination, if no cracks are visible, the specimen will probably not crack because of internal stresses.

An accelerated weathering or exposure test uses intermittent watering and drying to simulate the effect of sunshine and moisture in producing corrosion. Cabinets with suitable heating and watering units are needed for this test.

Microscopic Tests

Metals, as ordinarily fabricated, are composed of very small particles which are visible only with the microscope. Since H. C. Sorby's pioneer work with the microscope, the use of that instrument has contributed inestimably to the development of the metal industry. For the student of metallurgy the microscope is of utmost importance. Microscopic examination will show, for example, whether or not the most desirable structure has been produced, and if the material contains an excessive amount of dirt or nonmetallic inclusions. Such studies not only determine whether the best possible material has been produced, but they also teach us how to modify the manufacturing processes to obtain more desirable metals or alloys.

In order to present the true structures of metals, it is important that the samples be carefully prepared. For such examinations, small samples are selected, representing as nearly as possible the entire piece. One side of each sample is smoothed by means of a file or an emery wheel. This side, which is to be examined, is then ground with several emery papers of different coarseness, beginning with coarse and ending with a very fine paper. The sample is held so the scratches made by each emery paper are at right angles to the ones made in the preceding operation. The sample is then polished on a rotating moist cloth, using very fine alumina or rouge as a polishing agent. After this preparation, the sample should be nearly flat, free from pitting, and almost free from scratches when viewed through the microscope.

Several types of automatic polishing machines are now available for the grinding and polishing of specimens; one is shown in Fig. 7-13. Specimens to be ground and polished on this machine are first mounted in bakelite or other mounting epoxy, six mounts being used each time. These mountings are then placed in a fixture, and the fixture with the specimens is rotated in one direction, in con-

Fig. 7-13. Jarrett polishing machine. (Beuhler Ltd.)

tact, progressively, with stone, pitch, wax, and cloth laps, which rotate in the opposite direction to the fixture and specimens. Abrasive is fed to the face of the laps during each operation. After the last lap, which is a cloth disc with a finely ground alumina abrasive, the specimens are removed from the fixture. Examination will reveal a highly polished surface, free from scratches, and flatter than those usually obtained by hand grinding and polishing. Time is saved by keeping the specimens rather small. Too much care cannot be taken in the selection and preparation of specimens for microscopic examination. Each cutting, grinding, and polishing operation may alter the surface structure of the metal so as to destroy the evidence wanted in the analysis.

A microscope suitable for the examination of metallic specimens is shown in Fig. 7-14. This microscope differs from those used in viewing transparent substances in that a vertical illuminator is placed just above the objective, which is the lens next to the specimen. By means of this vertical illuminator, light from the small lamp is thrown onto the specimen through the objective. The vertical illuminator may consist either of a thin piece of glass or a half mirror. Both types are mounted so they may be rotated by a small disc; in examining specimens it is important to adjust the illuminator for the best view.

A microscope designed especially for metallographic work and equipped with an attachment for taking photomicrographs is shown in Fig. 7-15. In this machine the image of the specimen is projected on ground glass at the end of the bellows, just as the portrait camera projects the image of a person onto a ground glass. To record the magnified image, a photo-

Fig. 7-14. Table type metallurgical microscope. (Bausch & Lomb)

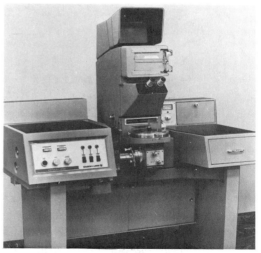

Fig. 7-15. Bench type research metallograph. (Bausch & Lomb)

Metallurgy

graphic plate or film is substituted for the ground glass and exposed for the proper length of time. With such a machine, magnifications ranging from 10 to 10,000 may be obtained. The magnifications most used are 75, 100, 200, and 500, although it is frequently necessary to use higher magnifications to see the finer structures.

When a polished but otherwise untreated sample is examined under the microscope, a few accidental scratches may be seen, together with some dirt or nonmetallic impurities, such as slag and manganese sulfide in iron. If an alloy is examined which consists of two or more constituents, one of which is appreciably softer than the other, the softer element will have eroded away somewhat in polishing, and the eroded space will appear black. An example of this is shown in Fig. 7-16, a photomicrograph of a polished but

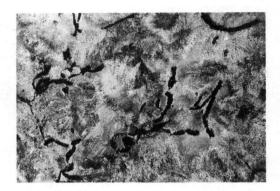

Fig. 7-17. Same as Fig. 7-16, etched; magnified 200 times.

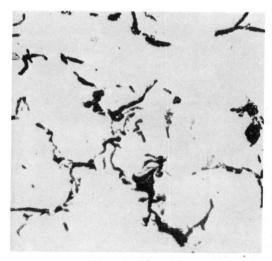

Fig. 7-16. Photomicrograph of gray cast iron, unetched, showing the flakes of graphitic carbon; magnified 200 times.

otherwise untreated sample of gray cast iron magnified 200 times. Gray cast iron consists of a mixture of iron, throughout which there are flakes of graphite. These graphite flakes are eroded away on polishing, and the areas these flakes occupied show black in the micrograph.

To reveal the structure, it is necessary in most cases to etch the polished samples. Fig. 7-17 is the same specimen as shown in Fig. 7-16 except that it has been etched in dilute nitric acid. The light structural background in Fig. 7-17 has now darkened and is revealed as a structural component called pearlite. This pearlitic structure is a very fine lamellar structure consisting of layers of ferrite and cementite (Fe_3C), as shown at higher magnifications in Fig. 7-18. Fig. 7-19 is a photomicrograph of malleable cast iron magnified 500 times. In etching, the sample is immersed in a dilute acid or other reagent which will dissolve a portion of the surface or tarnish the different constituents selectively. Fig. 7-20 shows a sample of iron

Testing of Materials

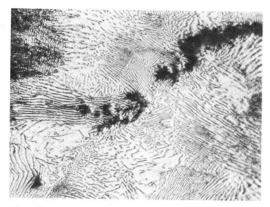

Fig. 7-18. Same as Fig. 7-17, but magnified 500 times.

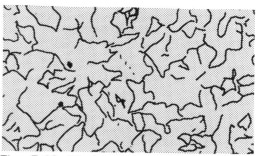

Fig. 7-20. Photomicrograph and typical chemical analysis of pure open-hearth iron. Carbon 0.015 percent; manganese 0.020 percent; phosphorus 0.005 percent sulfur 0.035 percent; silicon 0.005 percent; total metalloids 0.080 percent.

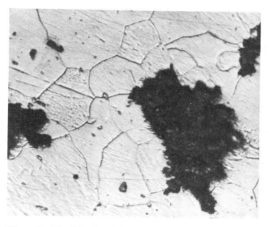

Fig. 7-19. Malleable cast iron magnified 500 times. Graphite nodules (dark) in a matrix of ferrite (etched).

Fig. 7-21. Photomicrograph of cartridge brass annealed at 1200° F (650° C), magnified 100 times.

which has been etched with a dilute solution of nitric acid in alcohol. As the micrograph shows, the material consists of minute and regularly shaped grains. The nitric acid etch has attacked the crystalline structure of the iron in such a manner as to develop a difference in level of the individual grains, so that the crystals or grains are easily seen with the aid of the microscope.

Investigations have indicated that the size of these grains—for other metals as well as for iron—has a profound influence on the physical properties of the material. In some instances, therefore, the grain size of metals is determined as a routine

Metallurgy

test. An example of a structure made visible through the staining or tarnishing of some parts of the specimen is shown in Fig. 7-21 which is of ordinary soft sheet brass. In this instance the etching solution of ammonia and hydrogen peroxide has stained the grains in varying amounts, depending upon their orientation.

Fracture Test

The fracture test is made simply by notching a suitable specimen and then breaking it. If the specimen is relatively small in cross section, and not very ductile, it may be fractured without notching. Although no concrete values of the properties of a metal are obtained through this test, it does reveal many characteristics, such as grain size, lack of soundness, case depth in casehardened steel, etc. The fracture test is probably the oldest of the methods used in the inspection and testing of metals. The earliest specifications for metals probably included a requirement for soundness, and doubtless the check for soundness, or the examination for defects that would contribute to unsoundness, was made by the fracture test. A fracture will usually follow the path of least resistance and thus uncover any previous cracks or defects in the metal, which would make it unsound. Fig. 7-22 shows a standard grain-size test as developed by Shepherd in which the standard

Fig. 7-22. Standard Fracture Grain Size Chart as developed by Shepherd. (Visual Examination of Steel by Enos, American Society for Metals)

Testing of Materials

consists of a series of ten steels of varying grain sizes. The test is made by comparing fractures with these actual standards.

From the examination of fractures, it is possible to tell whether a bar is high or low-carbon steel. A fractured casting will also show whether it is grey cast iron, white cast iron, wrought iron, or malleable cast iron.

When fracturing, *low-carbon steel* will bend and be quite tough. The fracture will be light grey, and often the grain size can be noted if the fracture has not been distorted badly. If heated to a bright red heat, about 1500°F (815°C) and quenched in water, not much change in properties can be noted upon fracturing after this treatment.

High-carbon steel will fracture more easily than low-carbon steel since it is more brittle and so will stand less bending. The fracture will glisten more, and the grain size will be readily seen. If heated to a red heat, 1425°F (about 775°C) and quenched in water, marked changes in properties occur. The steel becomes very hard and brittle, and the fracture will be very fine and silky in appearance after this treatment.

Gray cast iron cannot be forged, but sometimes it becomes necessary to distinguish it from white or malleable cast iron. The gray iron will fracture very easily, and the metal will be soft and brittle. The fracture is coarse grained and dark gray in appearance.

White cast iron will fracture very easily and is very hard and brittle. The fracture will show the grain size and be very white in color. White cast iron cannot be forged or machined, except by grinding.

Malleable cast iron cannot be forged or worked cold, but it will be found hard to fracture, being much tougher and more ductile than either white or gray cast iron. The fracture has the appearance of a case and core. The case at the edge of the fracture has the appearance of low-carbon steel, and the core or interior of the fracture is similar to gray cast iron but somewhat darker in color.

Wrought iron is very tough and hard to fracture. Heating and quick cooling, as of low-carbon steel, have little or no effect. Due to streaks of slag that often can be seen on the surface after cleaning off the scale, the fracture seems fibrous and darker than low-carbon steel. This fibrous condition and streaks of slag are the outstanding characteristics of this material.

Testing By Macro-Etching

In testing metals, certain characteristics which may not be apparent to the naked eye often can be brought out by deep etching of the specimen. Although a good surface finish is desired for this type of test, it is possible to deep etch a specimen that has been machined or ground without resorting to the fine grinding and polishing necessary in preparing a specimen for microscopic examination. Deep etching with Nytal, a 10 percent solution of nitric acid in alcohol, will be found useful in revealing soft spots in hardened steels, case depth in casehardened steel, welds, and small cracks. A drastic deep etching may be obtained with a 50 percent solution of hydrochloric acid in water. This solution, which is heated to 160° to 175°F (71 to

Metallurgy

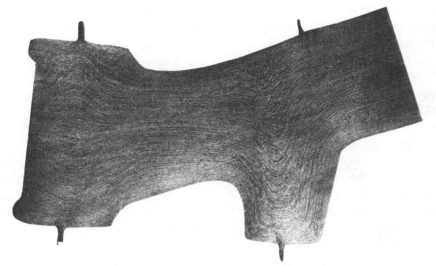

Fig. 7-23. Axle forging cross-section.

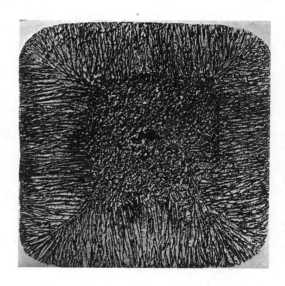

Fig. 7-24. Deep etched transverse fracture of steel ingot showing dendritic outer surface and normal free crystal inner structure. Note cleavage planes at corners and central pipe. (Gathmann Engineering Co., Baltimore, Md.)

79°C), will disclose the fiber or flow lines in a forging, as well as segregation, blowholes, cracks, seams, pipes, dendritic structures, etc. A deep etched forging showing the flow lines or direction of forging is illustrated in Fig. 7-23 while Fig. 7-24 shows the effect of deep etching a sectioned cast ingot of steel.

NON-DESTRUCTIVE TESTING

Magnetic Testing

Two types of magnetic test are used: one to discover defects such as cracks, seams, blowholes, etc., and the other to determine variations in structure and properties of a metal as compared with a standard specimen that is known to be satisfactory.

This test method may be used with the magnetic metals, such as iron, nickel, cobalt, and their alloys. The test may be used to determine the presence of any defects that would affect the soundness and mechanical properties of the metal. This type of test, being nondestructive, can be applied to all the material of many work pieces instead of to a sample specimen.

In the *Sperry test*, cracks and other flaws in rails may be uncovered by the passage of a heavy direct current longitudinally through both rails, by means of a test car that moves along the rail at about six miles an hour. Any flaw greater than 1 percent of the cross-section of the railhead causes an abrupt change in the magnetic field induced by the current. Both rails may be completely inspected at the same time. Periodic inspection also is possible by this method.

A powder method of magnetic particle inspection *(Magnaflux)* is used a great deal to detect defects on or just below the surface of steel forgings. Such defects might be revealed by the macro-etching method, but at the same time they will be destroyed by the deep etching; whereas, by the magnetic particle inspection method, indications of defects remain intact and may be examined under the microscope to identify their character and extent. Parts to be examined must first be magnetized, then coated with finely divided iron powder or magnetic iron oxide. The powder may be dusted onto the magnetized specimen or applied by immersion in a solution of kerosene and carbon tetrachloride containing the magnetic iron oxide powder. A deposit of the magnetic powder will adhere to any defects which may be present; see Fig. 7-25. This test has been used successfully to reveal fatigue cracks developing in machinery in service. Where safety is all important, as in aircraft, this test has become very popular.

In the use of magnetic tests for checking variations in structure and mechanical properties of metal parts, it is assumed that any two identical metal parts having identical mechanical properties should show identical magnetic properties. Unfortunately there appear to be many factors that upset this assumption, so this phase of the test has not been so successful. The simplest form of the test consists in comparing the characteristics of the unknown material with those of an acceptable or standard test sample, on the

Metallurgy

Fig. 7-25. Magnetic particle inspection on defective forging; forging crack before magnaflux test (left); after (right). (Wyman-Gordon Co.)

for this purpose, a photographic plate is placed on one side of the specimen or part to be examined, and the source of radiation on the other side. The X-rays or γ-rays penetrating the sample are registered on the photographic plate. If the sample has blowholes or large inclusions, for example, such defects will allow more rays to pass through and will be identified by dark spots on the photographic plate. These methods have been used to find defects in small castings, Fig. 7-26.

X-rays have also been used in studying the crystal structure of metals and alloys. In this use, the atoms deflect the rays in such a way that it is possible to determine just how the individual atoms are built up to form metallic crystals. Such studies have conclusively demonstrated the crystalline arrangement of metals.

theory that comparison of the magnetic properties of each one will reveal any differences between them. This method is used to check large quantities of similar materials, such as razor blades, bar stock, wire, and tubing.

Radiography Tests

X-rays and γ rays are rays of much shorter wave length than ordinary light rays. The fact that they will pass through materials impervious to ordinary light, such as metals, makes them valuable in searching for internal defects. When used

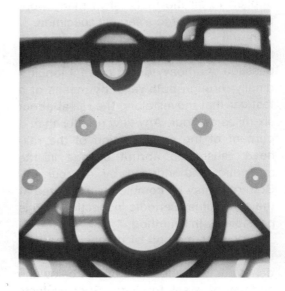

Fig. 7-26. An X-ray radiograph of aluminum casting. (Conductron Corp.)

Testing of Materials

Ultrasonic Hardness Testing

A device called Sonodur, Fig. 7-27, is an indentation type microhardness tester with an electronic read-out. It employs a diamond tipped magnetorestrictive rod which is electrically excited to its resonant frequency. As the vibrating rod is pressed onto the surface of the metal to be tested, the tip penetration is a function of hardness. The resonant frequency of the rod increases with penetration and the frequency change is displayed in the read-out meter in terms of hardness. The instrument is portable and readings can be made in less than three seconds. The extremely small penetration presents a practically non-destructive test.

There are several other hardness-testing machines that are satisfactory for hardness testing of metals. However, they all yield comparative values for hardness only. There is no definite relationship between hardness as measured by the different methods, or between hardness and strength. However, tests have been carried out and various conversion tables have been constructed which are useful when approximate conversion values are needed. See Table 7-1.

Fig. 7-27. Sonodur ultrasonic hardness tester. (Branson Instruments)

Metallurgy

TABLE 7-1. HARDNESS CONVERSION TABLE
FOR ALLOY CONSTRUCTIONAL STEELS—Approximate

VICKERS OR FIRTH DIAMOND HARDNESS NUMBER	BRINELL		ROCKWELL HARDNESS		SHORE	TENSILE STRENGTH
	IMPRESSION 10 mm ball 3000 kg load	HARDNESS NUMBER	C SCALE 150 kg, 120° DIAMOND	B SCALE 100 kg, 1/16" BALL		
	dia in mm		CONE			(psi)
1220	2.20	780	68	..	96	...
1114	2.25	745	67	..	94	...
1021	2.30	712	65	..	92	354
940	2.35	682	63	..	89	341
867	2.40	653	62	..	86	329
803	2.45	627	60	..	84	317
746	2.50	601	58	..	81	305
694	2.55	578	56	..	78	295
649	2.60	555	55	..	75	284
608	2.65	534	53	..	73	273
587	2.70	514	51	..	71	263
551	2.75	495	50	..	68	253
534	2.80	477	48	..	66	242
502	2.85	461	47	..	64	233
474	2.90	444	46	..	62	221
460	2.95	429	44	..	60	211
435	3.00	415	43	..	58	202
423	3.05	401	42	..	56	193
401	3.10	388	41	..	54	185
390	3.15	375	39	..	52	178
380	3.20	363	38	..	51	171
361	3.25	352	37	..	49	165
344	3.30	341	36	..	48	159
335	3.35	331	35	..	46	154
320	3.40	321	34	..	45	148
312	3.45	311	32	..	43	143
305	3.50	302	31	..	42	139
291	3.55	293	30	..	41	135
285	3.60	285	29	..	40	131
278	3.65	277	28	..	38	127
272	3.70	269	27	..	37	124
261	3.75	262	26	..	36	121
255	3.80	255	25	..	35	117
250	3.85	248	24	100	34	115
240	3.90	241	23	99	33	112
235	3.95	235	22	99	32	109
226	4.00	229	21	98	32	107
221	4.05	223	20	97	31	105
217	4.10	217	18	96	30	103
213	4.15	212	17	95	30	100
209	4.20	207	16	95	29	98
197	4.30	197	14	93	28	95
186	4.40	187	12	91	27	91
177	4.50	179	10	89	25	87
171	4.60	170	8	87	24	84
162	4.70	163	6	85	23	81
154	4.80	156	4	83	23	78
149	4.90	149	2	81	22	76
144	5.00	143	0	79	21	74
136	5.10	137	-3	77	20	71

The International Nickel Company, Inc.

Testing of Materials

Measuring Temperatures

In metallurgical work, temperatures are usually measured with a thermoelectric *pyrometer*, Fig. 7-28. Such an instrument may consist of two dissimilar wires welded together at one end, called the *thermocouple*, with the opposite ends connected to a millivoltmeter. The welded ends of the thermocouple are placed in the furnace where temperature is to be measured. As this end becomes heated, an electromotive force (emf) or voltage is generated and a current flows through the thermocouple wires. The voltage generated depends upon the temperature reached by the hot end of the thermocouple; a millivoltmeter may be used to read the voltage generated and the temperature may be determined from that voltage. Instruments (usually termed recorders) are available, which register the temperature on a roll of slowly moving paper.

The metals and alloys most used for thermocouples are iron with constantan, chromel with alumel, and platinum with a platinum-rhodium alloy. Platinum and platinum alloys, due to their high cost, are used only for measuring very high temperatures.

For measuring temperatures higher than those for which thermocouples are adapted, such as steel furnace temperatures, a pyrometer of the optical or radiation type may be used, Fig. 7-29. In the optical type of pyrometer, a lamp filament located in the instrument is heated by means of a variable current to the same brightness as the hot substance being measured for temperature. Temperature is then determined from the current necessary to heat the filament to the same brightness as the hot object.

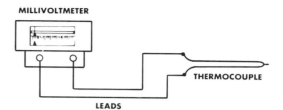

Fig. 7-28. Schematic for a thermoelectric pyrometer.

Fig. 7-29. Infrared radiation pyrometer. (William Wahl Co.)

While the optical pyrometer measures the intensity of a narrow spectral band of radiation emitted by a glowing object, the radiation pyrometer measures the intensity of all wave lengths, light and heat waves combined. Actually the energy of all wave lengths radiated by the source is forced to fall upon the hot junction of a small thermocouple. The temperature which this thermocouple reaches is approximately proportional to the rate at which energy falls upon it.

Metallurgy

REVIEW

1. What are the two main divisions of physical testing?

2. What properties does tensile testing determine?

3. Describe briefly how a tensile test is made on a specimen.

4. Describe fatigue failure.

5. Describe the hardness test as performed by a Brinell tester.

6. Describe a Vickers hardness tester and its advantage over a Brinell tester.

7. Describe the hardness test as performed by a Rockwell tester and its advantages over the Brinell and Vickers tests.

8. Describe the operation of a Shores scleroscope test.

9. Why is the salt spray corrosion test a common test used by the automobile industry?

10. Describe briefly the steps taken in the preparation of a specimen for microscopic analysis. Why is etching necessary?

11. From the fracture tests describe the following grain structure:
   low-carbon steel
   high-carbon steel
   wrought iron
   grey cast iron
   white cast iron
   malleable cast iron

12. How is the fracture method used in testing for grain size?

13. What etching fluid is used for deep etching in preparing specimens for testing by macro-etching? What is the fluid's composition?

14. Describe briefly Magnaflux testing.

15. Describe a thermocouple as used in a thermo-electric pyrometer.

16. Describe an optical pyrometer.

CHAPTER 8  THEORY OF ALLOYS

STUDY OF METALS

From the study of the nature of pure metals, we next come to the subject of alloys. We are concerned here with a study of the properties and structural characteristics of a pure metal when atoms of a second metal or a nonmetal are combined with it.

Increased strength, durability, and other qualities have made alloys of much greater industrial importance than pure metals. There are many combinations of properties that can be obtained by alloying metals with each other, or with nonmetals, in various proportions. Knowledge of the properties of alloys makes it possible to determine the uses to which a specific alloy is best adapted, and to make reliable predictions as to the behavior in service of that alloy. Today is indeed the age of metals.

Alloys are much more complex in their behavior and treatment than are the nearly pure metals. A great many aspects of the behavior of our common alloys are still not clearly understood or explained.

Most of the alloys were made by the "cut-and-try" method, and we have learned *how* before we have fully learned *why*.

Metallurgy may be one of the oldest practices of man; but until the latter part of the nineteenth century we were concerned chiefly with the extraction and production of the metals and alloys, rather than with the reasons for the behaviors they exhibited. During all this early work, it was believed that nearly all the characteristics found in the metals were created during the refining and producing stages, and that if the manufacturing methods

Metallurgy

could be controlled, a metal or an alloy could be produced that would possess the desired properties.

Shortly after the beginning of the present century, however, it became increasingly apparent that the properties of the common metals and their alloys were affected by many factors aside from methods and practices used in their manufacture. It became clearly evident that the properties of metals and their alloys were markedly affected by heat treatment, or heating followed by rapid cooling, and by the methods used in forming, such as rolling, drawing, etc. In an effort to understand and ultimately control these changes, metallurgists have engaged in active research in metals. Their work in chemical research, microscopic analysis, and study of the changes taking place during the heating and cooling of metals has resulted in a vast amount of useful knowledge. Consideration of the changes during heating and cooling, referred to as thermic-analysis, has been especially helpful. Information gathered by this method can be assembled to form the *alloy diagram.*

The difficulties encountered in the study of the characteristics of any series of alloys are lessened by use of the alloy diagram. The purpose of this study based on the alloy diagram is to simplify a difficult subject. The alloy diagram can be used by the student to predict what the normal structure of certain alloys will be. From this knowledge of the structural characteristics of any alloy, certain effects on the physical and chemical properties can be predicted also. Therefore, knowledge of these alloy diagrams, representing a system that will predict the structural and physical characteristics of any alloy, is of great value to workers in the metallurgical field. Let us then consider briefly how these alloy diagrams are constructed and interpreted.

Thermal curves. The alloy diagram is constructed from a series of cooling curves, which are obtained from a study of the cooling characteristics of any metal or alloy. These curves are obtainable because of the fact that bodies, while undergoing certain changes, liberate or absorb heat faster than their surroundings. Thermocouples placed in contact with the alloy while it is being heated or cooled uniformly within a certain temperature range, provide a means for studying the changes occurring during the heating and cooling cycle.

Time-temperature curve. The simplest type of curve is obtained by taking temperature readings at fixed intervals of time, and then plotting the results with temperatures as ordinates and time readings as abscissae. So long as the metal is being raised or lowered in temperature at a steady rate, this curve follows a smooth course. Any departure from this smoothness indicates that there has been either an abnormal absorption or evolution of heat within the specimen. Fig. 8-1 illustrates the common types of curves using the simple time-temperature method. Alloys when cooled from the liquid state to room temperature do not all undergo the same characteristic evolutions of heat; thus their thermal curves differ in appearance. The two most common types of alloys having different appearing thermal curves are: alloys that form a solid solu-

Theory of Alloys

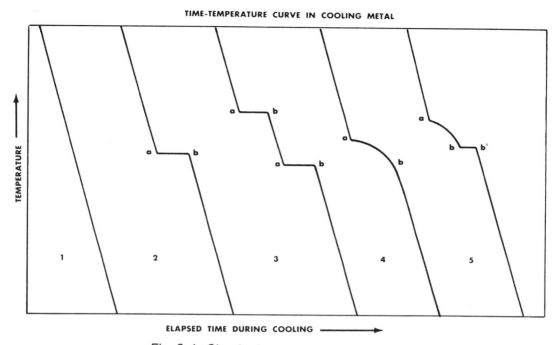

Fig. 8-1. Simple time-temperature curves.

tion, and alloys that form eutectic or eutectoid mixtures.

Curve 1, Fig. 8-1, is the cooling curve of a metal that undergoes no change within the temperature range of the cooling curve.

Curve 2, Fig. 8-1, is the cooling curve of three possible metals as follows: Case 1: cooling curve of a pure metal. A pure metal changes at a constant temperature from liquid to solid, or from one form of crystal structure to another (allotropic change) at a constant temperature. Case 2: cooling curve of a compound which solidifies at a constant temperature, similar to a pure metal. Case 3: cooling curve of a particular mixture of two metals known as the eutectic or eutectoid mixture, which transforms or freezes at a constant temperature.

Curve 3, Fig. 8-1, is that of a mixture of two elements that do not blend or influence each other in any way, so that the solidification of each takes place at its particular freezing temperature.

Curve 4, Fig. 8-1, is the cooling curve of an alloy of two metals which are mutually soluble in each other in both the liquid and solid states. The freezing or transformation of such an alloy takes place over a range of temperatures, *a* to *b* on the curve.

Curve 5, Fig. 8-1, is a cooling curve applicable to two cases. Case 1: when two metals are soluble in each other above *a,*

Metallurgy

but have zero solubility for each other below bb^1. With this condition, a eutectic or eutectoid mixture is formed, as indicated by the horizontal arrest in the cooling curve at bb^1, the excess of either metal being separated from solution along the curve from *a* to *b*—the resultant alloy being composed of either a eutectic or eutectoid and a pure metal. Case 2: where two elements are soluble in each other above *a*, but where the solubility becomes limited below bb^1. This condition results in the separation of a eutectic or a eutectoid mixture, similar to Case 1, but instead of the excess or primary metal to freeze being pure, it consists of a saturated solid solution.

Curves similar to those in Fig. 8-1 can be plotted by several different methods. Other methods of study of these changes also can be used. However, the time-temperature method will serve as an elementary illustration.

THE FREEZING OF ALLOYS

The alloy diagram. It is difficult to discuss the changes taking place during the cooling of an alloy from the liquid state to normal temperature without the aid of a diagram. The alloy diagram is to the metallurgist what the blueprint is to the toolmaker. The alloy diagram is made up from a number of cooling curves of a series of alloys of the metals used in the investigation. The coordinates for the diagram are temperature as the ordinate, and concentration or analysis as the abscissae. If temperature-time curves are used, time is visualized as in the plane perpendicular to the paper and is not shown on the diagram. Taking the simplest possible examples, the freezing of alloys is classified in five ways, as follows:

Type I alloys: the solid solution alloy system. The single cooling curve of this type of alloy is shown in Fig. 8-1, *curve 4*. Fig. 8-2 illustrates this alloy system by a diagram of the copper-nickel series. The following is a brief discussion of this diagram.

The upper line is known as the *liquidus,* for the alloy is completely liquid above this line. This liquidus line indicates the lowest temperature to which a given liquid composition can be lowered without freezing. It also indicates the compositon of the liquid alloy on the verge of freezing at any known temperature. The lower line of the diagram is known as the *solidus,* for all compositions in the area of the diagram below this line are in the solid state. The solidus line indicates the composition of the alloy which freezes at the given temperatures within the limits represented by the solidus curve. The area between the liquidus and solidus lines represents the mushy state of the alloy, partly liquid and partly solid.

Now, let us consider briefly the changes

Theory of Alloys

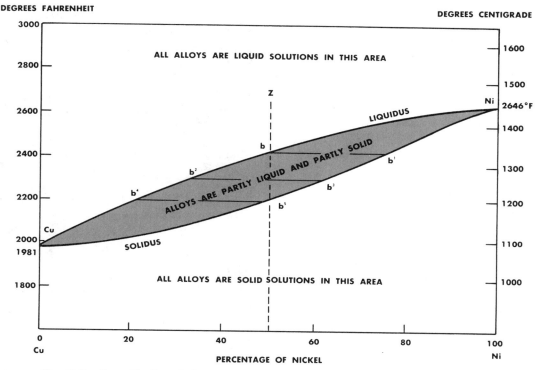

Fig. 8-2. Constitutional diagram of the copper-nickel alloys, Type I alloys.

that take place during cooling, when an alloy of 50 percent copper and 50 percent nickel is at a temperature indicated by Z, Fig. 8-2, and the temperature is lowered.

The alloy remains a uniform or *homogeneous* liquid solution until the temperature drops to a value indicated by the intersection of the liquidus line at *b*. Here the alloy begins to solidify, forming not pure crystals, but solid solution crystals. It is to be remembered that crystals form by progressive solidification along primary, secondary and ternary axes which form a skeleton onto which the remaining liquid solidifies. The primary axes of the crystals which form from a 50-50 liquid are not pure, but consist of a solid solution, the composition of which is found on the solidus line at b^1, 78 percent nickel and 22 percent copper.

As the mass cools, the composition of the growing crystals changes along the solidus line from b^1 to b^5, while the remaining liquid alloy varies in composition along the liquidus line from *b* to b^4. The solid that first solidifies from the liquid contains less copper than the liquid as a whole. The remaining liquid is thus left

Metallurgy

richer in copper than it was originally, and it therefore possesses a lower freezing point. As solidification proceeds, metal progressively poorer in nickel is deposited around the primary solid, that which solidifies last being richest in copper. The solidification, therefore, is not that of a single solution, but of an infinite number of solid solutions, and the solutions formed have a corresponding number of solidification temperatures, the result being a number of solid solutions of different chemical composition.

Hence, the structure of solid cast alloy differs from that of cast pure copper or nickel; it consists of dendritic crystals, originally so called from their branched tree-like appearance. These vary in composition, being rich in nickel at the center and rich in copper on the outside. Their lack of homogeneity can be corrected by diffusion if the rate of cooling is sufficiently slow, or if the alloy is rendered homogeneous by annealing at temperatures below the melting point. The rate of diffusion varies greatly for different metals, and the *heterogeneous* (non-uniform) cored structure persists longest in alloys having the slowest rate of diffusion. It is persistent in nickel alloys. Figs. 8-3 and 8-4 illustrate *heterogeneity* in a Monel metal, 67 percent nickel and 28 percent copper. Uniformity of composition of the crystal grains is assisted by mechanical work, i.e., by rolling, forging, etc., followed by annealing. All copper-nickel alloys in the rolled and annealed condition have a similar structure—that of a homogeneous solid solution, as indicated in Fig. 8-5.

Type II, eutectic alloys. These alloys are made up of two metals which are en-

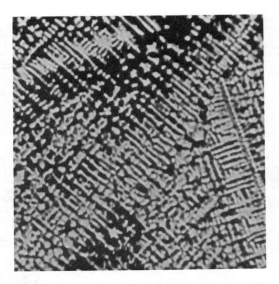

Fig. 8-3. Cast monel metal showing the dendritic or segregated structure; magnified 25 times.

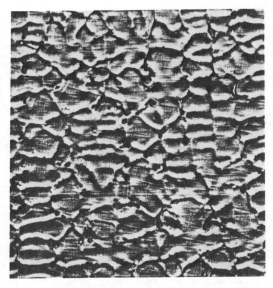

Fig. 8-4. Same as Fig. 8-3, but magnified 100 times; the rough cored structure is very apparent.

Theory of Alloys

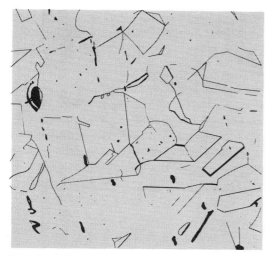

Fig. 8-5. Rolled and annealed monel metal; magnified 100 times. Compare this with Fig. 8-4.

tirely soluble in each other when liquid, but insoluble in each other when solid. When solid, they form two kinds of crystals or grains. Therefore, when they freeze, crystals of individual metals form; there is no solid solution. Only a few metals form alloys of this type, as a metal will rarely separate from a liquid solution in the pure state. Cadmium and bismuth, cadmium and zinc, aluminum and silicon, silver and lead are believed to represent this type of alloy.

A freezing-temperature-curve representing this Type II alloy is shown in Fig. 8-1, *curve 5*, and the diagram of cadmium and bismuth, Fig. 8-6 can be used to describe this type of alloy.

The liquidus of the diagram in Fig. 8-6 is *AEB*, and the solidus is *ACEDB*. The line *CED* is often referred to as the *eutectic line*, and point *E* as the *eutectic point*.

If a small amount of cadmium is added to molten bismuth, the freezing point of the bismuth is lowered. If, instead, a small amount of bismuth is added to pure cadmium, the freezing point of the cadmium is lowered. It is apparent, since each metal lowers the freezing point of the other, that the lines connecting these freezing points must intersect at some point as shown by point *E* in Fig. 8-6. This point of intersection, sometimes called the *eutectic point*, is of the greatest importance. The composition freezing at this point is called the *eutectic alloy*.

The alloy of this eutectic composition melts and freezes at a constant temperature, in this respect behaving as a pure metal. The eutectic alloy has the lowest melting point of any composition in the series. The eutectic is not a homogeneous alloy, but consists of crystals of nearly pure cadmium and crystals of nearly pure bismuth, approximately 40 percent cadmium and 60 percent bismuth. Point *E* of Fig. 8-6 informs us as to the chemical composition of the eutectic, and the extreme ends of the eutectic line, *C* and *D*, point out that the eutectic is made up of nearly pure cadmium and nearly pure bismuth crystals.

If the liquid solution contains less than 60 percent bismuth (the eutectic composition), when the temperature is lowered until the *AE* line is intersected, nearly pure cadmium begins to separate from the liquid solution. The separation of cadmium and the lowering of the freezing point go along the curve *AE*, Fig. 8-6, until the remaining liquid solution contains 40 per-

Metallurgy

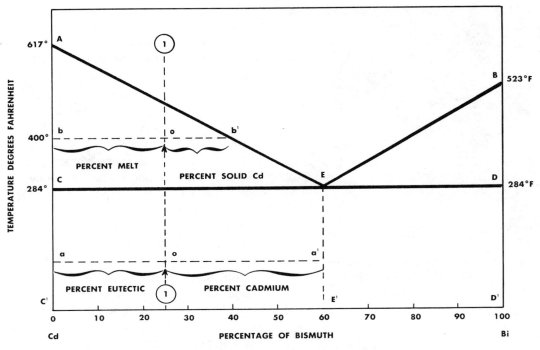

Fig. 8-6. Cadmium-bismuth diagram, Type II alloys.

cent cadmium and 60 percent bismuth. If the solution contains more than 60 percent bismuth, then nearly pure bismuth separates from the liquid, and the separation of bismuth and the lowering of the freezing point go along the curve *BE,* Fig. 8-6, until the remaining liquid solution contains the eutectic composition.

No matter what the original analysis was, the liquid solution of cadmium and bismuth at temperature 284°F (140°C) (the eutectic temperature) always contains 40 percent cadmium and 60 percent bismuth. This so-called eutectic mixture freezes at a constant temperature, forming crystals of nearly pure cadmium and nearly pure bismuth.

The photomicrographs in Figs. 8-7 to 8-10 show the structures developed in *Type II, eutectic alloys,* as illustrated by the cadmium-bismuth alloys.

You will note that in this discussion we say that the crystals which separate are nearly pure cadmium or nearly pure bismuth. Careful analysis will show that the cadmium crystals actually contain some bismuth, and the bismuth crystals some cadmium. Only in a very few instances are the crystals which separate really pure metal. In most cases the separating crys-

Theory of Alloys

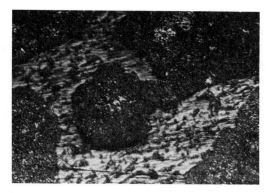

Fig. 8-7. The structure obtained if an alloy containing 75 percent cadmium and 25 percent bismuth is cooled from the molten state to room temperature. This alloy contains approximately 42 percent eutectic (wavy, flake-like structure) and approximately 58 percent nearly pure cadmium (dark structure). (Principles of Metallography by Williams and Homerberg, McGraw-Hill Book Co., Inc.)

Fig. 8-9. The structure developed with an alloy containing 20 percent cadmium and 80 percent bismuth. The structural composition consists of approximately 50 percent eutectic and 50 percent nearly pure bismuth (light constituent). (Principles of Metallography by Williams and Homerberg, McGraw-Hill Book Co., Inc.)

Fig. 8-8. An alloy containing the eutectic composition 40 percent cadmium and 60 percent bismuth. The structure of this alloy is 100 percent eutectic (wavy, flake-like structure). The dark constituent of the eutectic is nearly pure cadmium and the light constituent is nearly pure bismuth. (Principles of Metallography by Williams and Homerberg, McGraw-Hill Book Co., Inc.)

Fig. 8-10. The structure developed with an alloy containing 10 percent cadmium and 90 percent bismuth. This alloy contains approximately 25 percent eutectic and 75 percent nearly pure bismuth (light constituent). (Principles of Metallography by Williams and Homerberg, McGraw-Hill Book Co., Inc.)

Metallurgy

tals contain more or less of the other constituent. Such crystals are called mixed crystals or *solid solutions,* indicating that the crystals contain more than one metal. The term *solid solutions* is to be preferred to that of *mixed crystals,* since the crystals are not a mixture but a solution, in that the metals which are present in them cannot be separated by mechanical means.

As an example of this solid solution formation, let us consider the lead-antimony alloys rich in lead, which are represented in Fig. 8-11. If we consider the alloy represented by the point *o,* an alloy containing 3.2 percent antimony and held at a temperature of 660°F (332°C), it is entirely liquid. If we allow it to cool to the point C and skim the separated crystals, we will find on analyzing them that they are not pure lead but a solid solution of antimony in lead, containing the amount of antimony which is represented by the point G. Now as we cool and skim, we find that the amount of antimony in the skimmings increases up to a definite limit, which is called the limit of solubility, represented at M. The area Pb-M-N then represents a temperature and composition range in

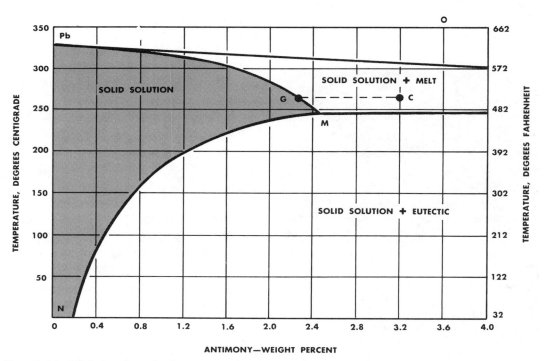

Fig. 8-11. High-lead end of antimony-lead equilibrium diagram. (American Society for Metals)

which the entire alloy will be in the solid solution range.

It will be noted that the limit of solubility of antimony in lead is reached at the point M, and that as we lower the temperature this solubility decreases along M-N as the temperature falls. This is analogous to the separation of salt from the liquid solution of it in water as a hot solution is cooled below its solubility limit.

Now this solubility limit and the change of solubility with temperature vary with every pair of metals. In some cases the limit of solubility becomes infinite, that is, the two solid metals form solutions in all proportions, as water and alcohol form liquid solutions. Such a case is copper-nickel (Fig. 8-2).

Type III alloys. Very few alloys which form perfect molten solutions crystallize into two pure metals as described in Type II solidification. Nevertheless, a large majority of the alloys which form molten solutions separate during freezing to form two kinds of crystals. These crystals are composed of the metals which are not pure, but which carry with them some part of the alternate metal in the state of solid solution. In other words, the two metals crystallize and separate, but the crystals are not pure; each contains some of the other metal as a solid solution.

The behavior of these alloys during freezing, and the shape of their diagrams is of an intermediate nature when compared to the diagrams of the alloys of Type I and Type II. A diagram of this type is shown in Fig. 8-12, the copper-silver system. Silver is capable of retaining 8.8 percent copper in solid solution, and alloys in this series will solidify, if slowly cooled, in a manner typical of solid solutions, Type I, and will exhibit the characteristic cooling curves of the Type I alloys.

On the other hand, those alloys of this series having compositions which place them beyond the limit of solid solubility of the metals, at either end of the series, behave in most respects very like the alloys which are entirely insoluble in one another in the solid state. The only difference is that the solid which first crystalizes is not pure metal, but a *saturated solid solution* of one metal in the other, similar to the saturated solid solution formed in the lead-antimony alloys. A complete diagram of such a system therefore, is made up of three distinct parts: a central portion BC, resembling the diagram of Type II alloys, and portions AB and CD at each end, typical of the Type I alloys.

The areas AB and CD of Fig. 8-12, as well as the Type I alloys (Fig. 8-2) represent the same characterisitic behavior of the alloys during solidification. Both form solid solutions upon solidification, which in the cast state are heterogeneous, with the crystals having a cored or dendritic structure. The solubility or *miscibility* of one metal in another in Types I, II and III alloy systems, then, are as follows:

Type I. The metals are *miscible* in both the liquid and solid states; the crystals separating from the liquid contain both metals in solution.

Type II. The two metals are miscible in the liquid state but separate as nearly pure crystals of the two metals when the melt solidifies.

Type III. The two metals are miscible in the liquid state, but are only partially miscible in the solid state.

Metallurgy

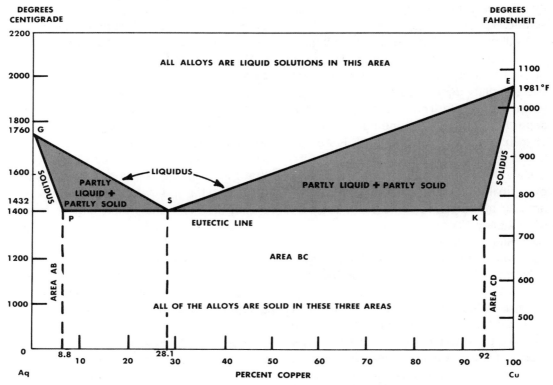

Fig. 8-12. Copper-silver diagram, Type III alloys.

Type IV alloys. Not all metals are miscible in the liquid state. Thus lead and zinc or lead and copper do not mix when melted together, but form two layers, as do water and oil. This immiscibility is of commercial importance in making bearing bronze in which droplets of lead are mixed through molten copper like an emulsion of oil in water, and the whole is solidified. Two metals, however, are rarely completely insoluble. Each one dissolves some of the other, and as the temperature is increased, eventually a point is reached where any metal becomes miscible. Fig. 8-13 shows an equilibrium diagram for lead-zinc.

Type V compound-forming alloys. In the study of alloys, we find that one of the most important of all possible behaviors takes place when two metals are added to each other and they combine to form a compound. Such a compound is called an *intermetallic compound* and is similar to the common compounds or combinations such as water or ice, H_2O, or sodium chloride, NaCl, etc.

Theory of Alloys

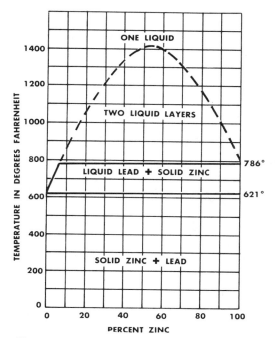

Fig. 8-13. Lead-zinc equilibrium diagram, Type IV alloy.

The alloy diagram of the compound-forming metals may take on many very different appearances. For discussion and introduction to this type of alloy behavior, Fig. 8-14 may be used, and the alloys of cadmium-nickel, iron-carbon, magnesium-tin, and many others may be used as illustrations of this type of behavior among the industrial alloys.

The compound formed in the alloy diagram presented in Fig. 8-14 is designated as *AmBn*, and has a composition of 50 percent metal *A*, and 50 percent metal *B*. A melt of 50 percent *A* and 50 percent *B* will solidify at temperature *T* to form a homogeneous solid, an intermetallic compound. It will be seen from a study of this diagram that two Type II eutectic alloys are formed by this behavior.

The components of the eutectic E_1, Fig. 8-14, are nearly pure metal *A* and the compound *AmBn*; whereas, the components of the eutectic formed at E_2 are nearly pure metal *B*, and the compound *AmBn*. We actually have two separate eutectic diagrams in one, and the discussion of such behavior is similar to that of a simple eutectic-forming alloy system with the intermetallic compound *AmBn* considered as a pure substance and as one of the components of the system. With this in mind, the significance of the various areas of the diagram becomes evident.

The alloy diagram of peritectic reactions. Under certain conditions of alloy behavior, some metals undergo a reaction between the solid and liquid phases during the process of solidification, which results in the disappearance of the two phases, and in their place a new solid phase is born. The new solid phase

In general, these intermetallic compounds are the principal hardeners in our industrial alloys, the compounds being hard and brittle with little or no strength. Such compounds or hardeners are found in alloys such as steel, aluminum alloys, and bearing metals, in which the compounds Fe_3C, $CuAl_2$, $NiCd_7$, $CuSn$, $SbSn$, etc., are found. Some of the hardest substances known to man are formed from compounds such as WC (tungsten carbide) and B_4C (boron carbide), both of which are harder than sapphire and nearly as hard as the hardest substance known, diamond.

Metallurgy

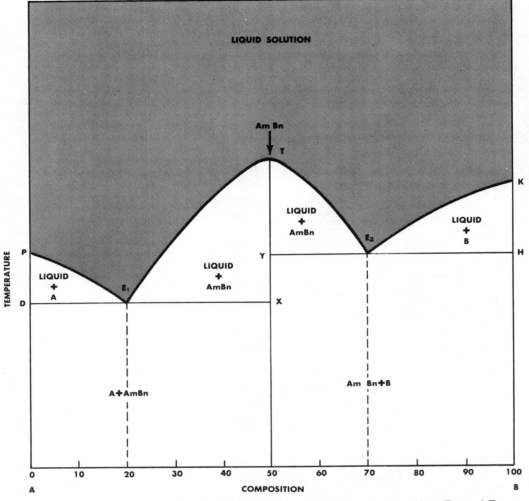

Fig. 8-14. Type V alloy behavior. Alloys that form two separate eutectics, E_1 and E_2, and a compound AmBn.

formed from a reaction between the liquid and solid phases may be a solid solution or an intermetallic compound. This reaction, resulting in the disappearance of two phases and the creation of a new one, has been called a *peritectic reaction*.

Such alloy behavior is illustrated by the diagram Fig. 8-15, where H corresponds to the formation of a compound *AmBn*. This compound *AmBn* forms along the

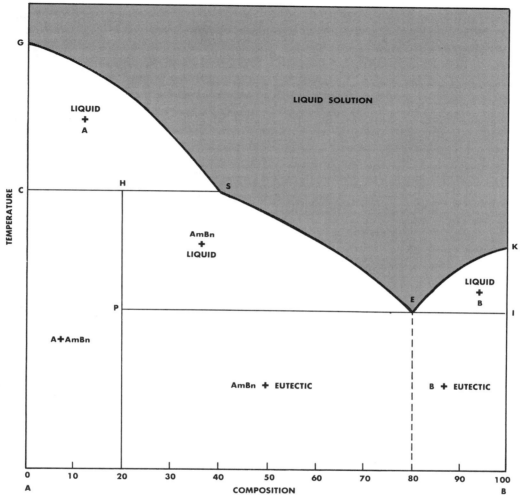

Fig. 8-15. Diagram of alloy behavior that results in a peritectic reaction, which causes the formation of a compound AmBn.

CHS line of the diagram from the reaction of a nearly-pure solid (metal A), shown by point C, and the liquid solution of composition S. If the original melt contained just 80 percent metal A and 20 percent metal B, the result of this peritectic reaction gives birth to a pure compound AmBn.

In the discussion of this diagram, we discover that nearly pure metal A separates from the liquid along the line GS,

145

Metallurgy

and upon cooling, from *GS* to *CHS*, we have a solid *A* and a liquid solution which may vary in composition from *G* to *S*. Upon cooling to the *CHS* line, the solid *A* reacts with the liquid solution of composition *S*, forming the compound *AmBn*. In case the original liquid contained less than 20 percent metal, *B*, the final structural composition would contain metal *A* plus the constituent *AmBn*.

With an original composition between 20 and 40 percent metal *B*, the reaction along the *CHS* line would be evidenced by the formation of the compound *AmBn* and a liquid. Cooling from the *HS* line to the *PE* line would result in more compound forming from the liquid solution and the remaining liquid becoming richer and richer in metal *B* until the temperature has dropped to the *PE* line, at which time the remaining melt would have a composition as shown by point *E*, or a eutectic composition. The remaining liquid at this time would solidify to form a eutectic structure. The final alloy would consist of crystals of *AmBn* and the eutectic.

The eutectic of this alloy solidifies along the *PEI* line, Fig. 8-15, with a eutectic composition which contains 20 percent *A* and 80 percent *B*, and consists of crystals of the compound *AmBn* and crystals of nearly pure metal *B*. The eutectic structure of this alloy would be similar in appearance to that of any common binary alloys solidifying with a eutectic structure.

Changes in the solid state. In dealing with the alloy diagram, we have discussed changes occurring during the process of solidification. The study of the alloy behavior is complicated by the fact that solids, during heating and cooling undergo certain changes in their structural makeup that may completely alter their properties. Such changes can be shown by the alloy-diagram principle. Some of the most common changes that occur include:

1. Changes in solid solubility upon cooling.
2. Changes from one allotropic form to another.
3. Formation of eutectoid structures.
4. Grain growth.
5. Recrystallization.

The first two changes, changes in *solid solubility* and changes *from one allotropic form to another,* are common changes that may be illustrated by the alloy-diagram method.

The iron-carbon alloys, which will be discussed in more detail later on, may be used as an illustration of these two important changes taking place during the cooling of a solid. Fig. 8-16 is a simplified iron-carbon diagram showing the changes that take place within a solid solution, which exists in area *AESGA,* when the solid solution is cooled to normal temperatures. The maximum solubility of Fe_3C in gamma (γ) iron is shown by point *E*. As cooling of the solid solution takes place, the solubility of gamma iron for Fe_3C follows along the *ES* line to point *S*.

At point *S,* all iron becomes alpha (α) iron, changing from the gamma form, and the solubility of the new alpha iron phase becomes practically zero, represented by point *P*. Point *G* represents the change of pure iron from one allotropic form to another, the change involving a rearrangement of a face-centered cubic atomic lattice form of iron to a body-centered cubic form of iron. The structure born along the

Theory of Alloys

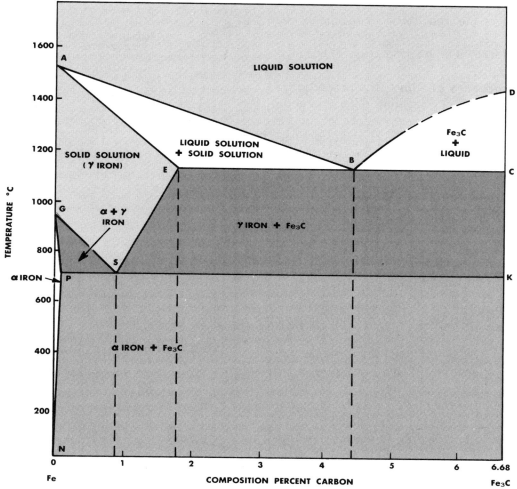

Fig. 8-16. Simplified iron-carbon diagram.

PSK line, from a solid solution, is similar to that formed along the *EBC* line from the liquid solution. However, one is born from a solid, and the other from a liquid.

The structure formed at *B* is called the *eutectic;* whereas the structure born at *S* is called the *eutectoid.* The student should note the similarity in appearance of the diagram in the area of solidification and the changes that follow during the cooling of the solid solution alloy. Line *AB* is similar to *GS, BD* similar to *SE, PSK* similar to *EBC,* and *AES* similar to *GPN*. The two changes, i.e., one of changing solubility

Metallurgy

and the other a change in the allotropic form of the iron, result in reactions that may be measured and account for the lines formed in the solid areas of this alloy diagram.

Properties of alloys. The properties of alloys may or may not differ markedly from their constituent metals. Much can be predicted about the properties of an alloy from a knowledge of the equilibrium diagram. The following rules will be useful:

1. If the two elements do not form solid solutions or compounds, their alloys will have properties intermediate between the two elements as would be expected of a mechanical mixture. See Fig. 8-17. This applies to hardness, electrical conductivity, color, and magnetic properties.

2. If the two elements form a solid solution, the alloy will be harder and stronger and have greater electrical resistance than would be obtained with a simple mixture of its constituent metals. See Fig. 8-18. Its color and magnetic properties cannot be predicted. In general, it may be stated that the less the solubility of one metal in another, the greater will be the hardening and the more difficult it will be to dissolve a given per cent of it in the second metal. Thus nickel, which is soluble in all proportions in copper, does not affect the properties of copper to anywhere near the extent of phosphorus, which is only slightly soluble.

3. If the two metals form a compound, nothing can be predicted of its properties from those of its constituents.

Finally, the properties of some alloys may be profoundly altered by heat treatment. These are the solid solution alloys which show a change of solid solubility

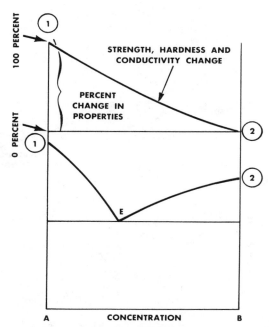

Fig. 8-17. Possible effect of behavior of Type II alloy upon the strength, hardness, and conductivity.

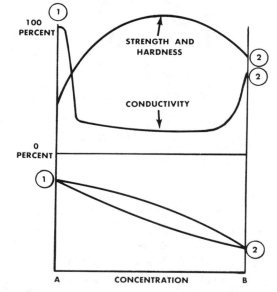

Fig. 8-18. Behavior that might be expected from a system of Type I alloy.

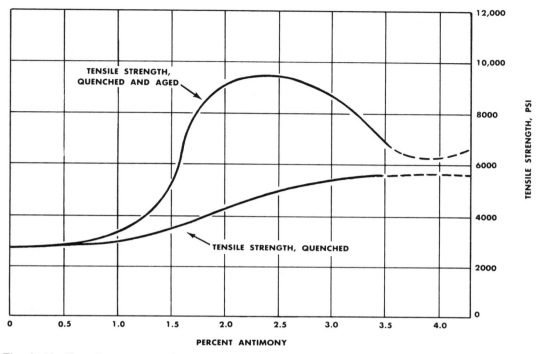

Fig. 8-19. Tensile strengh of lead antimony alloys, after quenching from 450° F (232° C), and after aging at room temperature.

with temperature, as described for lead-antimony. Such alloys may be greatly hardened by heating to the solid solution range and cooling suddenly, as by quenching in water, and subsequent aging at room or elevated temperature. The quenching preserves the solid solution which by the aging process becomes finely divided particles of one metal in the other. Such a finely divided mixture of alloy crystals is much harder and stronger than the same mixture in coarser particles. The tensile strengths of quenched and aged lead-antimony alloys are shown graphically in Fig. 8-19.

THE IRON-CARBON DIAGRAM

Alloys of iron and carbon are the most widely used and least expensive of all metal alloys. The importance of iron-carbon alloys is due to the fact that their

Metallurgy

properties may be very decidedly changed by heating and cooling under controlled conditions.

Iron, as an elemental metal, is relatively soft with low tensile strength, incapable of being hardened and of little use in industry. One important discovery about iron was that it is *allotropic,* meaning that it can exist in more than one type of crystalline structure, depending on temperature.

The term *critical point* is a name for a particular temperature at which iron or steel will undergo a molecular change which alters the metal's crystalline structure and physical characteristics. An examination of the critical points of a common substance, water, will give a clearer view of the temperature-related physical changes as steam is cooled.

The cooling curve for water. A cooling curve for water is shown in Fig. 8-20. At 212°F (100°C) steam condenses into water. This is a critical point and is shown as a plateau in the cooling curve. Again at 32°F (0°C) the water changes to ice. This is the second critical point with a second plateau. Steam, water and ice are all forms of water (H_2O) but with much different physical characteristics.

Notice also in Fig. 8-20 that a certain amount of time is required for steam to change to water and for water to change to ice. While these changes are taking place, the temperature remains constant (*isothermal*) until the complete transformation has taken place. These plateaus on the chart are typical of allotropic change and result from internal energy

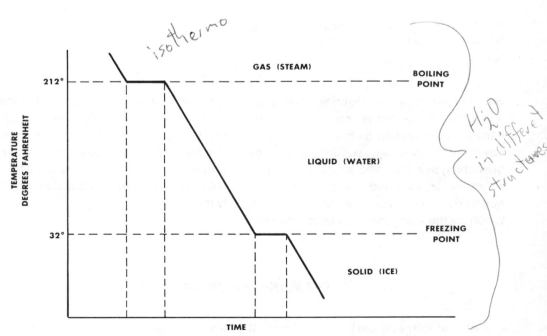

Fig. 8-20. Time-temperature cooling curve for H_2O.

Theory of Alloys

dissipation (a measurable amount of energy is given off). This is related to atomic rearrangement within the molecules.

The cooling curve for pure iron. Iron and steel behave in much the same way as water. Fig. 8-21 shows the cooling curve for pure iron with four critical points (plateaus) shown. When liquid, pure iron at about 3,000°F (1640°C) is cooled below its melting point of 2795°F (1535°C) it crystallizes into what is called a *body-centered cubic crystal* and is known as *delta* iron.

During the process of solidification, the atoms of liquid metal arrange or group themselves into an orderly pattern. This pattern is known as the space lattice. Fig. 8-22 represents a *body-centered cubic* (B.C.C.) structure comprised of 9 atoms; 8 on each corner and one in the center of the space lattice.

The second critical point at about 2535°F (1390°C) sees an allotropic change from B.C.C. crystals to *face-centered cubic* crystals known as *gamma* iron. Fig. 8-23 represents a *face-centered cubic* (F.C.C.) comprising of 14 atoms; 8 on each corner and one in the center of each face (6). A very important physical characteristic of gamma iron is that it is non-magnetic.

At about 1670°F (910°C) a transitional form of B.C.C. crystals are formed and are called *beta iron*. This phase is still non-magnetic but at 1414°F (768°C) the familiar magnetic *alpha* iron is formed.

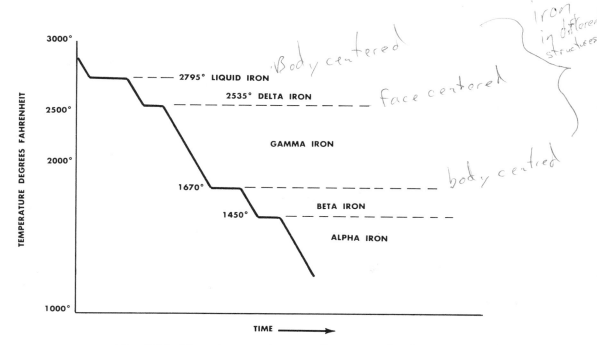

Fig. 8-21. Time-temperature cooling curve for iron.

Metallurgy

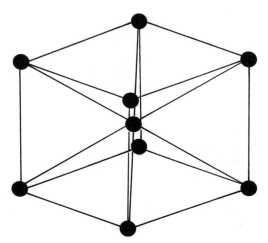

Fig. 8-22. Body-centered cubic crystal.

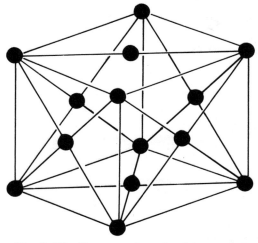

Fig. 8-23. Face-centered cubic crystal.

The cooling curve for carbon steels. Steel is an alloy of iron and carbon and is properly referred to as *carbon steel* or *plain steel* to differentiate this alloy from the multi-alloyed steels called *alloy steels*.

When carbon is added to iron, the resulting alloy, steel, may be strengthened and given hardenability in proportion to the percentage of carbon added. Steel is defined as having a carbon range of from 0.008 to 1.7 percent carbon. Irons with carbon percentages above 1.7 percent are known as cast irons; these are discussed in a later chapter.

The hot, liquid metal produced by a blast furnace is known as *pig iron* and is a solution of carbon dissolved in the liquid iron, much like sugar dissolved in water. The percentage of carbon in pig iron is from 3 percent to 5 percent but never more than 6.68 percent which is the solubility limit of carbon in iron. Subsequent refining processes reduce the carbon content into the usable ranges for steels, 0.02 to 1.7 percent and cast irons, 1.7 to 4.5 percent.

Plain carbon steels are classified by carbon content into three groups; (1) low-carbon steel often called *mild steel,* with 0.02 to 0.30 percent carbon, and non-hardenable, (2) medium-carbon steel with 0.30 to 0.60 percent carbon, and moderately hardenable, and (3) high-carbon steel with 0.60 to 1.7 percent carbon and greatly hardenable. The steel men refer to percentages of carbon as points so that 0.30 percent is called 30 points, etc.

As hot liquid steel cools below 2800°F (about 1540°C) and begins to crystallize, the carbon drops out of solution (precipitates) and chemically combines with the iron to form a very hard compound, iron carbide (Fe_3C), which is called cementite. The greater the percentage of carbon present, the greater the amount of cementite formed.

If a series of samples of iron-carbon al-

Theory of Alloys

loy with varying percentages of carbon from 0 to 6.68 percent are made and their critical points determined in the laboratory, a composite cooling curve called the *iron-carbon diagram* may be charted. This chart is shown in Fig. 8-24.

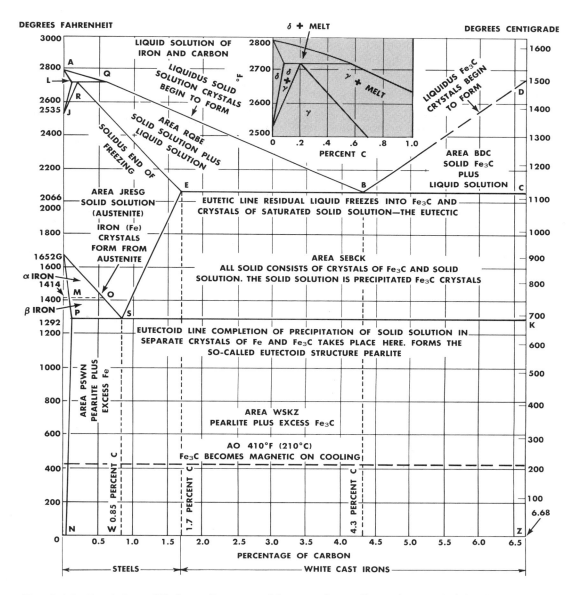

Fig. 8-24. Partial equilibrium diagram of iron-carbon alloys. Inset of delta iron region. (After Frank Adcock)

153

Metallurgy

TEMPERATURE RELATED STRUCTURES

Solidification. Pure iron, represented by the point A, and the iron-carbon eutectic represented by the point B, melt and solidify at a constant temperature. All other alloys represented in the diagram melt and freeze over a range of temperatures.

Alloys containing 0.0 to 4.3 percent carbon begin to solidify with the separation of solid solution (austenite) crystals from the liquid.

Alloys containing more than 4.3 percent carbon begin to solidify with the separation of Fe_3C (cementite) from the liquid, on cooling to the *BD* line.

In the case of alloys containing 1.7 percent or less, solid solution (austenite) begins to freeze out on cooling from *AB* to *ALRE*, Fig. 8-24. At *ALRE* all the alloy is solid, and consists of dendritic crystals. These dendritic crystals are formed, not of a single solid solution, but of an infinite number of solid solutions. The first to freeze is a solid solution relatively low in carbon content; the last to freeze is relatively rich in carbon concentration. It is to be recalled that this lack of homogeneity can be corrected by diffusion if the rate of cooling is sufficiently slow, or if the alloy is rendered homogeneous by hot working or by annealing at temperatures below the melting point. For the present it can be assumed that the area *JRESG* of the diagram represents an alloy in one homogeneous phase, *austenite.*

In alloys containing from 1.7 to 4.3 percent carbon, solid solution (austenite) freezes out of the liquid, beginning at *AB* and continuing as the temperature is lowered to *EB*, Fig. 8-24. At *EB*, the eutectic line, there is some residual liquid, which is of the *eutectic composition*. This liquid then solidifies at a constant temperature, forming the eutectic mixture.

The eutectic of the iron-carbon system consists of alternating layers of Fe_3C (cementite) and saturated solid solution (austenite) crystallizes simultaneously at *EB* to form the eutectic structure. Therefore, alloys containing from 1.7 to 4.3 percent carbon, just below the *EB* line, consist of mixed crystals of two constituents, austenite and eutectic. The amounts of each vary from 100 percent austenite and 0.0 eutectic with 1.7 percent carbon alloy, to 100 percent eutectic and 0.0 austenite with 4.3 percent carbon iron alloy.

In alloys containing more than 4.3 percent carbon, Fe_3C (cementite) freezes out on cooling from *BD* to *BC,* Fig. 8-24; the residual liquid at *BC* is of eutectic composition and solidifies at a constant temperature, forming the eutectic constituent. These alloys just below BC consist of mixed crystals of cementite and eutectic.

In the discussion of the other diagrams, Types I, II, and III alloys, it was considered that the structure of the solid alloy just below the solidus line and the eutectic line, remained unchanged during the cooling to room temperature, except for diffusion taking place when dealing with a heterogeneous solid solution phase. The structure at room temperature was that of the cast alloy. However, with iron-carbon alloys, very marked changes occur during the cooling from the solidus *ALRE* and

Theory of Alloys

eutectic *EBC*, Fig. 8-24, to room temperature. These changes are brought about principally because iron has the ability to exist in several allotropic forms, and during the cooling the iron changes from one of these forms to another.

Allotropy. In the *JRESG* area and the *SEBCK* area, Fig. 8-24, iron exists wholly in the form known as *gamma*. This gamma iron has its atoms arranged in the face-centered cubic lattice. Gamma iron dissolves carbon; its grain size depends on temperature, time and working; and it is nonmagnetic. Gamma iron is denser than alpha iron. At normal temperatures, and in the areas *PSWN* and *WSKZ*, Fig. 8-24, iron exists wholly in the form known as *alpha*. Alpha iron has a body-centered cubic lattice, is found to be magnetic, dissolves little carbon, and its grains do not change under normal conditions. Forging of the alpha iron at room temperature causes a distortion of the grains.

Other allotropes of iron. It is reported that at 2535°F (1390°C) gamma iron changes to a different form called *delta* iron. This delta iron is said to have a body-centered atomic arrangement similar to alpha iron.

Pure iron undergoes certain discontinuous changes in physical properties at 1414°F (768°C). These are small changes in the internal energy, volume, electrical conductivity, etc. There is a marked change in the magnetic properties. At temperatures a little below 1414°F (768°C) iron is strongly magnetic, while at higher temperatures it is not.

Because of these changes, it has been held that at temperatures between 1414°F (768°C) and 1652°F (900°C), iron exists in the form of a distinct allotropic modification which has been called *beta* iron. As the structure of both alpha and beta iron is body-centered, this structure remains unchanged in passing from alpha to beta iron. It is stated that the marked change in the physical properties which takes place at 1414°F (768°C) is caused by a change within the atom of the iron.

Ferrite solubility curve. Pure gamma iron exists on cooling from *J* to *G*, Fig. 8-24. At *G* all the pure gamma iron changes to beta iron. This beta iron exists during cooling to *M*, changing at M to alpha iron, and this alpha form remains unchanged below *M*. No change takes place at *P* with a pure iron. This alpha iron is capable of holding in solution considerable amounts of various elements such as nickel, silicon, phosphorus, etc. The solubility for carbon is very slight and is not definitely known, being perhaps less than 0.015 percent carbon at normal temperature. The term *ferrite* is applied to solid solutions in which alpha iron is the solvent.

The temperature at which alpha iron forms from gamma iron on cooling is lowered by the presence of carbon in solid solution in the gamma iron, that is, in austenite. This is represented in the diagram by the line *GOS*, Fig. 8-24, which shows that the temperature of formation of ferrite decreases from 1652°F (900°C) with pure iron, to 1292°F (700°C) for an alloy containing 0.85 percent carbon. The formation of ferrite from the solid solution austenite is analogous to the precipitation of cadmium from the liquid solution of cadmium and bismuth (Fig. 8-6). The line *GOS*, Fig. 8-24, expresses the solubility of ferrite in austenite, and may be referred to

Metallurgy

as the ferrite solubility curve. The straight line *GOS* is often represented by a curved or broken line.

Cementite solubility curve. At the eutectic temperature 2066°F (1130°C) austenite will hold 1.7 percent carbon in solid solution. When austenite of this composition is cooled, Fe_3C (cementite) precipitates from solution. The solubility of carbon in austenite at various temperatures is shown by the line *SE*, Fig. 8-24, which may be called the cementite solubility curve, as cementite is the phase which is separated as the temperature drops below this line. The maximum solubility at 2066°F (1130°C) is 1.7 percent carbon; at 1292°F (700°C) the maximum solubility becomes 0.85 percent carbon.

The eutectoid. It is thus seen that, no matter what carbon content is started with, austenite reaches a temperature of 1292°F (700°C) on *slow cooling* with a carbon content of 0.85 percent carbon. This situation will exist only so long as the metal remains at 1292°F (700°C). It cannot cool below this point without a complete separation of all the solid solution austenite into crystals of ferrite and cementite.

On cooling through this temperature, austenite containing 0.85 percent carbon deposits simultaneously a mixture of ferrite and cementite. This mixture is called the *eutectoid,* from analogy to the eutectic. The eutectoid structure formed from a saturated solid solution austenite at 1292°F (700°C) is often referred to as *pearlite,* because its appearance when viewed with the aid of a microscope is similar to that of mother-of-pearl, see Fig. 8-25.

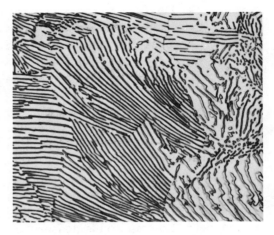

Fig. 8-25. 0.85 percent carbon steel, all pearlite. Magnified 2800 times.

Cooling of a structural steel. If a structural steel containing 99.75 percent iron and 0.25 percent carbon is cooled slowly from above the liquidus *AB* to normal temperature, it will solidify in the usual manner of a solid solution of its components, carbon and iron. But with a slow rate of cooling, this solid solution will not remain in the state of chemical solution below the *GOS* line, Fig. 8-24. When it reaches the lowest temperature at which the solid solution austenite can retain all the iron in solid solution, approximately 1550°F (840°C), any further cooling will result in the separation of beta iron crystals out of the austenitic crystals. The structure of the steel will then consist of beta iron crystals formed around the solid solution austenitic crystals.

As cooling continues, more beta iron crystals are precipitated from the austenite, there being a constantly increasing number of grains of beta iron among the solid solution crystals.

When the temperature drops to the *MO* line, Fig. 8-24, all the beta iron crystals change to alpha iron crystals; no other change takes place at *MO*. Any further cooling results in more iron crystals separating from the remaining austenite, the iron crystals now being the alpha iron form. This separation continues until the line *PS* is reached, Fig. 8-24. At this temperature the steel consists of a mixture of solid solution (austenite) crystals containing 0.85 percent carbon and crystals of alpha iron (ferrite). Practically all of the carbon is now concentrated in the solid solution; the alpha iron (ferrite) grains will be almost free from it. The steel cannot be cooled below this *PS* line, 1292°F (700°C), without a complete separation of all the solid solution into crystals of alpha iron (ferrite) and Fe_3C (cementite). Therefore at a constant temperature of 1292°F (700°C), two changes in the austenite take place simultaneously: first, the gamma solid solution iron changes to alpha iron; second, the carbon that was dissolved in the solid solution austenite is separated from it to form separate crystals of Fe_3C (cementite). Further cooling to room temperature takes place without change of structure.

Cooling of a tool steel. When a tool steel containing 1.5 percent carbon is first frozen, it is composed structurally of a solid solution of gamma iron and carbon. As it cools, however, the alloy soon reaches a temperature where the gamma iron cannot hold this amount of carbon in solid solution, which results in a separation of Fe_3C (cementite) from the solid solution

austenite. This temperature corresponds to the intersection of the *SE* line, Fig. 8-24, which is at approximately 1900°F (about 1040°C). The precipitation of Fe_3C cementite crystals continues during the cooling.

When 1292°F (700°C), indicated by line *SK*, Fig. 8-24, is reached, the steel is composed of part solid solution austenite, containing 0.85 percent carbon, and part Fe_3C (cementite), containing 6.67 percent carbon. Finally, at *SK*, Fig. 8-24, the remaining austenite separates into crystals of alpha iron and cementite, forming the structure pearlite. The ultimate structure consists of crystals of free cementite (approximately 10 percent) forming a network around the crystals of eutectoid pearlite, see Figs. 8-26 and 8-27. This steel is classified as a *hypereutectoid* steel.

Slow heating and slow cooling of steel. It should be understood that the description of changes taking place during

Fig. 8-26. 1.43 percent carbon steel; the dark is pearlitic and the light area is free cementite. Magnified 50 times.

Metallurgy

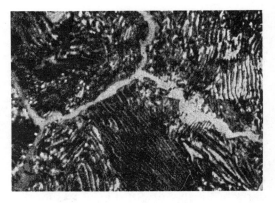

Fig. 8-27. 1.43 percent carbon steel; same as Fig. 8-26. Magnified 500 times.

the cooling of steel refers to cooling at a rate slow enough to allow the changes to occur in a normal manner. Each change requires some time, perhaps several minutes, and normal change is not obtained unless the proper rate of cooling is employed. It should also be understood that changes occurring on slow cooling are reversed on heating.

Decomposition of cementite. If alloys which contain more than 1.7 percent carbon are slowly cooled, the cementite decomposes into iron and graphite. The graphite, in the usual form of fine soft flakes, tends to weaken and embrittle the alloy. Most iron castings are made in sand molds which allow the iron to cool comparatively slowly; this permits the formation of graphite. The graphite as it occurs in slowly cooled or grey cast iron is shown in Fig. 7-16. If the iron is cast into metal molds, it cools so rapidly that the cementite does not decompose and come out of solution in flakes of graphite. This cementite is responsible for the hardness of white or *chilled* cast iron.

The decomposition of the cementite into graphite and iron is influenced not only by the rate of cooling, but by the other elements present in the iron. A high silicon content tends to increase graphitization; thus when thin sections of soft iron are desired, a high silicon iron is used to insure the decomposition of the graphite. Manganese has the opposite effect; and irons having a high manganese content may have hard portions because the manganese prevents the formation of graphite.

Crystalline grain. The grain size changes that occur during the heating and cooling of metals and alloys are very important. The changes can be classified as follows: crystallization, crystal growth, recrystallization.

Crystallization of the liquid into a solid usually results in relatively large dendritic grains. The best way to break down these large dendritic grains is by hot working.

Crystal growth takes place in most metals and alloys at relatively high temperatures, but below the solidus. Crystal growth results as an individual grain steals particles from grains surrounding it, thus becoming larger in itself. Ultimately, the number of grains is diminished, and the average grain size is increased. This grain growth is rapid and is accelerated as the temperature is increased. The tendency of the structure is to become *single grained*. The term single grain is used to describe the structure of a piece of metal in which grain growth has advanced to such a degree that there is only one grain in the cross-section of the piece.

Theory of Alloys

No grain growth will occur with the alpha form of iron, unless it has been cold worked and subjected to temperatures around 1200°F (650°C). However, gamma iron grains grow rapidly, so that heating above the *GOSK* line, Fig. 8-24, causes grain growth; and heating to near the solidus line results in marked grain growth.

Recrystallization is the changing of crystalline grains into other crystalline grains without the aid of fusion. It can result from two causes: first, allotropic transformations, such as changing alpha iron grains to gamma iron grains; second, by mechanical straining, such as working of the metal. When gamma iron (*austenite*) changes to alpha iron (*ferrite*) and Fe_3C (cementite), as on cooling from *GOS* to below *PSK,* Fig. 8-24, complete recrystallization takes place. Grains which crystallize out (recrystallize) above *PS* are relatively coarse; grains which crystallize out at *PS* 1292°F (700°C) are relatively fine. If a piece of steel is heated through *PS* 1292°F (700°C), to *GOS,* recrystallization completely obliterates the original grain structure, and new, small grains of gamma iron (austenite) are formed.

Grain size changes on heating. At normal temperature a 0.25 percent carbon steel consists of pearlite and free ferrite grains. No change of grain occurs during the heating below the *PS* line, Fig. 8-24. On reaching the *PS* line, however, the grain structure of pearlite recrystallizes to form a very fine-grained austenite, the alpha iron changing to gamma iron. This change affects only the original pearlite, approximately 29 percent of the structure, the 71 percent of free ferrite remaining unchanged.

During the raising of the temperature from *PS* to *GOS*, Fig. 8-24, the free ferrite recrystallizes from alpha to gamma iron and is absorbed by the austenite. The recrystallized structure of gamma iron remains fine grained during this change because it is being disturbed and grain growth is being prevented by diffusion and recrystallization of the free ferrite. After the temperature has been raised to above the GOS line, all the iron has changed from alpha to gamma and gone into solid solution with the austenite; grain growth then sets in and continues up to the *ALRE* solidus line, Fig. 8-24.

The changes taking place in grain size during the heating cycle are illustrated in Fig. 8-28 which shows the relative grain size changes which may be expected during the slow heating of a 0.25 percent carbon steel to 1800°F (about 980°C) and the slow cooling back to normal temperatures.

At 1800°F (about 980°C) the grain size of the austenite would be relatively coarse. Upon slow cooling, no marked change takes place until the *GOS* line, Fig. 8-24, is reached, at 1475°F (about 800°C). Here the austenite precipitates fine ferrite crystals to the boundaries of the austenite crystals. During the slow cooling from *GOS* to *PS*, this precipitation continues with gradual growth in volume and size of the newly formed ferrite grains. The grain size of the austenite is decreasing, and the grain size of the ferrite is increasing. Upon reaching the *PS* line, 1290°F (695°C), the remaining austenite recrystallizes to form pearlite. This pearlite is a very fine mixture of crystals of ferrite and cementite (Fe_3C). From slightly

Metallurgy

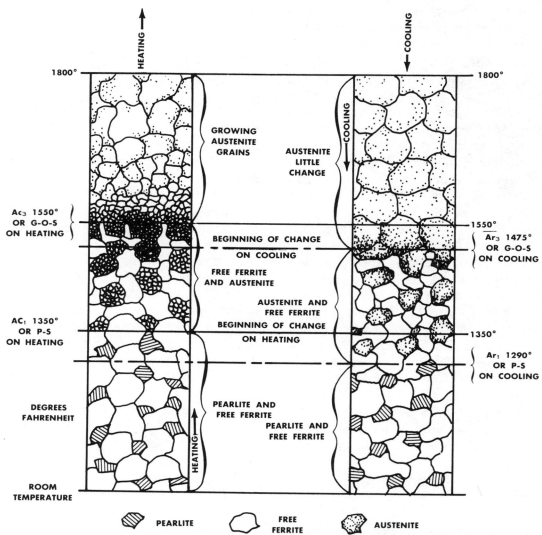

Fig. 8-28. Changes in grain structure during heating and cooling of 0.25 percent carbon steel.

below *PS* no further changes take place; however, if very slow cooling is employed, further growth of the ferrite grains may take place and a spheroidizing of the layers of cementite in the pearlite may occur.

From this it may be seen that the size

Theory of Alloys

and shape of the original austenitic grains influence to a marked degree the resultant grain size and distribution of free ferrite and pearlite. A very fine grain size in the original austenite will result in a very fine aggregation of free ferrite and pearlite.

If a 1.5 percent carbon steel containing mixed grains of pearlite and free cementite is heated, it is found that no change of grain structure occurs below the *SK* line, Fig. 8-24. At the *SK* line the pearlite changes to austenite (alpha iron to gamma iron and dissolving of the Fe_3C crystals), resulting in a very fine-grained austenite. The excess cementite is unaffected at this temperature.

On further heating, up to the *SE* line, Fig. 8-24, this excess cementite is dissolved by the fine-grained austenite. During this heating, marked grain growth of the fine austenite is prevented by the absorption and diffusion of the excess cementite. However, some grain growth takes place, which results in a coarser-grained austenite, and when the *SE* line is reached and all the excess cementite is dissolved, marked growth sets in and continues up to the *ALRE* line.

Critical temperature diagram. The diagram shown in Fig. 8-29 is seen to be part of the iron-carbon diagram and is often referred to as the *critical temperature diagram*. It is through this temperature range during heating and cooling that steels undergo the marked structural changes that have been described. It is apparent from Fig. 8-29 that the line *PS* in Fig. 8-24 has formed two lines *Ac1* and *Ar1*. Also, what is true of the *PS* line will be true of all other lines in this section of Fig. 8-24.

The splitting up of the temperature lines in Fig. 8-24 is due to a change in the temperature of transformation *during cooling* of iron-carbon alloys, as compared with the temperature of transformation *upon heating*. That is, the transformation of an 0.85 percent carbon steel from austenite to pearlite during cooling occurs at 1292°F (700°C) (point *S* in Fig. 8-24), but the change from pearlite to austenite, upon heating, occurs at 1350°F (730°C). This difference between the heating and cooling transformations, Fig. 8-29, is approximately 50°F (10°C), if very slow heating and cooling are employed. In reading this critical temperature diagram the *Ac* temperature lines are used if the iron-carbon alloys are being heated, and the *Ar* temperature lines are used if cooling is taking place.

The method of designating the various points, which is borrowed from the French, needs a word of explanation. The halt or *arrest* in cooling or heating is designated by the letter *A* standing for the French word *arret*. An arrest on cooling is referred to by the letters *Ar*, the *r* standing for *refroidissement*-cooling. An arrest on heating is designated by the letters *Ac*, the *c* standing for *chauffage*-heating. The various arrests are distinguished from each other by numbers after the letters, being numbered in the order in which they occur as the temperature increases.

Lag. The temperatures mentioned and given in the diagram refer to conditions of equilibrium. Under practical conditions there is a delay or lag in the attainment of equilibrium, and the critical points are found at lower temperatures on heating than those given. That is, there is a differ-

Metallurgy

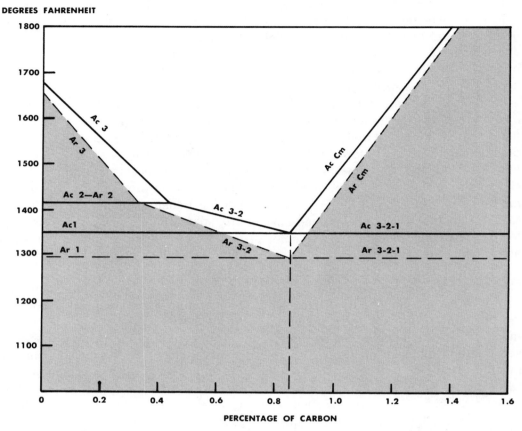

Fig. 8-29. Simplified iron-carbon equilibrium diagram showing the critical temperature and lag between the heating and the cooling temperature.

ence between the *Ar* points and the *Ac* points. This difference increases with the rate of cooling or heating. For instance, the *Ac* temperatures can be raised about 300°F (about 150°C) from their normal occurrence by fast heating, and the *Ar* temperatures can be lowered about 1000°F (540°C), in fact, can be forced below room temperature by rapid quenching of thin sections.

The equilibrium temperatures represent very slow cooling or heating.

Effect of impurities on alloys. Impurities other than carbon, when added to iron, will modify or change the temperature of occurrence of critical points, or the transformation from alpha to beta to gamma iron. For instance, the addition of nickel will lower the *Ac* critical temperatures about 30°F (-1°C) for each 1.00 percent.

Theory of Alloys

Critical range. The critical points, considered collectively, are known as the critical range. For instance, the critical range of a 0.45 percent carbon steel, on heating, extends from the *Ac1,* 1340°F (about 730°C), to 1410°F (765°C). See Fig. 8-29.

REVIEW

1. What is thermic analysis?
2. What information is needed to construct an alloy diagram?
3. In the study of a diagram covering the freezing of alloys; what does the *liquidus* line represent?, the *solidus* line?
4. What characteristic is common to Type II eutectic alloys?
5. Define the term solid solution when applied to a crystal structure.
6. Name the three classes of solubility relations in metals.
7. What is meant by the term *two-layer alloys.*?
8. Define the term *binary alloy.*
9. What are intermetallic compounds? Of what importance are they in connection with industrial alloys?
10. What is the meaning of a *peritetic reaction*?
11. List some of the most common changes in structural make-up which occur in an alloy during heating and cooling.
12. Define the term *allotropic.*
13. Define the term *critical point.*
14. Define the term *isothermal.*
15. How many atoms are there in a B.C.C. crystal?
16. How many atoms are there in a F.C.C. crystal?
17. What is an important physical characteristic of gamma iron?
18. What is mild steel?
19. What is iron carbide called and what is its chemical symbol?
20. What is ferrite?
21. What is the eutectoid and what is the name of its structure?
22. How does heat affect the magnetic properties of iron and steel?
23. When steel cools to a temperature of 1292°F (700°C), what changes take place?
24. What effect does hot working have upon grain size?
25. Name the two causes for recrystallization.
26. What is the effect of impurities on alloys?

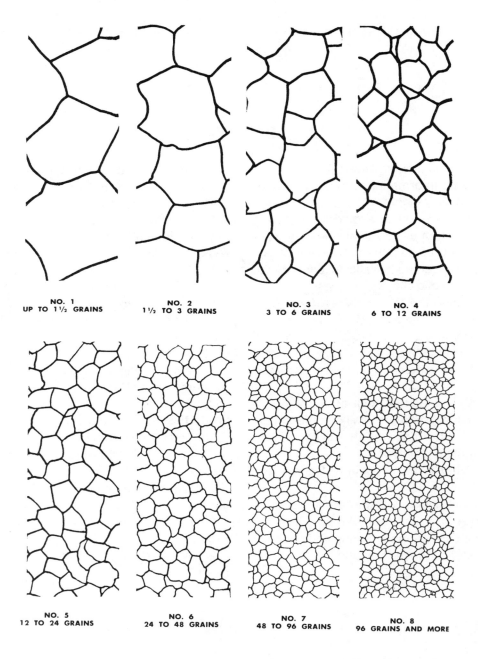

ASTM grain size chart. Grain size measured at 100 magnifications.

CHAPTER 9 — HEAT TREATMENT OF STEEL

The properties of all steels may be changed greatly by heating and cooling under definite conditions. *Heat treatment* is a broad term that may have reference to any one of a number of heating and cooling operations that will provide desired properties in a metal or alloy. It is necessary to specifically describe the type of heat treatment such as normalizing, annealing, or hardening.

NOTES:
1. Heating should be slow and uniform
2. For best results never heat higher than necessary
3. Normalizing=Air cooling
4. Annealing=Furnace or pack cooling
5. Hardening=Quick cooling

The object of heat treatment is to make the steel better suited, structurally and physically, for some specific industrial application. Among the many purposes of heat treatment is to soften the steel for easier machining operations, to improve the ductility and to increase the hardness and strength. Other heat treatments are used to relieve internal stresses in forgings and castings to eliminate the effects of cold working of the metal and to give a more uniform grain size. The first heat treatment to be considered is normalizing.

NORMALIZING

If a normalizing heat treatment is specified, it usually refers to a heat treatment that involves heating steel to above its critical range, Fig. 9-1, followed by cooling in air. The maximum temperature, the time at that temperature, and the cooling

Metallurgy

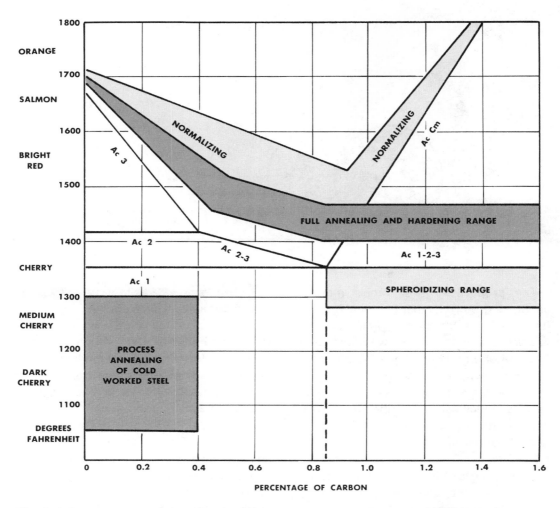

Fig. 9-1. Iron-carbon equilibrium diagram showing the critical temperature and the visual normalizing, annealing, and hardening ranges.

conditions are very influential, and in any application should be carefully worked out. The normalizing heat treatment is applied to steels that are known to have poor structures, structures that are uneven in grain size and segregated.

Normalizing is used for forgings that have a banded and laminated type of structure, see Fig. 9-2. Fig. 9-3 illustrates the improvement of structure obtained from a normalizing heat treatment applied to a forging of low-alloy steel, carbon 0.30 to 0.04 percent. Normalizing also is used as a treatment to improve the uniformity

Heat Treatment of Steel

and refine the structure of castings as well as forgings. It improves steel from the viewpoint of better results obtained in other treatments, including annealing and hardening. Also, some steels develop satisfactory physical properties from a single normalizing heat treatment, so that they are put into service without additional heat treatments.

Steel forgings may vary in physical characteristics even when made out of the same lot of steel. This is due to the different handling each may receive while being forged and while cooling from the forging heat. The temperature of the forging at the finishing operation and the cooling rate from the finishing temperature play an important part and influence the structure of the finished forging. Any variations in these factors account, in a large measure, for the variations found in forgings and constitute one of the main reasons for specifying a normalizing heat treatment.

Occasionally, in high-carbon steels, hot-worked steel may be treated so as to develop a structure as illustrated in Figs. 9-4

Fig. 9-2. Structure of hot-forging, showing badly banded structure and mixed grain size, magnified 75 times. Undesirable structure.

Fig. 9-3. Same steel as shown in Fig. 9-2 after correcting the undesirable structure by a normalizing treatment, magnified 75 times.

Fig. 9-4. Network, or eggshell type structure in hypereutectoid steel, magnified 100 times.

Metallurgy

Fig. 9-5. Same as Fig. 9-4, but magnified 500 times.

and 9-5. This cementite surrounding pearlite is a network structure which may be described as an eggshell-like structure, causing a very poor physical condition if subjected to a hardening heat treatment.

To break up this type of eggshell structure of cementite and distribute the cementite more uniformly throughout the entire structure, we might specify a normalizing heat treatment. Heating to within the normalizing temperature range, Fig 9-1, followed by air cooling would cure this condition.

Occasionally very much higher temperatures than those recommended in Fig. 9-1 are used in normalizing. This is true when the structures are difficult to break up. Using higher temperatures increases the solubility and diffusion rate, thus aiding in the operation.

Cooling in air means removing the steel from the heating furnace and exposing it to the cooling action of air in a room. The best practice consists in suspending the steel objects in air, or placing them on a special cooling bed so that they become surrounded by the cooling action of the air. Careless practice, such as dumping the steel objects into a pile on the floor of the shop, may result in very poor treatment and very non-uniform structure.

ANNEALING

The usual annealing heat treatment requires heating steel to above the *Ac 1-2-3* critical temperatures, Fig. 9-1 followed by suitable cooling. The heating is carried out slowly and uniformly, usually in regular oven-type furnaces. The time at heat should be long enough to insure heating to the selected temperature and to allow the structure to become completely refined.

The rate of cooling from the annealing temperature varies with the analyses of the different steels and with the properties most desired in the annealed steel; however, in general, the steel is cooled very slowly, either with the furnace (furnace an-

Heat Treatment of Steel

nealed), or cooled in a container surrounded by heat-insulating material (box or pot annealed).

The principal reason for specifying annealing is to soften the metal so as to make the steel easier to cold-form, machine, etc. Annealing may be carried out to develop mechanical properties that are required and obtained only by this treatment; for example, high ductility.

Annealing for Free Machining

Metal machining problems are very complex, and to put any steel into the best structural condition for easy machining operations may require special considerations based on the type of steel, type of cutting operations, and the degree of surface finish desired. Annealing for free machining should be carried out to a specified type of structure and Brinell hardness.

There are two basic types of structures from annealing heat treatments: (1) the laminated pearlite type and (2) the spheroidized cementite type of structure. To obtain the laminated type of structure, the steel is heated to within the annealing range, Fig. 9-1, and slowly cooled, at a rate of approximately 100°F (56°C) per hour down to about 1000°F (540°C) or below. The spheroidized type of structure is obtained by heating to the critical temperature, *Ac 1-2-3*, or slightly below this temperature for several hours, followed by cooling slowly to about 1000°F (540°C). The degree of spheroidizing depends largely upon the time of soaking at the spheroidizing temperature. Cementite that

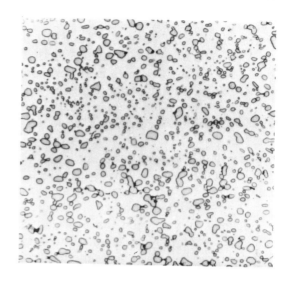

Fig. 9-6. Results of spheroidized heat treatment of SAE-3250 steel; 3 percent Nital etched; magnified 750 times.
DESIRED MICROSTRUCTURE:
C = 0.45 to 0.50   Ni = 1.50 to 2.00
Mn = 0.30 to 0.60   P = 0.025 max
Cr = 0.90 to 1.25   S = 0.025 max

is out of solution and red hot will ball up or form into spheres; upon slow cooling, the cementite that is precipitated will add to this spheroidized condition of cementite. Fig. 9-6 illustrates the degree of spheroidization desired in an SAE-3250 type of steel that is to be machined in automatic screw machines.

Many of the carbon tool steels and low-alloy steels are annealed to a completely spheroidized structure usually requiring a long annealing time cycle. The steel to be annealed is heated to the austenitic temperature range and held or soaked at this temperature generally from 1 to 4 hours. The cooling cycle is controlled from the

Metallurgy

annealing temperature so as to cool at a maximum rate of 50°F per hour (about 30°C) to a temperature of 1000°F (540°C) or below. The cooling rate is largely controlled by the furnace characteristics and the size of the load in the furnace. A large well-insulated furnace with a large tonnage charge will cool very slowly and result in a well-spheroidized structure. A small furnace with a light load will cool relatively fast and may cool too fast to obtain the desired degree of spheroidization.

Recently, it has been discovered that spheroidization may be effected during a shorter cycle than the usual time required by the slow-cool method. This new process requires accurate control of temperatures and relatively fast cooling from a high to a low spheroidizing temperature. In this process, the steel is first heated to slightly above the lower critical $AC1$ of the steel and held there for about two hours. The heat is then lowered relatively rapidly to a temperature of 1200° to 1300°F (650 to 700°C), the cooling rate being as fast as is possible to obtain.

The steel is then held at the lower spheroidizing temperature in the furnace at 1200° to 1300°F (650 to 700°C), for a period of from 4 to 15 hours depending upon the degree of spheroidization and softness required. After the soaking period at 1200° to 1300°F (650° to 700°C), the steel may be removed from the furnace, or furnace-cooled if softer and stress-free steel is required. The hardness or softness resulting from a spheroidization cycle depends largely upon the degree of spheroidization and upon the size of the spheroidized cementite particles.

In general, continous cutting, as in lathe and screw machines, requires a spheroidized structure in the high-carbon steels, but a laminated structure in steels of lower carbon steel, below 0.40 percent carbon. Laminated pearlitic types of structures are best for cutting operations, including gear cutting, milling, facing, broaching and splining, regardless of the carbon content of the steel being treated.

Annealing of Cold-Worked Steel

Cold-worked steels may have their original structure and ductility restored by a simple annealing operation consisting of heating to within the process annealing range shown in Fig. 9-1 followed by slow cooling. However, the rate of cooling may not affect the structure or properties greatly. This operation may be carried out in regular oven-type furnaces, although frequently it becomes necessary to anneal the cold-worked steel without scaling or ruining the surface finish, i.e., *bright annealing.* Bright annealing may be carried out in specially designed muffle furnaces containing reducing gases, or through the use of containers or boxes. The steel to be annealed is placed in a box and sealed. Gases that will prevent oxidation are forced into the box to displace any air present, and the box with the charge of steel and gas is heated to the annealing temperature.

Control of grain size in steels having lower carbon content, and control of the size and shape of cementite particles in steels of higher carbon content constitute the most important factors governing the results of any annealing operation following cold working.

Stress-Relief Treatment

Forgings, castings, and welded structures may retain a high residual internal stress or load due to uneven cooling and shrinkage effects. These internal stresses may be the cause of serious warping and even failure by rupture. Due to internal stresses, castings that have been accurately machined and assembled into a complicated machine tool may warp (after going into service) and cause the finished machine to become inaccurate or prevent its proper operation. These internal stresses may be removed in any annealing operation. However, if the parts to be stress-relieved do not need an annealing operation, they may be treated by heating to a much lower temperature than the annealing range.

In fact, internal stresses may be partly removed at room temperature, but this may require months; whereas heating to above room temperature but below the critical temperature of steel or cast iron will remove the internal stresses in a few minutes. Frequently a temperature of 1200°F (650°C) is recommended for the stress relief of cast steel, cast iron, and welded structures. Cooling from the stress-relieving temperature should be carried out so as to prevent any uneven cooling. This usually requires furnace cooling to at least 800°F (about 430°C).

HARDENING OF CARBON STEEL

Hardened steel weapons replaced the bronze swords used by the armies of Greece and Rome. While the swordsmith had the knowledge to produce a strong sword with a keen edge, no one in the early iron age actually knew what steel was nor why it hardened. In fact, only in the last century has the science of metallurgy discovered the importance of the carbon content in the ferrous alloy called steel and the scientific principles of hardening.

There are three main requirements for the satisfactory hardening of steel by the quench hardening method. These requirements are:

1. An adequate carbon content in the steel to form the crystalline structure called *martensite*.
2. The proper hardening temperatures, together with the optimum length of time at this temperature to form the nonmagnetic structure called *austenite*.
3. Controlled rapid cooling (quenching) to convert the red-hot austenite to the hard structure, martensite.

Hot-rolled or forged steel that has been slowly cooled from above red heat consists structurally of pearlite with either free ferrite or free cementite, depending upon the amount of carbon present in the steel.

Metallurgy

This slowly cooled steel has a relatively soft and plastic condition.

If the soft steel is to be hardened, two operations are performed:

1. The steel is heated to a temperature in excess of the critical temperature or the *Ac 1-2-3* line of Fig. 9-1.

2. The red-hot steel must be cooled rapidly (quenched) to bring the metal back to room temperature.

The object of the heating operation is to charge the steel from the normal, soft pearlitic structure to a solid structure of austenite. Many find it hard to imagine a solid solution. This would occur, for example, if sea water, which is a salt solution, were subjected to a very low temperature and frozen. The conversion of the pearlitic structure to an austenitic structure is accomplished by heating any carbon steel to within the temperature range suggested for the hardening of steel as shown in Fig. 9-1.

Steels with less than 1.0 percent carbon, when heated into this range will become 100 percent austenitic; whereas steels with greater than 1.0 percent carbon will contain an excess of free cementite (iron carbide, Fe_3C) in the austenite. The amount of free cementite in the higher carbon steels will depend upon the total amount of carbon contained in the steel and on the maximum temperatures to which the steel is heated.

It will be recalled that the AcCm line in Fig. 9-1 indicates the solubility of carbon in austenite and can be used to determine the amount of free cementite present in the steel at any selected temperature. When any carbon steel is heated through the temperature zone, *Ac 1-2-3* of Fig. 9-1, the alpha iron in the steel changes to gamma iron. The newly formed gamma iron dissolves the cementite and greatly refines the grain structure of the steel. When the grain structure has been refined and the cementite dissolved to form the austenitic structure, the first operation in the hardening of steel has been accomplished. It is with the austenitic type of structure that the second and equally important operation in the hardening of steel can be carried out, the rapid cooling or quenching of the austenite.

The object of the second operation in the hardening of steel is to undercool or prevent any change in the austenite other than that of cooling until a temperature of approximately 200°F (93°C) is reached. If the austenite is successfully undercooled to 200°F (93°C) it will rapidly change to a hard and relatively brittle structure known as *martensite,* the structure of fully hardened steel. It is born from austenite at approximately 200°F (93°C) by the rapid change of gamma iron of austenite to the alpha iron found in pearlite. However, this is the only resemblence of martensite to pearlite, as martensite is hard and pearlite is soft. Pearlite contains cementite in rather coarse particles whereas martensite apparently retains the carbon or cementite in a dissolved or nearly dissolved state. At least the microscope does not reveal the form of carbon in martensite.

The second operation in the hardening of steel, that of quenching or rapid cooling of red hot steel transforms a relatively soft, plastic steel to a hard brittle steel. The techniques of quenching and the various quenching media, as well as close control of maximum heating tempera-

Heat Treatment of Steel

tures, and the time a part is held at those temperatures, are most important and must be understood and carefully observed for optimum hardening.

Martensite

While martensite can be recognized from its hardness and its structural appearance which is that of a needle-like structure, Fig. 9-7, the reasons for its high hardness and strength are not fully known. Its hardness is in all probability due to several causes. The reduction of grain size, the presence of a fine precipitation of iron carbide in the slip planes of the crystals, and also the possible distortion of the atomic lattice or a disturbed atomic orientation due to thermal strain could account for its hardness. The martensitic state of steel is one of the most famous and common structures found in steels.

Why Quenching is Necessary

There is only one way to fully harden a piece of steel and that consists of undercooling austenite to approximately 200°F (93°C). This results in the formation of martensite, the structure of fully hardened steel. Fig. 9-8 illustrates the rate of quenching of austenite necessary in order to produce martensite in 0.85 percent (eutectoid) steel.

It is apparent from Fig. 9-8 that only two types of structures can be obtained from the usual method used in quenching of steel in any hardening operation—that is, pearlite and martensite. A slow-quench as in line *B*, Fig. 9-8, results in the transformation of austenite to pearlite. However the pearlite formed by this slow quench is a very fine pearlite with a relatively high hardness. During the slow quench, the transformation to pearlite, or the change from gamma iron to alpha iron takes place at approximately 1000°F (538°C) in about two seconds. Slower cooling, as in lines *C* and *D*, Fig. 9-8, cools austenite to a lesser degree and allows a more complete transformation to a coarser and softer pearlite.

If a faster cooling rate is employed, faster than line *B*, no change of the austenite takes place until 200°F (93°C) is reached. At this temperature martensite is formed. The rate of quenching (rapidity of cooling) necessary to undercool the austenite is referred to as the *critical rate* of quenching. A study of Fig. 9-8 will show that the rate necessary to cool the austenite to below

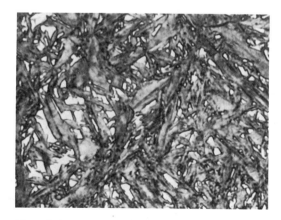

Fig. 9-7. Martensitic structure of hardened tool steel; very coarse needle-like structure is due to quenching from a course grained austenitic condition. Magnified 1500 times.

173

Metallurgy

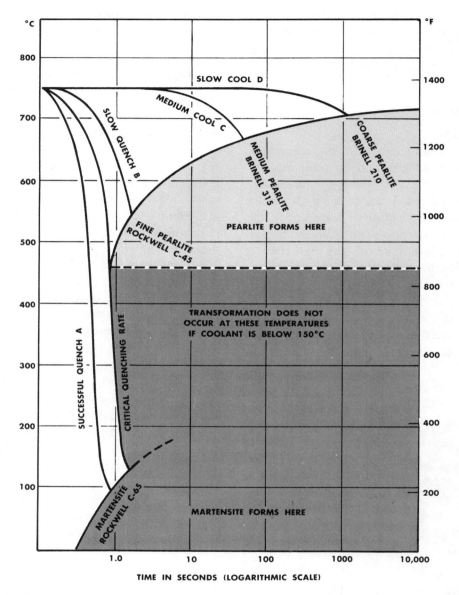

Fig. 9-8. Chart showing schematically the time and temperature curve of initial transformation of a eutectoid steel as cooled at various uniform rates. Logarithmic abscissa. (Transactions of the American Society for Steel Treating, Vol. 20, American Society for Metals).

1000°F (583°C) and thereby prevent formation of the softer pearlite is approximately a one-second cooling time. Faster cooling will enhance the opportunity to

Heat Treatment of Steel

Fig. 9-9. Fine pearlite (primary troosite) in martensite as a result of quenching at slower than the critical rate. Magnified 1500 times

form martensite but slower cooling than the critical rate will result in the formation of soft spots on the surface of the steel or an all-soft steel. Such soft spots are due to the formation of a type of pearlite that forms at about 1000°F (538°C). This fine pearlite structure was called *primary troosite* in early texts, Fig. 9-9.

Not all of the austenite transforms to martensite by being undercooled. Some of the austenite may be retained in the hardened steel when it reaches room temperature. While the retained austenite is hard, it is not as hard as the martensite, which can cause problems at the sharpened edge of a tool. The amount of retained austenite may vary from 1 to 30 percent.

The temperature at which austenite begins to transform to martensite has been designated by the letter *M* and is referred to as the *M-point*. A schematic diagram illustrating the M-point for an 0.80 percent carbon steel is shown in Fig. 9-10. In this illustration it will be noted that the M-point, or the temperature of the beginning of the austenite to martensite transformation, occurs at approximately 500°F (260°C) and the transformation to martensite is 80 percent complete at room temperature. This is to say that at room temperature the structure of the quenched steel consists of 80 percent martensite and 20 percent retained austenite.

Cryogenic Treatment

It has been discovered that any austenite retained in a hardened tool steel at room temperature may be transformed into martensite by cold treating, submitting the steel to sub-zero temperatures. Cold treating is carried out by cooling the workpieces in dry ice (solid CO_2) or in special refrigerating units. A temperature range of from -70°F to -110°F (-58 to -79°C) is usually employed in this hardening process.

Cold treating to transform retained austenite will not result in any loss of the original hardness of martensite and will result in maximum hardness being obtained for the part. Conventional tempering operations always result in loss of some hardness. In order to lessen the danger of cracking, the tool or workpiece should be given a low-temperature, 300 to 350°F (149 to 177°C), tempering operation before the cold treatment. The treatment should be followed by a second low-temperature tempering treatment.

Metallurgy

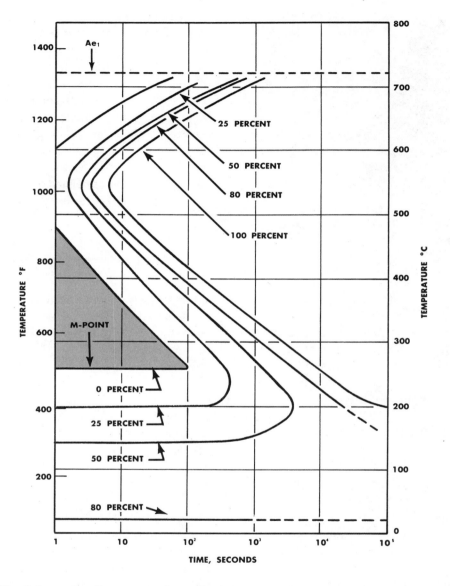

Fig. 9-10. Schematic diagram illustrating the M-point for an 0.80 percent carbon steel. (Tool Steels by Rose, Roberts, Johnston, and George. American Society for Metals)

Cold treatments are now being used for increasing the strength and hardness levels of carburized and hardened gear teeth, ball bearings, and other products having high surface-carbon content and which are subjected to high working stresses.

Heat Treatment of Steel

Quenching Media

Various compositions of steels do not require the same rate of cooling to transform austenite to hard martensite or to prevent the formation of a softer structure. However they all require a rate of quenching that will undercool the austenite to somewhere near room temperature, regardless of whether it is fast or slow. Carbon steels in general require a fast rate of cooling, such as can be obtained by a water quench. These steels are classified as *water hardening* steels.

Many of the low-alloy steels may be fully hardened with a slower rate of cooling than obtained in a water quench and are often quenched in oil. Such steels may be classified as oil hardening steels.

Some of the high-alloy steels have a very slow rate of transformation from austenite and upon cooling in still air from the red-hot austenitic temperatures will develop, when cold, a fully hardened martensitic structure. Such steels are referred to as *air hardening* steels.

The three common quenching media used in the hardening of steel are water, oil and air. The common action of the two liquid quenches may be described as having three stages. These are:

The *first stage,* which occurs when a piece of red-hot steel is immersed in the liquid bath and the liquid comes in contact with the hot metal. At the moment of contact, the liquid is converted into a vapor—steam in the case of water—which may completely envelop the hot metal. When this happens, the vapor acts as an insulating blanket around the metal and stops the cooling by preventing any further con-

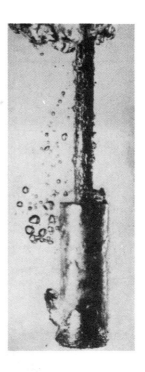

Fig. 9-11. First water quench stage of cooling of a steel cylinder—vapor blanket stage. At start of quenching from 1550° F (about 845° C), the vapor moves wave-like toward the top of the specimen. (Metal Progress)

tact with the cooling liquid. Unless this vapor film is broken, the cooling of the hot metal takes place by slow radiation and conduction through the film which results in a relatively slow cooling rate. If the film is allowed to remain, the metal will fail to harden by forming little or no martensite. See Fig. 9-11.

The *second stage* occurs when the film formed by the vapor and surrounding the hot metal collapses and breaks away from the surface of the hot metal. This action

Metallurgy

allows the hot steel to come in direct contact with the cooling liquid. During this stage, as the cooling liquid wets the hot metal surfaces, violent boiling occurs. This greatly activates the cooling liquid and creates the fastest stage of cooling. This action was referred by the old-time smith as when *"The water bites the steel."* See Fig. 9-12.

The *third stage* of cooling occurs when the hot metal cools down to approach the boiling point of the liquid quenching medium. Cooling continues by the thermal process of liquid conduction and convection. The results in a relatively slow cooling rate. However during this stage of cooling, a fast rate of quenching is not as necessary as during the earlier stages. Fast quenching is particularly necessary during the second stage when the steel is passing the temperature range where the transformation of austenite is at a maximum. See Fig. 9-13.

The vapor film formed during the first stage of quenching may prevent the successful hardening of steel. Soft steel, soft spots, warping and cracking may be the result of the vapor state. A great majority of the common hardening difficulties occur during this stage. Vigorous agitation

Fig. 9-12. Second water quench phase—vapor transport stage. Violent action of bubbles tears scale from work. (Metal Progress)

Fig. 9-13. Liquid cooling stage. Vapor no longer forms, and cooling is by conduction and convection. (Metal Progress)

of the steel by the smith holding the steel in his tongs and thrusting it through and around in the water was a common sight in the old blacksmith shop. What he was doing was breaking down the vapor film.

Water is the most commonly used quenching medium and is used in the hardening of many of the common carbon and low-alloy steels. Steels that have a very rapid transformation rate require water or some aqueous solution as a quenching medium. In order that the quenching rate may be great enough to consistently harden carbon steel, water should be kept at room temperature or below and be continuously agitated during the quenching operation.

As the heat from the steel is absorbed by the water, the water must be cooled during continuous quenching operations. In small shops, ice is added to the quench tanks to cool the water. In larger installations, artificial cooling by cold water or refrigerant that is circulated through coils in the quench tank must be employed. Agitation to break down the phase one vapor film can be accomplished by using underwater jets that spray streams of cooled water onto the steel. This action is necessary to permit more uniform and faster cooling action.

A brine solution is often employed to a better quench to carbon steels. A 5 percent sodium-chloride solution is commonly used. The salt acts to lower the freezing temperature and give the solution a greater capacity to absorb heat. As a result, the brine gives a faster and more uniform quenching action and is less affected by an increase in the quenching bath temperature. A 3 to 5 percent sodium hydroxide (caustic soda) quenching bath is also recommended as a good quenching medium for carbon steels. This bath cools even faster than the sodium chloride bath.

Oil is frequently used as a quenching medium and may be used when the operation involves the hardening of carbon steels of such a thin section that quenching in oil results in the formation of martensite. Thin sections such as knives, razor blades and wire may be successfully hardened in an oil-quenching medium.

Oil is recommended as a quenching medium in preference to water whenever it can be used because there is less danger of cracking, with less distortion of internal quenching stress. The action of oil is quite different from that of water as it has a more rapid quenching rate during the first or vapor stage but a much slower rate during the second stage when the oil is wetting the hot metal surfaces. Oil can be used only with light sections of shallow hardening steels and for heavy sections when a hardening operation is applied to steels that have a slow transformation rate from austenite and are deep hardening as is true of many alloy steels.

Oil cools steel much more slowly during the last cooling state. This is desirable because it results in much less danger of severe internal stresses, warping and cracking. Also, as this last stage is usually below the M-point for most steels, no undesirable results occur. Oils differ in quenching characteristics and should be carefully selected. Such properties as flash point, boiling point, density and specific heat should be considered.

Control of the temperature of the oil quenching bath is very important as is the

Metallurgy

volumetric capacity of the bath in relation to the tonnage of hot steel that will pass through the bath. In general it requires approximately one gallon of oil for every pound of steel quenched per hour. If 100 pounds of steel are to be hardened every hour, a tank of 100 gallons capacity is required.

Some method must be used to maintain the temperature of the oil in the tank between 80° and 150°F (27° and 66°C). Keeping the oil between 120° and 130°F (49° and 54°C) is advantageous. It will give a good quenching rate and reduce the danger of cracking and warping of the steel by lowering the residual quenching stresses.

Air cooling is employed with some high-alloy steels of the air-hardening type. Steels to be hardened are removed from

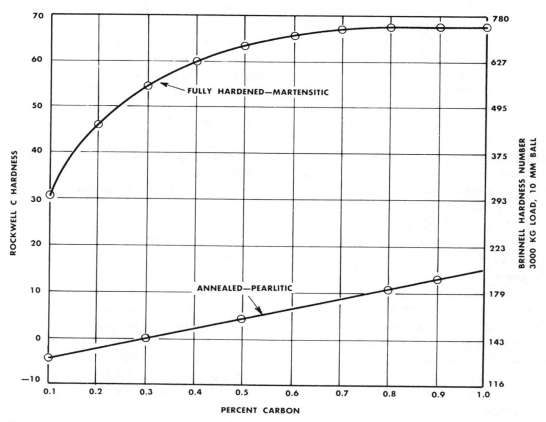

Fig. 9-14. Effect of carbon content upon hardness of carbon steel in martensitic and pearlitic conditions.

Heat Treatment of Steel

the heating furnace and exposed to still air. The rate of cooling in air may be modified by the use of an air blast.

Hot-quenching treatments are used in an attempt to combine the principles of hardening and tempering in the same operation. This method consists of quenching the hot steel in a molten salt bath held at a constant temperature (isothermal) usually between 350° and 800°F (177° and 427°C). The steel is quenched in the hot bath and held long enough to equalize at the temperature of the bath, or long enough to transform from austenite into martensite isothermally, or to some softer type of structure.

Effect of Carbon

The hardening capacity of steel clearly increases with the increase of carbon up to about 0.60 percent carbon. Carbon plays two roles in the hardening of steel:

1. Up to about 0.60 percent carbon it is easier to undercool austenite therefore making it easier to harden steel. Apparently, in this case carbon acts as a retardant, slowing up the rate of transformation.

2. The carbon present in the fresh formed martensite increases the hardness. Carbon, either dissolved or partly precipitated in the form of cementite, is probably the greatest single factor in causing the martensite to be hard.

Fig. 9-14 illustrates the effect of carbon upon the hardness of carbon steels in both the hardened and normal pearlite condition. An interesting fact is that the maximum hardness obtained in steel by making it martensitic rarely exceeds a Rockwell hardness of C-67 or Brinell hardness of 745, regardless of the maximum carbon content or the addition of alloys. (A nitrided case may exceed C-74. See page 216.)

Increasing the carbon content beyond the eutectoid point, 0.85 percent increases the hardness of the steel in both the pearlitic and martensitic condition. In addition, the high-carbon content increases the rate of transformation of austenite to pearlite and makes it more difficult to successfully harden this steel to a full martensitic condition. Although the curve for fully hardened steel in Fig. 9-14 shows that it is possible to obtain an appreciable increase in hardness in low-carbon steels by heating and quenching, it becomes increasingly difficult to trap 100 percent alpha martensite in lower carbon steels. Steels with less than 0.30 percent carbon are seldom hardened because of this difficulty.

The curve in Fig. 9-14 shows a possible hardness of Rockwell C-55 with a 0.30 percent carbon steel. This hardness cannot be obtained unless the section is very thin and is drastically quenched in a water spray of severely agitated brine or caustic solution. With a carbon content of 0.20 percent, it is not practical to expect maximum hardness of Rockwell C-45 from an ordinary quench-hardening operation. When the carbon content of steel exceeds 0.30 percent carbon it becomes easier to harden the steel to a fully martensitic condition and obtain the maximum hardness indicated in Fig. 9-8. In the hardened condition, 1.00 percent carbon steel consists of martensite with some free cementite particles as shown in Fig. 9-15. The great-

Metallurgy

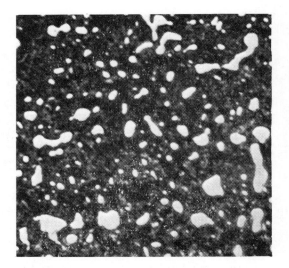

Fig. 9-15. Undissolved cementite in a tempered martensitic background. High-carbon tool steel, magnified 1500 times.

er hardness of this type of steel is due to the presence of cementite and not to a greater ability to harden.

Shallow hardening of steels. It has been pointed out that in order to harden a piece of steel it must be rapidly cooled from the proper red-hot hardening temperature. It is to be expected that the action of the liquid quenching media in the rapid cooling of the steel acts only on the surface of the steel object. The interior of any steel object must of necessity cool at a slower rate than the surface. It is natural to expect from this behavior of the cooling medium that the interior of the steel object may cool slower than the rate required to produce martensite, and thus remain relatively soft and tough.

Due to the fast cooling rate needed in the hardening of carbon steels, the center of any steel object may remain soft and tough, only the surface attaining maximum hardness and becoming martensitic. This leaves the steel object with a hard surface and a tough core; this characteristic of carbon steel has made it a favorite where a combination of hardness and toughness is needed.

The very property that makes carbon steels difficult to harden, its rapid transformation rate, allows us to produce a hardened tool or steel object with a hard, wear-resistant surface and a tough core to guard against breakage.

The shallow hardening effect of carbon steels makes it difficult to harden a piece of carbon steel with a heavy section. The tremendous amount of heat energy stored in a heavy section may even prevent the steel from cooling fast enough to produce a fully hard surface. In any event, the penetration of hardness in a heavy section would be very slight, and perhaps not enough to be effective.

To avoid this mass effect, heavy sections may be cored or hollow bored, thereby cutting down the maximum section. If the section cannot be reduced, a special or alloy steel may be employed with a deeper hardening effect, thus overcoming the effect of the heavy section. Also, in recent years, the steelmaker has discovered that grain size plays an important part in controlling the hardening ability of steel. A coarse-grained steel will harden with a deeper penetration of martensite than will a fine-grained steel of the same chemical composition.

Effect of grain size. For many years plants in the consumer industry have endured continued variation in the quench-

Heat Treatment of Steel

hardening results of carbon steels. Steels with the same chemical composition, given identical heat treatments, yield variations during hardening; some steels harden satisfactorily, others harden with a very deep penetration of hardness, and still other steel objects harden to a shallow depth, and in some cases with soft spots.

This condition is the reason that the different steel heats are kept separate. (As each heat of steel is kept separate, it soon became evident that individual heats vary in such important properties as hardening ability and distortion during hardening.) It has been determined that one of the most important factors contributing to this condition is the inherent grain size with which a steel is first made. The inherent grain size refers to the grain size that a given steel develops when heated to the proper hardening temperature. The steelmaker, during the steelmaking operations, can control the inherent grain size of a steel and therefore can control, within certain limits, the grain size the steel will develop upon heating to above the critical temperature, as in hardening of the steel.

It has been determined that fine-grained steels have a fast rate of transformation from the austenitic condition and are therefore shallow hardening; whereas the coarse-grained steels are deep hardening and more easily hardened, with a slower rate of transformation. The standard grain-size chart as determined by the American Society for Testing Materials is

Fig. 9-16. No. 1 coarse to No. 8 fine grained steels, show effects of penetration of hardness. Hardened steels have been sectioned and etched to bring out depth of hardening. White is hard; dark core is soft. (Transactions of the American Society for Steel Treating, Vol. 20, American Society for Metals)

Metallurgy

shown on the page facing this chapter. With this added control of grain size, carbon steels are treated with much greater uniformity and receive much favor, as they are inherently better suited for many applications than special or alloy steels. Fig. 9-16 illustrates the effect of inherent grain size upon the depth of hardness of one inch rounds of 1.0 percent carbon tool steels.

Effect of varying hardening temperature. The penetration of hardness in any standard carbon tool steel may be affected to a large degree by control of the maximum temperature selected by the hardener. It will be recalled that the grain size of austenite is a minimum at a temperature close to the *Ac 1-2-3* line of Fig. 9-1. Heating just over the *Ac 1-2-3* temperature results in a shallow hardening.

Heating beyond the critical temperature and developing a coarser-grained austenite will produce a deeper hardening effect. Larger pieces to be hardened, forging dies, etc., which require a deep penetration of martensite are usually heated to a much higher temperature in the hardening operation to insure uniform surface hardness and rather deep penetration of hardness. Small dies and thin sections may be hardened from a temperature close to the critical temperature with satisfactory results.

Warping and cracking. Warping and particularly cracking are serious menaces in nearly every hardening operation. In fact, one of the most severe tests a piece of steel receives is that of quenching in water or brine from the hardening temperature. The quenching operation may ruin the steel object by causing severe distortion or even breakage of the piece. Warping or cracking is caused by severe stresses or load, set up by uneven contraction and expansion that takes place during the hardening operation. Some of the sections of the steel object being hardened may be expanding while other parts of the same steel may be contracting. At least the contraction that takes place during the severe cooling as in quenching will never be an even contraction; therefore, some distortion always occurs.

Distortion or warping will always take place, and if the stresses become severe enough and the steel is made martensitic and brittle, danger of breakage by cracking is always present. The distortion that takes place as a result of the hardening operation leads to the costly operations involved in straightening and grinding the hardened steel in an attempt to remove the distortions and make the steel parts true to shape. A study of the part to be hardened may often save a great deal of future trouble and cost.

The heat treater may exhibit real skill and craftsmanship in the manner he selects to treat a given steel object to be hardened. Some of the many factors influencing warping and cracking may include: (1) unbalanced and abrupt changes in section; (2) sharp corners and deep tool marks, which act as crack formers; (3) intricate shapes, with many cut out sections; (4) defects in the steel, such as seams, dirty steel, segregation, and coarse grains, which may weaken the steel and cause trouble. Fig. 9-17 illustrates a slag stringer in steel that caused failure in the hardening operation. Fig. 9-18 illustrates a common type of failure in shallow harden-

Heat Treatment of Steel

Fig. 9-17. Slag stringer in steel is source of trouble in hardening operation.

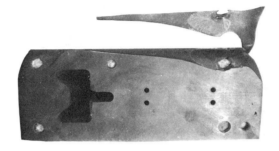

Fig. 9-18. Hardening die of carbon steel with corners and edges spalled off. Shallow hardening steel.

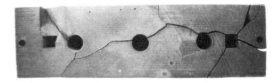

Fig. 9-19. Die that was split in hardening. Deep hardening steel.

ing carbon tool steel, that of spalling failure. Corners or surfaces of the steel spall from the body of the object. Fig. 9-19 illustrates the bursting or splitting type of failure occurring in deep hardening type of steels, steels that harden nearly all the way through the object. Fortunately, these failures may be avoided in the majority of the hardening operations.

Soft skin or surface. Soft skin or surface on hardened steel is the result of burning off the surface carbon or decarburization of the steel. Decarburization of steel causes a low-carbon or nearly pure iron layer to form on the surface of steel, leaving this decarburized surface soft after the hardening operation. Fig. 9-20 illustrates a badly decarburized carbon tool steel.

Decarburization may be prevented by proper control of the atmosphere surrounding the steel during the heating operation. The proper use of lead or salt baths, or the control of the gaseous atmosphere in an oven type of hardening furnace, may prevent any decarburization from taking place. However, all steels as received from the steel mill retain a decarburized skin which will cause trouble unless it is removed prior to the hardening

Metallurgy

Fig. 9-20. Decarburized tool steel, magnified 200 times. Surface of steel is white against black; The decarburized zone is the middle-section; and the high-carbon zone is at bottom, dark.

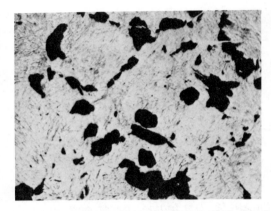

Fig. 9-21. Soft spots in hardened steel. Soft spots, black in light martensite, magnified 500 times.

operation. All steel should be bought oversize so that it may be ground or machined, removing the decarburized surface, before the hardening operation.

Soft spots. Soft spots or soft areas may occur on the surface of hardened tools and dies. These soft spots are caused by the transformation of austenite to pearlite at 1000°F (538°C) in the quenching operation. Allowing the quenching medium to get too hot or dirty, poor agitation of the quenching medium, scale or dirt on the steel surface, interference to the quenching action by the tongs or holders used in handling the steel, or anything else that will retard the quenching action may contribute to soft spots. Deep recesses in the steel surfaces where pockets of vapor may collect will prevent proper quenching action and cause soft areas. To avoid soft spots, a spray quenching practice, which allows the cooling medium to flow rapidly past the parts to be hardened may be necessary. Fig. 9-21 illustrates soft areas in martensite.

EFFECT OF ALLOYS

In general, alloy steels, steels containing special alloying elements, are easier to harden than plain carbon steels. Alloy steels are less sensitive, and allow more latitude in the selection of a hardening temperature and in the manner of quenching. This makes it easier for the heat treater to carry out the hardening operations and obtain the desired results. In general, alloy steels may be quenched in oil to produce the hardened state. Oil, being a slower cooling medium than water or brine, causes less distortion and reduces danger of cracking.

Heat Treating of High-Speed Steels

Heating for hardening may be carried out in controlled-atmosphere furnaces or in salt baths, or the tools, before heating, may be dipped in a saturated solution of borax and water that is heated to between 150° and 180°F (66 to 82°C). Tools dipped in this warm solution will dry when removed and will be covered with a thin powdery film of borax that will not rub off if the tools are handled with moderate care. If too much borax is used, blisters may form on the surface of the hardened tools. After hardening, the tools will be found coated with a thin layer of borax, which can be removed with a wire brush, or a 10 percent solution of acetic acid.

Salt baths offer one of the best methods available for the heat treating of high-speed steels, particularly when the maximum temperature for hardening does not exceed 2300°F (1260°C). Some compositions of high-speed steel required a hardening temperature in excess of 2300°F (1260°C), and therefore it may be found inadvisable to use a salt bath. Some other heating medium may prove more satisfactory. The straight tungsten types and those containing cobalt require higher hardening temperatures than the molybdenum and molybdenum-tungsten steels.

Salt bath furnaces are usually complete in themselves, consisting of a preheat, a high heat, and a quench bath. The preheat salt operates at about 1500°F (815°C), and the steel is placed in this bath for a period of time that allows it to reach the heat of the bath. From the preheat salt, the work is placed in the high-heat bath and held in this bath for a predetermined length of time.

The time in the high heat is determined by the usual trial and error method, although with the use of hardness tests and microscopic examination of the treated steel, the correct time can be determined accurately. The time and temperature of the high heat are governed by the amount of carbide solution and the grain size developed. Fig. 9-22 illustrates the type of microstructure obtained from heating a piece of 6-6-2 type (molybdenum-tungsten) high-speed steel to 2250°F (1230°C) for 5 minutes and quenching in hot salt.

The hot salt quenching bath acts as a quenching medium, undercooling the austenitic structure formed at the high heat and also helping in the removal of the salt that adheres to the work from the high-heat bath.

Under ideal operating conditions, the salt-bath method of heat treating for a high-speed steel furnishes full protection against any surface skin such as decarburization or carburization. Fig. 9-23 illustrates a slight skin of about 0.0003" that was formed in a salt-bath treatment for hardening of high-speed steel. This slight

Metallurgy

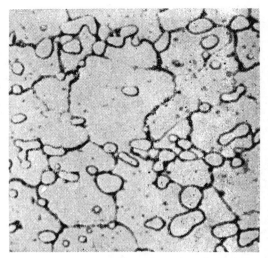

Fig. 9-22. Molybdenum-tungsten (6-6-2 type) high-speed steel in as-quench-hardened condition. Undissolved carbides are light; austenitic grain boundaries are dark. Magnified 1000 times.

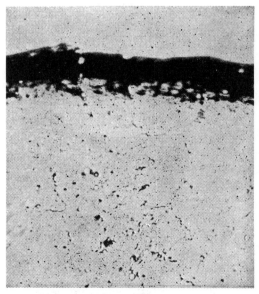

Fig. 9-23. Decarburized skin (dark) on molybdenum-tungsten 6-6-2 type high-speed steel, hardened from salt-bath furnace. Magnified 1000 times.

skin is negligible insofar as it influences the useful properties of the steel for tools.

Frequently, high-speed steel is heated for hardening in a standard type of oven furnace with protection offered by a suitable gaseous atmosphere. The atmosphere commonly used is rich in carbon monoxide, running about 34 percent and with only a trace of CO_2. With the proper controlled atmosphere, high-speed steels of any composition can be successfully hardened without any detrimental effects from either carburizing action. Hardening of the tungsten type (18-4-1) or cobalt high-speed steels can be carried out successfully in gas or electric-fired furnaces with a suitable gaseous atmosphere.

Carburization of the surfaces of high-speed steels, particularly of the tungsten type, may occur when a rich CO atmosphere is used as protection. Carburized surfaces will cause the austenitic structure to become quite stable and prevent its breaking down to martensite in the drawing or tempering operation.

Fig. 9-24 illustrates a soft skin formed on the surface of the 18-4-1 type of high-speed steel by carburization during the hardening operation. The austenitic layer (light in photomicrograph) resists transformation to martensite during the tempering operation and remains relatively soft. This condition is undesirable and should be avoided by reducing the CO content in the atmosphere surrounding the steel during the hardening operation.

Preheating for hardening. It is recommended that high-speed steels be preheated to a temperature range from 1400° to 1500°F (760° to 815°C) before placing them in the hardening furnace.

Heat Treatment of Steel

This preheating should be carried out in a furnace separate from the one used for the hardening heat. The molybdenum high-speed steels can be preheated to a lower temperature than the tungsten high-speed steels. This apparently helps to avoid decarburization. Small tools of light section, under ¼" (6.35 mm) can be hardened without preheating.

Hardening temperatures for high-speed steels. The straight tungsten type of high-speed steel, such as the 18-4-1, can be hardened by heating to a temperature range from 2300° to 2375°F (1260° to 1300°C), followed by quenching in an oil bath. The time required to heat will depend upon the size of the section and upon the type of heating medium. The timing should be such as to allow the steel to come up to the temperature of the furnace, or nearly so. For light sections, this may take only a few minutes, whereas for a section of 1 to 2 inches, it may require a heating time of 20 minutes. Careful observation of the steel during the heating cycle will allow the hardener to judge the time needed for heating.

Anyone who has hardened the 18-4-1 type of tungsten high-speed steel can harden molybdenum high-speed steel. The hardening is carried out in a similar way, preheating at 1200°F to 1500°F (650° to 815°C) and transferring the steel to the high-temperature furnace for hardening. Quenching and drawing are carried out as with the 18-4-1 type. The steel should never be soaked excessively at the hardening temperature but should be re-

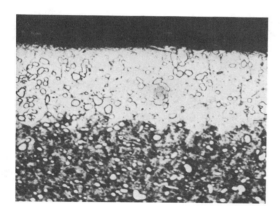

Fig. 9-24. High-speed steel of the 18-4-1 type, hardened and double-drawn to 1050° F (about 565° C). White surface layer is retained austenite, resulting from heating in a carburizing atmosphere during hardening, magnified 500 times. Surface hardness is Rockwell C-56.

Fig. 9-25. High-speed steel severely overheated in hardening, magnified 500 times. Note the coarse, needle-like martensitic markings with a network structure resulting from slight fusion. Note freedom from carbides.

Metallurgy

moved from the furnace and quenched shortly after it is up to heat. The temperatures recommended for the hardening of the molybdenum type of steels are somewhat lower than for the tungsten steels, 2150° to 2250°F (1175°-1230°C). Overheating of the molybdenum type will put it in a somewhat brittle condition, whereas the 18-4-1 type may be overheated 75°F (24°C) and still make a reasonably satisfactory tool.

Overheating of high-speed steel ruins the steel for further use, making it weak and relatively soft. The structure resulting from severe overheating is illustrated in Fig. 9-25. This structure shows large austenitic grains with fused grain boundaries (dark), and coarse needle-like martensitic markings. Practically all of the carbides are dissolved when high-speed steels are overheated.

Quenching of high-speed steels. The quenching may be done in oil, salt bath or air for both the tungsten and molybdenum steels. Oil is the most common quenching medium. However, if warping and cracking become a problem, it is recommended that *hot quenching* be used, quenching in molten salts held at a temperature range from 1100° to 1200°F (about 595° to 650°C), or quenching in air if scaling is not serious and the section hardens satisfactorily. A practice that may be tried is to quench in oil for a short time, then to place the tool in the pre-heat furnace held at 1200°F (650°C) until the temperature becomes equalized, and finally to air-cool to room temperature. The time in the oil quench should be such as to cool the steel from the high heat to around 1200°F (650°C), or a very dull heat.

Tempering or drawing of high-speed steels. High-speed steels, after hardening, do not develop their full hardness and toughness without tempering or drawing. Drawing may be carried out successfully by heating the hardened steel to a temperature range extending from 1000° to 1100°F (about 540° to 595°C) for at least 1 hour, preferably 2 hours. The steel, before tempering, should be cooled to at least 200°F (93°C) from the quench and reheated slowly to the temperature used in drawing. The cooling from the drawing temperature may be done in air but not in water. Multiple tempering is recommended when maximum toughness is required i.e., the tempering operation should be repeated.

The microstructure of high-speed steels. Microscopic examination of annealed high-speed steels shows the structure to consist of a large quantity of carbides in a matrix of ferrite. The structure of annealed high-speed steel is shown in Fig. 9-26. The carbides (light particles) are made up of a complex mixture of carbon, tungsten, molybdenum, chromium, vanadium, etc. The complex carbides found in high-speed steels give these steels outstanding properties.

In the hardening process, the steel is heated to a relatively high temperature in order to dissolve a large percentage of carbides. However, the austenitic structure formed at the high temperature does not dissolve all of them, and it is not desirable to do so. The carbides remaining out of solution add to the wear resistance and the cutting ability of this steel. Upon quenching from the high heat to room temperature in either oil, air, or hot salt,

Heat Treatment of Steel

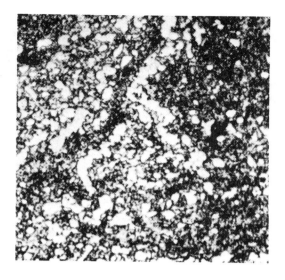

Fig. 9-26. Annealed high-speed steel of molybdenum-tungsten type 6-6-2, magnified 500 times. Note light carbides.

the austenite transforms to martensite, but not completely, so that the hardened high-speed steel always retains at room temperature some austenite from the quench. In fact, as much as 30 percent austenite may be retained in the quenched high-speed steel.

Microscopic examination of the quench-hardened steel shows a structure that appears to be austenitic with undissolved carbides (see Fig. 9-22). The original austenitic grain outlines can be plainly seen. The size of the austenitic grain is one of the factors that influence the properties of the finished tool or die and is a criterion of the correct temperature-time factor used in the hardening operation.

Because of the retained austenite in the hardened steel, it becomes necessary to temper or draw the steel after hardening. The objective of the drawing treatment is largely to transform the remaining austenite to martensite. This is accomplished by reheating the hardened steel between 1000° and 1100°F (about 540° to 595°C), usually 1050°F (565°C), followed by slow cooling to room temperature. By this treatment, the retained austenite of the quench-hardened steel is transformed to martensite, and the internal stresses in the hardened steel are removed. The newly developed martensite of the tempered or drawn steel should now be drawn to relieve any stresses set up by its growth. This is accomplished by repeating the drawing operation, i.e., heating to 1050°F (565°C), followed by slow cooling to room temperature. The steel is now in the best condition for use. The microscopic appearance of high-speed steel after a double draw following hardening is shown in Fig. 9-27. The structure consists of white

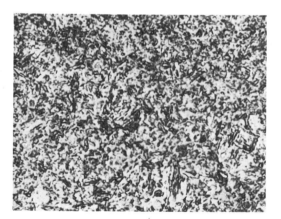

Fig. 9-27. Martensite tempered to 600° F (about 315° C), magnified 1500 times. Note cementite particles.

Metallurgy

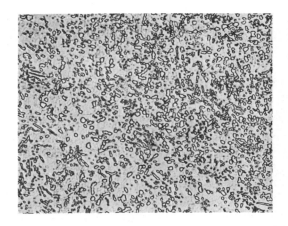

Fig. 9-28. Martensite tempered to 1200° F (about 650° C). White cementite particles have grown noticeably in size during spheroidizing.

carbides in a dark martensitic matrix, with the austenitic grain outlines, as seen in the hardened steel, completely gone. The grain size of the original austenite may be brought out in the drawn high-speed steel by means of a special etching treatment, as shown in Fig. 9-28, which shows the structure of hardened and drawn high-speed steel with the grain outlines of the original austenites appearing in the photomicrograph.

Hardness of high-speed steels. High-speed steel in the fully annealed condition will run between Rockwell C-10 and C-20 and is machinable. In the hardened condition, as quenched, the hardness should run between Rockwell C-63 and C-65 for satisfactory results. After drawing, the hardness should run between Rockwell C-64 and C-66.

Forging of high-speed steels. Forging should be carried out from 2100° to about 1800°F (1150° to 980°C) in order to obtain a good condition in the final forged shape. In most applications, the user of high-speed steel buys the steel in the forged condition, such as forged or rolled bars and forged blanks. However, if some shaping is required, it may be carried out successfully if the necessary precautions are taken. In the manufacture of high-speed cutters, where the outside diameter is greater than 4 inches, it is recommended that upset-forged blanks be used. The blanks may be made by buying a small-diameter bar and upset-forging a section of the bar to make the size cutter required. The upset-forging operation produces a better structure in the high-speed steel than can be obtained from a rolled or hammered bar of large diameter.

Annealing of high-speed steels. Annealing may be carried out by slowly heating the high-speed steel to a temperature from 1550° to 1675°F (about 845° to 915°C) for about 6 to 8 hours, followed by very slow cooling of the steel. The cooling cycle should require about 24 hours. The steel must be protected from oxidation and decarburization by packing in cast-iron chips or a suitable reducing agent such as charcoal and ashes mixed. As most high-speed steel is purchased in the annealed condition for machining, it is not necessary to anneal this type of steel unless reannealing and rehardening of a used tool are required. In the event that this is the case, the steel should be reannealed before hardening; otherwise a brittle condition of the rehardened tool will result.

Heat Treatment of Steel

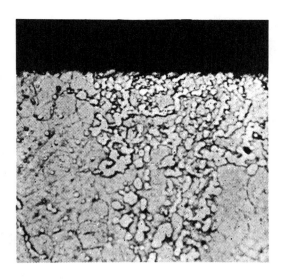

Fig. 9-29. Photomicrograph of the edge of a sample of hardened high-speed steel, showing practically no surface skin; magnified 500 times.

Decarburization or carburization in high-speed steels. Successfully hardened and tempered high-speed steels should not have more than 0.001" of soft skin of decarburization or carburization. However, if hardening of the molybdenum type is carried out in a standard oven-type furnace without the necessary controlled and protective atmosphere, decarburization may become a very serious factor. Decarburization and soft skin on the surface of tools hardened in this manner may extend to a depth of more than 0.010" and if the allowance for finishing or grinding does not remove this skin, a poor tool will result. Fig. 9-29 illustrates a satisfactory hardened piece of high-speed steel having no skin.

Tempering of Hardened Steel

The heat-treating operation known as tempering of steel is carried out following the hardening operation. Tempering of steel consists of heating quenched, hardened steel, steel in the martensitic condition, to some predetermined temperature between room temperature and the critical temperature of the steel for a certain length of time, followed by air cooling. The rate of cooling from the tempering temperature in most cases is without effect. There are very few applications where a fully hardened martensitic condition of steel may be successfully used, because of its brittle condition. The heat treater, by means of the tempering operation, conditions the hardened steel for successful performance.

The hardening operation imparts the following characteristics:
1. Smallest grain size
2. Maximum hardness
3. Minimum ductility
4. Internal stresses and strains

In this condition, steel is sometimes harder than necessary, although the most obvious deficiency is the lack of toughness in fully hardened steel. This lack of toughness, with the internally stressed condition present, makes hardened steel unsuitable for use. The hardness is needed in most applications, but the extremely brittle condition makes it necessary to subject the steel to further heat treatment, i.e., tempering. In order to relieve this stressed state and decrease the brittleness, while preserving sufficient hardness and strength, the steel is generally tempered.

Metallurgy

Reasons for tempering are:
1. To increase toughness
2. To decrease hardness
3. To relieve stresses
4. To stabilize structure
5. To change volume

Explanation of tempering. Quench-hardened steel is considered to be in an unstable condition, that is, the structure of the hardened steel (martensite) was obtained by severe quenching of the steel from the austenitic state, thereby undercooling the austenite to nearly room temperature before it transformed to the martensitic structure. Perhaps in most quenching operations some austenite is retained at room temperature.

The martensitic structure of the hardened steel is much different from the structure of normal pearlite formed in steels that have been slow-cooled from the austenitic state. The newly coated martensite is known as the alpha martensite and has a tetragonal atomic arrangement. Martensite in this condition tends to change to a more stable structure, more nearly pearlitic, and actually undergoes this change when given an opportunity, such as when the temperature is raised during the tempering operation.

Upon heating the alpha martensite formed in plain carbon steels to approximately 200°F (93°C), the tetragonal atomic arrangement is said to change to a body-centered cubic lattice similar to the pattern found in alpha ferrite. The martensite tempered to 200°F (93°C) with the body-centered cubic structure is known as beta martensite. Microscopic examination of the tempered beta martensite reveals a darkening of the needle-like martensite structure. This change is accompanied by a decrease in volume causing a shrinkage in the dimension of the hardened steel.

Further heating causes the beta martensite to precipitate carbon in the form of cementite (Fe_3C) which has been held in supersaturated solution in the martensite. The precipitated cementite particles are of minute size when this action starts, so tempering to low temperatures around 400°F (about 205°C) will not reveal their size by microscopic examination. However, if the tempering temperature is raised beyond 400°F (about 205°C), the minute cementite particles continue to grow in size, finally becoming microscopic.

The grains of iron also grow in size. The hotter the hardened steel is heated, the faster this change, the atomic structure becoming less rigid and more free to change. Finally, the martensite may be changed to a soft, coarse structure similar to pearlite. The microscope may be used to follow the structural changes taking place during the tempering of a martensitic structure, but the structural details, such as the action of precipitation, cannot be resolved as a result of the lower temperatures used in tempering. Apparently the cementite particles do not reach a size that becomes microscopic until a temperature of 600°F (315°C) is reached.

Figs. 9-27 and 9-28 earlier illustrate the change from martensite that might be expected upon tempering to 600°F and 1200°F (315° and 650°C), as seen with the aid of a microscope. Actually, the cementite has grown to a size that is easily seen with the microscope.

Any retained austenite found in hardened carbon or low-alloy steels may be transformed to martensite or a tempered form of martensite upon reheating to 500°F (260°C). The changing of austenite to martensite upon tempering is accompanied by an expansion which may be very marked. Such a change results in added internal stresses and should be taken into consideration by increasing the tempering time so as to at least partially remove them. Transforming the retained austenite to martensite by reheating to 500°F (260°C) will effect a change in the original martensite resulting in a loss of maximum hardness.

It has been discovered that any austenite retained at room temperature may be transformed to martensite by cold treating, i.e., cooling the hardened steel to subnormal temperatures [(70° to −110°F) (21 to −79°C)]. Cold treating to transform the austenite will not result in any loss of hardness of the original martensite, as that which occurs during tempering, and will result in maximum hardness being obtained. The cold treating of hardened steel may be followed by the usual tempering treatment and is sometimes used after a tempering treatment. There is some danger of cracking due to residual stress and added stresses during a cold treating operation; however, this may be avoided by first tempering the hardened steel and then subjecting it to the cold treatment, followed by another tempering treatment.

Effect of tempering upon hardness and toughness. Too little is known about the influence of tempering upon the hardness and toughness of hardened steels. It is assumed that softening of the martensite is progressive throughout the whole of the tempering range. Fig. 9-30 illustrates the change that might be expected when a hardened 1.10 percent carbon steel is tempered. It is certain that tempering of hardened plain carbon tool steel, at temperatures above 390°F (about 200°C), reduces its cutting hardness and its hardness as measured by the Rockwell and Brinell hardness testing methods.

Tempering of carbon tool steel at temperatures near room temperature has been reported as increasing the hardness of some quenched, hardened steels. It has been noted that quenched, hardened steels increase in hardness with time when aged or tempered at room temperature. This increase in hardness may be due to some retained austenite in the quenched steel slowly changing to martensite at room temperature.

It should be understood clearly that the purpose of tempering is not to lessen the hardness of quenched steel, but to increase the toughness so as to avoid breakage and failure of the heat-treated steel. Fully hardened steel with a carbon content greater than 0.60 percent carbon is brittle and therefore dangerous as a tool or structural part of a machine. Any slight overload or sudden shock load may cause failure.

The heat treater by means of tempering may reduce the brittleness and increase the plasticity of martensite, thereby imparting a greater degree of toughness to the steel. Measuring hardened steel for toughness and resistance to failure has been difficult, particularly in fully martensitic steels.

Tests have been made by impact testing and torsional impact testing methods,

Metallurgy

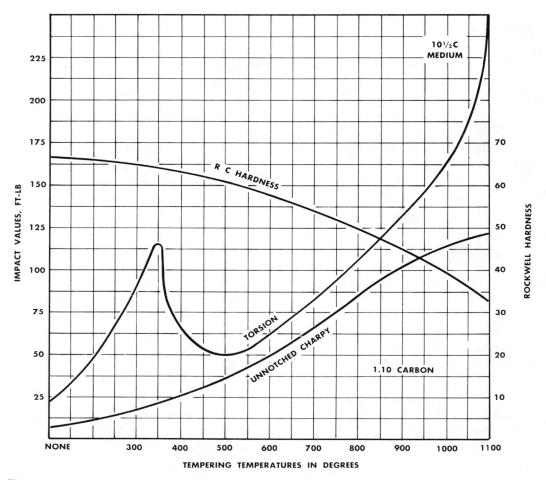

Fig. 9-30. Influence of tempering heat on fully hardened 1.10 percent carbon steel as represented by Rockwell hardness, torsional impact, and unnotched Charpy impact testing. (After Palmer)

and some very interesting results have been obtained. Fig. 9-30 illustrates the effect of tempering on a fully hardened carbon tool steel as measured by the Charpy impact testing methods, and the torsional impact method. If we assume that these methods of testing are measures of toughness, then it becomes apparent that the gain of toughness upon tempering is not progressive over the entire tempering range. In fact, tempering to within a temperature range of 500 to 600°F (260 to 315°C) may show a loss of toughness. Because of this effect, the region of 500 to 600°F (260 to 315°C) has been referred to as the *brittle tempering range*. Also some steels, particularly in the chromium-nickel steels of relatively high carbon-content, develop a seemingly brittle condition if tempered around 1150°F (620°C). The cause of this condition has not been clearly explained. Such steels may be cured of this brittle condition by quenching from the tempering temperature when the steel will have its expected toughness.

Effect of time in tempering. In general, the longer the time of treatment at a given tempering temperature the better the results from tempering. Longer time seems to release, to a greater extent, the locked-up stresses of the quenched steel, and to increase the plasticity and toughness without marked decrease in the hardness. Apparently, increasing the temperature during tempering increases the rate of any transformation, such as precipitation and growth of the cementite, but changes in the structure seemingly continue with time and allow a more complete change to take place. It is recommended that at least one hour be allowed at any temperature for satisfactory results. Some tempering operations require many hours.

Volume changes. It is known that when austenite changes to martensite during the hardening operation, a marked expansion takes place. Martensite has the greatest volume of all the structures in steel. Aging of freshly formed martensite at room temperature results in a contraction in volume, and upon tempering, the contraction continues. This volume change, and other changes taking place during tempering, are accompanied by an evolution of heat. This change in volume adds to the problem in controlling the size and shape of steel during heat-treating operations. Tempering usually will bring the steel back to more nearly the original volume; in some steels, the tempering temperature selected for the steel is somewhat determined by the amount of contraction that takes place, or at what temperature it returns to more nearly its original volume.

Tempering methods. Tempering of steel may be accomplished in liquid baths such as oil, salt or lead. These baths are heated to the correct temperature, and the steel is immersed in the bath for the determined length of time, after which the steel is removed and allowed to cool to room temperature. Tempering is very successfully carried out in air tempering furnaces. In these furnaces, air is heated by gas or electric means, and the hot air is circulated around parts to be tempered. These furnaces are fully automatic.

HEAT-TREATING FURNACES

Heating steel for the purpose of annealing, hardening, tempering, etc., may be carried out in a number of different types of furnaces. The choice of any furnace depends largely upon the type of heat-treating operation and upon the size and tonnage of the steel involved. Large-scale heat-treating operations are carried out in

Metallurgy

batch type or continuous type furnaces which are equipped with a means of automatic temperature control and mechanical devices for the handling of the steel in and out of the furnace and into the quenching baths. A great many of the furnaces are designed and constructed to carry out some specific heat-treating operation; accordingly, a great diversity of types and designs of furnaces are in use.

Furnaces are of the oil-fired, gas-fired, or electric type. Oil-fired or gas-fired furnaces may be designed as direct-fired, semi-muffle, or muffle. The direct-fired furnaces allow the steel to come into direct contact with the hot gases of combustion and are mainly used for heating steel for forging operations. The semi-muffle type is so designed that the gases of combustion are deflected and do not impinge directly upon the steel being heated, although they circulate around it. This affords better control of the heat, which results in a more uniform product. See Fig. 9-31.

In order to protect steel from oxidation during any heat-treating operation, furnaces of the full-muffle type are commonly used. These full-muffle furnaces are so constructed that they provide a chamber, known as the *muffle*, separate from the combustion chamber, into which the steel to be heated is placed. The gases of combustion circulate around the chamber or muffle and cannot come in contact with the steel being heated. See Fig. 9-32.

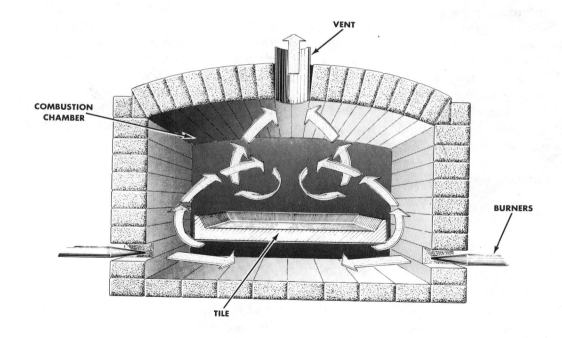

Fig. 9-31. The principle of the semi-muffle type furnace.

Heat Treatment of Steel

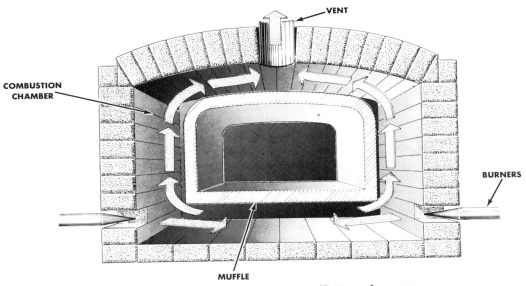

Fig. 9-31. The principle of the muffle type furnace.

Fig. 9-33. Electric muffle furnace with resistance and glo-bar hardening furnaces, air tempering furnace, endothermic atmosphere control, and oil and water quenching tanks. (Waltz Furnace Co.)

In the electric-type furnace, the resistors are usually placed around the outside of the muffle. The muffle is made of a special heat-resisting alloy or special refractory material. The hot gases of combustion, or in the case of an electric furnace, electric resistors, heat the muffle and the steel within the muffle. A neutral or reducing gas may be supplied to the interior of the muffle so as to greatly increase the degree of protection afforded the steel during the heat-treating cycle. The full-muffle type furnace is commonly employed with the so-called *atmosphere-type furnace* and is used when maximum protection from oxidation is required. See Fig. 9-33.

199

Metallurgy

Recent improvements in heat-treating furnaces have been made possible because of improved heat-resisting alloys for muffles and other parts, improved refractories, more efficient burner equipment, improvement in temperature control apparatus, and better electric resistor alloys.

Furnace Atmospheres

Heating steel in an electric furnace or in a muffle furnace in the presence of air results in the oxidation of the steel surfaces with the formation of scale and loss of surface carbon (decarburization). The ideal atmosphere is one which will not scale or burn out carbon (decarburize) or add carbon to the steel (carburize) during the heating operation. In order to obtain this ideal atmosphere, artificial gas atmospheres are added to the heating chamber in a muffle or electric furnace.

Bottled gases such as methane, propane, ammonia or city gas are frequently burned in endothermic (heat absorbing) gas generators. Such a gas generator is shown in Fig. 9-34. Endothermic gas generators are those in which heat is supplied in order to crack the gases and provide atmospheres generally composed of hydrogen, nitrogen, water vapor, methane, carbon monoxide and occasionally oxygen. With the exception of nitrogen, all of these gases react with steel, causing oxidation, carburization or de-carburization.

The selection of the ideal gas atmosphere for the heating chamber depends upon many factors, such as the type of steel being treated, the temperature of the operation, etc., and requires a careful

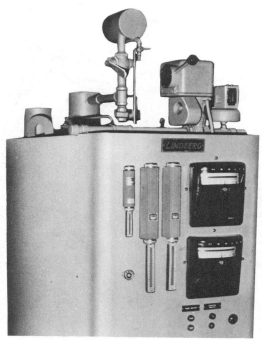

Fig. 9-34. Endothermic gas generator for preparation of protective furnace atmospheres. (Lindberg Engineering Co.)

study of the problem involved and an understanding of the reactions occurring at the temperature of operation. In general, this problem is solved by trying several gaseous mixtures and analyzing their effect upon the steel being treated. Some commonly employed gas atmospheres are shown in Table 9-1.

The most common gas mixtures contain a relatively high percentage of CO or H_2, both of which are reducing gases, and may be used to prevent the formation of surface scale or decarburization. The presence of CO_2, H_2O, or O_2 in the gaseous atmosphere may result in harmful and excessive scale or surface decarburi-

Heat Treatment of Steel

TABLE 9-1. PREPARED GAS ATMOSPHERES USED IN HEAT TREATING OPERATIONS.

TYPICAL APPLICATION	GAS CONSTITUENTS—PERCENT BY VOLUME					
	CO_2	CO	H_2	CH_4	H_2O	N_2
NO. 1 Case carburizing low or medium-carbon steels	- - - -	20.7	38.7	0.8	- - - -	39.8
NO. 2 Bright annealing copper	10.5	1.5	1.2	- - - -	0.8	86.0
NO. 3 Process annealing copper, low-carbon and stainless steels	0.05	0.05	3-10	- - - -	- - - -	Remainder
NO. 4 P/M sintering—bright annealing (cracked anhydrous ammonia)	- - - -	- - - -	75.0	- - - -	- - - -	25.0

zation. However, this will depend upon the type of gas used the type of steel, and the temperature of operation. In general, the atmospheres used in any heat-treating operation are such as to produce a light scale, thus preventing decarburization to a depth of more than 0.001'' all of which could normally be removed in any finishing operation.

Liquid Heating Baths

Another type of heating equipment commonly employed in heat treating operations is the pot type furnace used for liquid heating baths such as shown in Fig. 9-35.

Both molten lead and molten salt baths are used for the heating of steel in operations involving hardening, annealing, tempering, etc. Molten lead is a fast-heating medium and gives complete protection to the surfaces of the steel. With a melting point of 621°F (327°C), lead may be used

Fig. 9-35. Pot type electric furnace used for liquid heating, cyaniding, and martempering (Pacific Scientific Co.)

Metallurgy

successfully from 700° F to about 1600° F (370-870° C). Lead oxidizes readily with the formation of a dross, making a dirty bath, and gives trouble because of the sticking of lead and dross to the surfaces of the steel being heated. This can be overcome by suitable lead coverings such as wood charcoal, coke, carburizing compounds, salts, etc. The steel can be further protected from lead sticking to its surfaces by the use of a coating of a thin film of salt or some other material which is applied to the steel before it is placed in the lead bath. Dipping of warm steel into a saturated brine solution and allowing it to dry leaves a protective salt film on the steel. The use of a water emulsion containing bone charcoal, rye flour, potassium ferrocyanide, and soda has proved successful. The steel is dipped into the emulsion and allowed to dry before placing in the lead bath.

Molten salt baths have been found to be satisfactory for the heating of many steel parts that are to be given heat treatments. They transmit heat quickly and uniformly and afford a protection to the steel during the heating cycle. Upon removal of the steel from such a salt bath, a thin film of salt adheres to it, giving further protection from air prior to the quenching, etc. Although the ideal salt bath has not been discovered, if care is used in the selection of a salt and precautions are exerted in its use and maintenance, good results may be expected.

Salts may be used for low-temperature tempering. They usually consist of nitrates and may be used in a range of temperatures from 300°F to 1000°F (149 to 538°C). Salt baths used in temperature ranges of 1000°F to 1650°F (about 538 to 895°C) consist mainly of sodium carbonate, sodium chloride, sodium cyanide, and barium or calcium chloride. In a temperature range of 1800°F to 2400°F (980° to about 1315°C) salt baths are made from mixtures of barium chloride, borax, sodium fluoride, and silicates, Precautions should be observed in the use of salt baths in order to prevent violent reactions from the mixing of certain salts. One should also be careful to prevent moisture from coming into contact with the fused salts. With the cyanide salts, precautions should be observed because of their poisonous nature.

REVIEW

1. What is meant by the term heat treatment?

2. What is accomplished by the process of normalizing?

3. Describe the application of an annealing heat treatment.

4. What two basic types of structures are obtained by annealing?

Heat Treatment of Steel

5. What is meant by bright annealing?
6. What temperature is frequently recommended for stress relief of cast steel, cast iron and welded steel structures?
7. What are the three main requirements to satisfactorily harden steel?
8. Name the two operations necessary to harden steel.
9. What is austenite and how is it formed?
10. What is martensite and how is it formed?
11. What is the structural appearance of martensite?
12. With careful cooling, at what temperature does martensite form?
13. What is the M-point?
14. What method is used to transform retained austenite at room temperature to martensite?
15. Carbon steels can only be hardened in a water quench. Therefore these steels are classified as what type of steels?
16. Name the three common quenching media.
17. Why should water be agitated during quenching?
18. Below what temperature should the water for quenching be kept?
19. In what respects is brine superior to plain water for quenching?
20. What type of workpiece is most successfully hardened in oil as a quenching medium?
21. What two parts does carbon play in the hardening of steel?
22. What is the maximum Rockwell C hardness obtainable from any steel?
23. What effect does fast cooling rate needed in hardening carbon steels have upon the centers of the work-piece?
24. What means are used to aid the cooling of work having heavy sections?
25. What may be said about the rate of transformation of fine-grained steels? Of coarse-grained steels?
26. If deep penetration of hardness is desired in large pieces, how does this affect the temperature in the hardening operation?
27. What causes warping and cracking during hardening operations?
28. How is the distortion that takes place during hardening operations corrected?
29. Name a type of failure occuring in deep hardening operations.
30. How may decarburization during the heating operation be prevented?
31. How may soft spots on the surface of hardened tools be avoided?
32. What advantage has oil as a cooling medium in hardening alloy steels?
33. How may hardening of high-speed steels be carried out?
34. What are the reasons for tempering steel?
35. What protection does the salt bath method give to the surface of the steel?
36. How can a standard type of oven furnace be operated to heat high-speed steel for hardening?
37. What quenching media may be used in hardening of high-speed steels? Which is the most common?
38. Why is it necessary to temper or draw high-speed steel after hardening?
39. What four characteristics are furnished to steel by hardening?
40. Name five results of tempering hardened steel.

Metallurgy

41. What happens to the atomic structure of hardened steel when it is tempered?

42. What is the purpose of tempering?

43. What is the minimum time to be used at any tempering temperature for satisfactory results?

44. What effect does tempering of steel have upon its volume?

45. What liquid baths may be used in tempering?

46. What three energy sources are used to heat furnaces?

47. What is the ideal furnace atmosphere?

48. What is dross and why is it troublesome?

49. What simple operation will protect a part from dross?

50. Why must cyanide salts be used with extreme caution?

CHAPTER 10 — SURFACE TREATMENTS

Many metal objects made from either ferrous or nonferrous metals may be subjected to some form of surface treatment that affects only a thin layer of the outer surfaces. Such treatment may be carried out for the purpose of developing greater resistance to corrosion, for greater sales appeal from an appearance viewpoint, to increase surface hardness, or to obtain a combination of hardness and toughness in the same object that would be difficult or impossible by any other method. Surface treatments include hardening treatments, metallic, organic, oxide and ceramic coatings, and mechanical finishing.

SURFACE HARDENING

Carburizing

Carburizing is probably the oldest heat treatment in which the early smith made a steel by adding carbon to wrought iron. Carbon was likely added by the carbonaceous gases from the charcoal in the forge fire.

Carburizing is used when a hard steel surface coupled with a tough core is desired. This is done by heating low-carbon steel, plain or alloy, in contact with a rich carbon gas and allowing the low-carbon steel to dissolve the carbon from the gas. This produces a high-carbon surface layer, the *case*, that can be hardened by

Metallurgy

quenching. The balance of the interior of the steel is unaffected by the surface carbon absorbtion and remains low in carbon. This area is the core, which does not harden and remains strong and tough.

Carburizing may be achieved through either of three methods: (1) pack carburizing, (2) gas carburizing or (3) liquid carburizing. The pack carburizing process involves packing the low-carbon workpiece into a heat resisting metal box, usually cast iron, and completely surrounding the

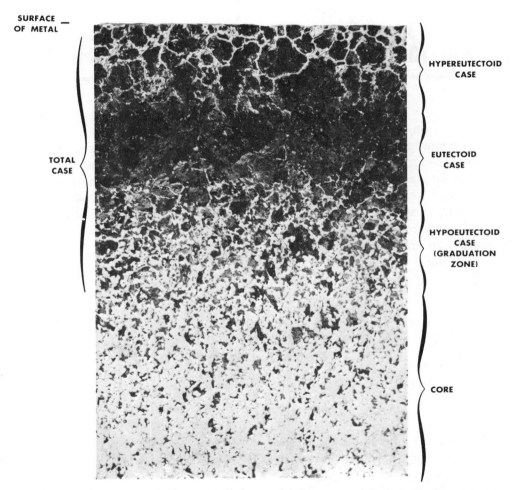

Fig. 10-1. Machine steel, 0.10 to 0.20 percent carbon, carburized. The unsatisfactory case produced of network or eggshell cementite is in hypereutectoid layer. Picral etched, magnified 75 times.

Surface Treatments

part with solid carburizers such as carbon-rich charcoal or fine ground charred bone. The container is sealed at the edge of the cover with clay and placed in a furnace, with the heat held at 1600 to 1700°F (about 870 to 925°C).

During the heating period, large quantities of carbon-rich gas evolve from the carburizing compound. This gas, primarily carbon monoxide, is absorbed by the austenitic steel and diffuses slowly into the interior of the part being carburized. The rate of carburizing is slow as shown in Fig. 10-1. A carbon pick-up of less than 1.15 percent carbon is considered good practice, although a carbon content of 0.90 percent is the best for maximum hardness and toughness of the case.

The microstructure of an unsatisfactory case of a pack carburized steel specimen is shown in Fig. 10-2 and a satisfactory case is shown in Fig. 10-3. Applications for the carburizing process include case hardening of gears, cams, crankshafts, firearm parts, roller bearings, ball bearings, conveyor chain and some inexpensive tools. Pack carburizing is relatively slow and dirty and has been supplanted to a great extent by gas and liquid carburizing.

Carbon will gradually diffuse into the steel, and by controlling the time at the carburizing temperature the penetration may be controlled. The rate of penetration may vary a great deal with different steels and carburizing practice; however, a penetration of approximately 0.006" per hour for shorter carburizing time should be expected. If a very deep or thick case is needed, over 0.100", an average hourly penetration of only 0.002" might be expected so as to require some 48 hours carburizing time.

The case depth may be measured by several different methods. In Fig. 10-4 the case thickness obtained from a carburizing cycle is revealed by deep-etching a section of the carburized part. Cooling slowly in the box or pot after carburizing will result in a high-carbon case on a low-carbon core, but both case and core will be in a soft and pearlitic condition.

Figs. 10-5 and 10-6 illustrate the condition of low-grade, low-carbon, screw-stock steel after carburizing and slow cooling. The banded and segregated condition of the original steel results in the uneven penetration of the carbon seen in this steel.

The carburized case is not machinable after hardening; therefore any parts to be

Fig. 10-2. Carburized and hardened surface of ball race illustrates defective case due to excess free cementite in the martensitic structure. Magnified 1000 times. The area of the free cementite is the white area at top, just below the surface.

Metallurgy

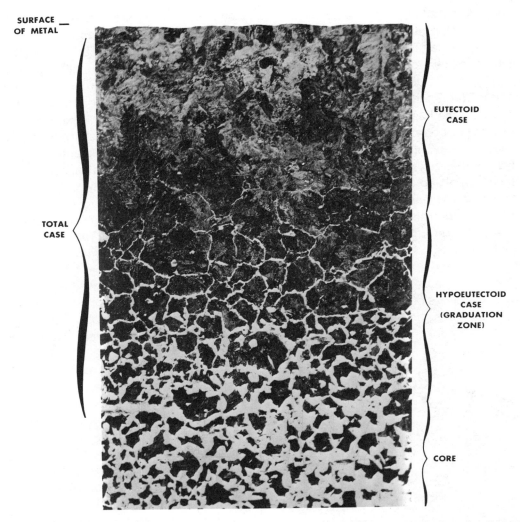

Fig. 10-3. SAE-1020 steel, carburized to produce a satisfactory eutectoid case. Picral etched, magnified 75 times.

machined or drilled after carburizing must be kept soft. The work may be cooled from the box or pot and any machine work done before hardening; however, this is not possible if the work is pulled out at the carburizing heat and quenched for hardening. To overcome such difficulties, inert materials are packed around places to be kept soft, and holes and other places are packed with fire clay and other suitable materials to keep them unaffected by the carburizing compounds.

Surface Treatments

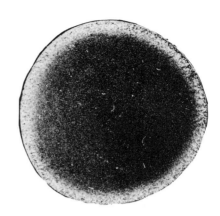

Fig. 10-4. Deep-etched section of case-hardened steel. The light outer layer illustrates case depth. Etchant 5 percent Nital, not magnified.

Fig. 10-5. Carburized screw stock, longitudinal section, magnified 100 times. The eutectoid case at top and the hypereutectoid just below it are the total case. The bottom half is the core.

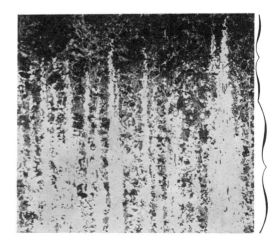

Fig. 10-6. Same as Fig. 10-5. End-section of rolled stock. Dark area of case is more penetrating into the core.

Metallurgy

Selective carburizing. Another method employed to isolate the work from the action of the carburizers is to copperplate them. The whole piece can be plated with copper and the copper machined off the particular portions to be hardened, or selective plating of the portions to be kept soft may be employed.

Case carburizing seeks to attain a certain level of surface hardness and wear resistance coupled with good adherence of this surface or case to the inner portion or core of the work. Carburizing media, carburizing temperatures and times, and alloying elements control the rate of absorption of carbon at the surface and the rate of diffusion of this carbon. Carburizing conditions and heat treatment are based on the depth of the case needed, the structure and hardness desired in the case, and the desired structure and properties of the core. A hypereutectoid case with the cementite in a network or in spines is brittle like an eggshell, (see Fig. 10-1) and is likely to spall off either on quenching in heat treatment, during grinding, or in service. Slow cooling of high-carbon austenite results in a tendency of the excess carbides to separate in embrittling forms. It is therefore desirable to have the carbide in a more or less spheroidized form, thus eliminating embrittlement of the case and adding to its wear resistance. This is obtained by controlling and using a relatively fast rate of cooling from the carburizing temperature to below the critical range—roughly below 1200°F (650°C). In industry this may be done by using separate cooling chambers either water jacketed or cooled by passing a cooled gas through this chamber.

Another method that may be selected in order to avoid a brittle eggshell case of free cementite is that of controlling the maximum carbon content of the case and keeping it around 0.90 percent. This control of the carbon content of the case may be had by either the liquid or gaseous carburizing process, but it is almost impossible to obtain by the pack-carburizing process using solid carburizers.

Gas carburizing. Both continous and noncontinuous, or batchtype, furnaces are used in the gas carburizing process. Carburizing may be accomplished by means of natural gas, propane, butane, and special gas mixtures produced from compounded oils that are fed into the retort containing the work to be cased. A popular gas carburizing machine of the batch type consists of a cylindrical drum or retort of special noncarburizing alloy steel within an outer heat-insulated cylinder. Heat is applied to the annular space between the inner and outer cylinders. The work is charged into the furnace on suitable fixtures, and the furnace is sealed. The carburizing gas or oil is introduced into the retort and circulated by means of a fan. This process has the advantage of being fast and offers the possibility of close control of the total case depth and the total carbon of the case. See Fig. 10-7 for rate of gas carburizing.

In controlling the total carbon of the case by this process, one method that has proved successful employs the use of a neutral gas during the heating-up cycle. After the carburizing retort has been purged of any oxygen in the form of CO_2, natural or propane carburizing gas is introduced into the retort. A supply of about

Surface Treatments

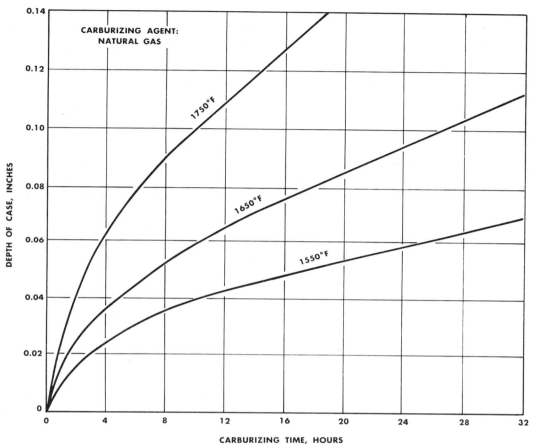

Fig. 10-7. Carburizing rates using natural gas as the carburizing agent.

30 cfh of carburizing gas will then yield an efficient carburizing atmosphere. The time and temperature are closely controlled with the carburizing temperature running between 1650° to 1700°F (895° to 930°C). After a predetermined time at the carburizing temperature, the carburizing gas is shut off, and the steel is allowed to cool in the neutral atmosphere. Cooling in the neutral atmosphere increases the case depth or penetration of the carbon and reduces the carbon concentration at the surface of the steel by diffusion of carbon from the outside toward the lower carbon core.

Liquid-bath carburizing and hardening. Many small parts, made from low-carbon steel, that require a high surface hardness with only a light or thin case, may be successfully carburized and hardened in the cyanide liquid baths. Steel to be casehardened is immersed in a molten

Metallurgy

bath containing more than 25 percent sodium cyanide held at a temperature usually around 1550°F (840°C). At this temperature, the steel will pick up both carbon and nitrogen from the bath, and in 15 minutes a penetration of approximately 0.005" is obtained. The nitrogen in cyanide-hardened cases is present as finely dispersed iron nitrides which impart high hardness and brittleness. If deeper penetration is required, a longer time is necessary. A treatment of one hour results in about 0.010" depth of case. To harden the case, quenching direct from the carburizing temperature into water or brine is practiced. If less distortion is needed than from the usual practice, lower carburizing temperatures may be employed, or it may be cooled to lower temperatures before quenching.

To keep the liquid carburizer active as a carburizing agent, the concentration of the sodium cyanide must be maintained above 25 percent concentration. This may be accomplished by frequent additions of new carburizer or by additions of a salt mixture rich in sodium cyanide.

The pots used in this process may be made from pressed steel, although special pots containing nickel and chromium are frequently used because of their much longer life and lower ultimate cost. Steel parts may be wired and suspended in the molten baths, and in many instances small parts are placed in wire baskets while in the carburizing bath. If a wire basket is used, care must be exercised in the quenching operation to insure even and fast cooling of all the parts treated to obtain uniform surface hardness.

All parts placed in the molten bath must be free from moisture, or danger from an explosive spattering of the bath may occur. The operator should be well protected by gloves and goggles or helmet. The fumes given off by the bath, because of their dangerous nature, should be removed by means of a hood over the bath and suitable exhaust system.

Liquid-bath carburizing and hardening result in distinct advantages over pot or pack carburizing. Lower temperatures are used, which reduce the depth of penetration because of the lower diffusion rate but also reduce distortion. The process allows for rapid heat transfer to the parts and also results in the elimination of oxidation during heating. The hardening action of the nitrogen is added to that of the carbon. When parts are removed from the cyanide bath, a film of cyanide adheres to the surfaces and acts as excellent protection from oxidation or decarburization during transfer to the quenching medium. These are a few advantages of cyanide hardening over pack hardening. It might be added that the clean, superficial hardness of surfaces resulting from cyanide hardening, done speedily and at low cost, makes this process more adaptable than pack hardening of small parts where service requirements permit the use of a thin case.

Chapmanizing

In developing thin cases harder than those from cyaniding and faster than those from nitriding, a process called *Chapmanizing* is used in which dissociated ammonia gas is bubbled through the molten cyanide bath thereby increasing the nitrogen content of the case.

Treatment after carburizing. The lowest-cost treatment after carburizing is to pull the steel out of the carburizing boxes while at the carburizing heat and quench in water or brine for plain carbon steels. Special alloy carburizing steels may be quenched in oil from the carburizing temperature. This treatment will result in a fully hardened case with a Rockwell hardness of over C-60. The core of the steel will also be hardened to its maximum, which may run less than Rockwell C-20.

The disadvantages of this treatment are due to the amount of warping and distortion resulting from the high quenching temperature, and perhaps a rather brittle condition of the steel due to excessive large grains in both case and core. It will be recalled that heating steel above its critical temperature results in grain growth, and due to the length of time the steel is held at the carburizing temperature, rather large grains may develop.

A practice that might be selected to avoid excessive distortion would involve cooling the carburized steel to a lower temperature before quenching. A quenching temperature of 1400° to 1500°F (760° to 815°C) may be selected; this will result in less distortion but will not improve the grain size. To refine the case and fully harden it requires cooling from the carburizing heat to below the critical temperature *Ac 1-2-3* and reheating to slightly above this temperature and quenching. This will result in a very fine and hard case. If the core is to be refined, cooling to below the critical temperature *Ac1* is required and then reheating to above the *Ac3* critical temperature followed by quenching, preferably in oil, to avoid excess distortion. Following this treatment, often referred to as the *regenerative quench*, the steel may be reheated to the proper temperature of the case (1400° to 1450°F [about 760 to 790°C]) followed by quenching to produce a martensitic case with its maximum hardness of Rockwell C-67. This double heat treatment, following carburizing, will result in the maximum combination of physical properties that can be developed by carburizing and hardening. Fig. 10-8 illustrates the fracture appearance of a steel treated by the double quench, resulting in a hard brittle case and a tough core.

Carburizing that produces a case with a carbon content in excess of 0.90 percent is one of the major causes of trouble and may be greatly aggravated by the method

Fig. 10-8. Fracture of case-hardened steel which has been given a double quench.

Metallurgy

Fig. 10-9. Carburized and hardened surface of ball race illustrates defective case due to excess free cementite in the martensite structure. Magnified 1000 times. The white portion of the area is free cementite.

used in the hardening operation following the carburizing cycle. Fig. 10-9 illustrates an undesirable condition found in a case hardened ball race treated by slow cooling from the carburizing temperature of 1700° to 1500° (930° to 815°C), followed by a direct quench in brine. The slow cool from 1700° to 1500°F (930° to 815°C) in a rich carburizing atmosphere served to segregate the excess cementite into a massive condition near the surface of the steel. This condition of massive cementite in the hard martensitic matrix of the fully hardened steel resulted in a brittle and cracked ball race.

Newer grain-controlled carbon and special alloy steels are available for carburizing, and due to a fine grain even at the higher carburizing temperatures, such steels may be quenched direct from the carburizing treatment without excessive distortion and brittleness. Even if these steels are slow-cooled to room temperature, they usually require only a reheating to the hardening temperature of the case, and upon quenching, the core properties are satisfactory and no regenerative quenching is required.

Tempering of the quench-hardened carburized steels may be done at a low temperature for the relief of stresses; and tempering at high temperatures may be carried out to increase the toughness of the hard and otherwise brittle case. Most casehardened steel is finished in the machine shop by grinding. If tempering has been carried out, there will be less danger of further warping and less danger of grinding cracks developing.

Nitriding

The nature of the nitriding process used to obtain a casehardened product is very different from that of the carburizing process. Nitrogen, instead of carbon, is added to the surface of the steel. Carbon does not play any part in the nitriding operation but influences the machinability of the steel and the properties of the core in the finished nitrided steel. The temperatures used in nitriding are much lower than those used in carburizing and below the critical temperature of the steel.

Simple carbon steels, which are often used for carburizing, are not used for nitriding. Steels used in the process are special alloy steels. With the nitriding de-

Surface Treatments

TABLE 10-1. NITRIDING STEELS[1]

ELEMENT	N 125 (Type H)	N 125N[2] (Type H with nickel)	N 135 (Type G)	N 135[2] (modified)	N 230
Carbon	0.20–0.30	0.20–0.27	0.30–0.40	0.38–0.45	0.25–0.35
Manganese	0.40–0.60	0.40–0.70	0.40–0.60	0.40–0.70	0.40–0.60
Silicon	0.20–0.30		0.20–0.30		0.20–0.30
Aluminum	0.90–1.40	1.10–1.40	0.90–1.40	0.95–1.35	1.00–1.50
Chromium	0.90–1.40	1.00–1.30	0.90–1.40	1.40–1.80	
Molybdenum	0.15–0.25	0.20–0.30	0.15–0.25	0.30–0.45	0.60–1.00
Nickel		3.25–3.75			

[1] N preceding a number indicates Nitralloy steels
[2] Aircraft specifications

veloping rather thin cases, a high core hardness is required to withstand any high crushing loads. High tempering temperatures call for a steel with a higher carbon content in order to develop this increase in core hardness. In addition to higher carbon content, various alloying elements are called for in the steel to bring about an increase in the formation of these nitrides. Aluminum seems to display the strongest tendency in the formation of these nitrides. The aluminum precipitates the compound AlN in a finely divided state, accounting for the extreme hardness of these nitriding steels. Chromium, molybdenum, vanadium, and tungsten, all being nitride formers, also are used in nitriding steels. Nickel in nitriding steels hardens and strengthens the core and toughens the case with but slight loss in its hardness.

In general, steels used in the nitriding process contain the elements within the percentage range given in Table 10-1.

Nitriding operation. In the nitriding process, nitrogen is introduced to the steel by passing ammonia gas through a muffle furnace containing the steel to be nitrided. The ammonia is purchased in tanks as a liquid and introduced into the furnace as a gas at slightly greater than atmospheric pressure. With the nitriding furnace operating at a temperature of 900° to 1000°F (480° to about 540°C), the ammonia gas partially dissociates into a nitrogen and hydrogen gas mixture. The dissociation of ammonia is shown by the following equation:

$$2NH_3 \rightleftarrows 2N + 3H_2$$

The operation of the nitriding cycle is usually controlled so that the dissociation of the ammonia gas is held to approximately 30 percent but may be varied from 15 to 95 percent depending upon operating conditions. The gas mixture leaving the furnace consists of hydrogen, nitrogen, and undissociated ammonia. The undissociated ammonia, which is soluble in

Metallurgy

water, is usually discharged into water and disposed of in this manner.

At these temperatures the free nitrogen formed by this dissociation is very active, uniting with the iron and other elements in the steel to form nitrides. These nitrides are more or less soluble in the iron and form a solid solution or are more likely in a fine state of dispersion, imparting hardness to the surface of the steel. From the surface the nitrides diffuse slowly, and the hardness decreases inwardly until the unaffected core is reached. The depth of penetration depends largely upon the length of time spent at the nitriding temperature. Diffusion of these nitrides is much slower than diffusion of carbon in the carburizing operation, so a much longer time is required to develop similar penetration. The case depth specified in nitriding is usually a very shallow one but requires from 18 to 90 hours to obtain. Nitriding at 960°F (about 515°C) for 18 hours results in about 0.010" case depth; 48 hours yields approximately 0.020" depth of case; 90 hours, approximately 0.030" depth of case.

Nitriding is usually carried out in muffle-type furnaces designed to operate within the temperature range of the nitriding cycle (800° to 1200°F [about 430 to 650°C]). The muffle or chamber is frequently made from a high alloy metal containing various amounts of chromium, 12 to 25 percent, with nickel from 20 to 80 percent. In order to obtain uniform dissociation of the ammonia gas in the nitriding chamber, the furnace should be designed to provide means of gas circulation. This may be accomplished by a fan built into the muffle or nitriding chamber.

The nitrided case. The depth hardness that may be expected from a nitriding operation as compared to a carburized and hardened steel is illustrated in Fig. 10-10. In general, the surface hardness is much greater after nitriding than it is after carburizing and hardening. The maximum hardness obtained from a carburized and hardened case runs around C-67 Rockwell; whereas, it is possible to obtain surface hardness values in excess of C-74 Rockwell by nitriding. The surface hardness of the nitrided case cannot be measured by the Rockwell C scale due to the extreme brittleness and shallowness of the case; it is usually measured by the Rockwell superficial scale such as the 15N or 30N scale. The hardness value on the Rockwell C scale may be estimated from a hardness conversion table.

The nitrided case is made up of three zones. The first zone is a thin white layer; the second zone, the effective nitrided case; and the third zone, a zone of gradation of hardness from that of the effective case to the core. The thin white layer, see Fig. 10-11, forms on the surface of the steel at the beginning of the nitriding cycle. This white layer is extremely brittle and is usually ground off after the nitriding operation so as to avoid failure by chipping which may occur in handling or during service, particularly where sharp corners exist in the work. The white layer has good corrosion-resistant characteristics, and from this standpoint it is desirable not to remove it.

Protection against nitriding. In order to obtain localized nitriding of parts or to stop nitriding on some surface areas, the best method is to use tin as a protecting

Surface Treatments

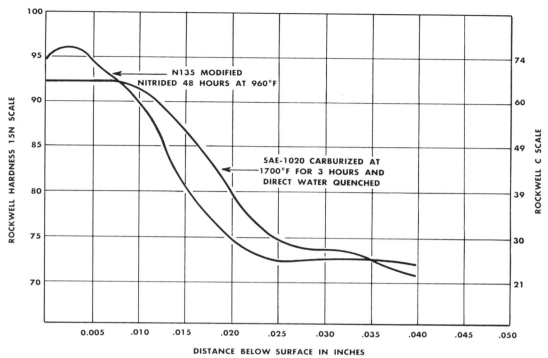

Fig. 10-10. Comparison of hardness of various depths below the surfaces of nitrided and carburized steels.

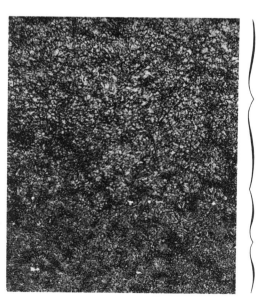

Fig. 10-11. Modified steel nitrided at 960°F (about 515°C) for 18 hours, illustrates white surface layer and case depth of approximately 0.011'' above core material at bottom.

217

Metallurgy

agent against the nitriding action. Tin, in the form of a paste or paint made from tin powder, or tin oxide mixed with glycerine or shellac, may be applied to the areas to be protected against nitriding. A thin tin electroplate may be applied as a satisfactory stop-off method. Although tin melts at a lower temperature than that used in nitriding, ample protection is provided by the thin layer of tin that is held to the surface by surface tension. Care should be exercised to apply the tin paint or electroplate to a clean surface and to avoid a thick layer of tin which may run or drip onto surfaces where protection is not desired. A tin plate of 0.0005" is sufficient to prevent nitriding.

Heat treatment in nitriding. All heat treatments, such as the quench hardening of steel in the nitriding process, are carried out before the nitriding operation. After a rough machining operation, the steel is heated to about 1750°F (955°C), held for the necessary length of time at this temperature, and then quenched in oil. This high hardening temperature is necessary in order to have the alloying elements go into solution in the austenite, thereby imparting core strength and toughness after quenching and tempering. Tempering is usually done slightly below 1100°F (about 595°C) before and after finish machining, to produce a sorbitic structure which has a tough case and eliminates any brittleness resulting from any free ferrite. This tempering also relieves internal stresses resulting from machining and hardening, thus reducing distortion during nitriding. After tempering, all oxide film and any traces of decarburization are removed.

Any decarburization left on the surfaces of the steel to be nitrided will usually result in failure of the nitrided surfaces by peeling or spalling off.

The steel is then nitrided and allowed to cool slowly to room temperature in the nitriding box or chamber. No quenching is required; the steel develops its maximum hardness without necessitating a further quenching operation.

The advantages of nitriding as a hard-surfacing operation are listed as follows:

1. Greater surface hardness
2. Greater resistance to wear and corrosion
3. Better retention of hardness at elevated temperatures
4. Less warping or distortion of parts treated
5. Higher endurance limit under bending stresses
6. Greater fatigue strength under corrosive conditions

Factors limiting its application include:

1. High furnace costs due to the long time of treatment
2. Necessity of using special alloy steels
3. Necessity of using high alloy containers to resist the nitriding action
4. Expense of medium used

Other Diffusion Processes

There are several other processes in addition to carburizing and nitriding which are designed to improve wear resistance by the diffusion of some element into the surface layers of the part. They are:

1. Carbonitriding
2. Chromizing

Surface Treatments

3. Siliconizing
4. Aluminizing

Carbonitriding is a process for case hardening a steel part in a carburizing gas atmosphere containing ammonia in controlled percentages. Both carbon and nitrogen are added to steel, the nitrogen serving chiefly to reduce the critical cooling rate of the case and allow hardening with oil or forced air cooling. Carbonitriding is normally carried out above the $Ac1$ temperature for steel. The penetration rate is quite close to that obtained in carburizing.

The widest application for carbonitriding is as a low-cost substitute for cyaniding. Operating costs may be as low as one quarter the cost of cyaniding.

Chromizing involves the introduction of chromium into the surface of the metal to improve corrosion and heat resistance. This process is not restricted to ferrous metals but may be applied to nickel, cobalt, molybdenum and tungsten. When it is applied to iron or steel, the surface is converted into a stainless steel case. Coatings about 0.0005" thick may be obtained after 3 hours at 1650°F (900°C). Increased case thickness may be obtained at higher temperatures.

A new development in chromizing is the process of ion bombardment. Chromium, the ore of which is imported from Rhodesia, the U.S.S.R. and the Union of South Africa is costly and becoming increasingly scarce. Corrosion-resistant steel can be produced by using only one-millionth the amount of chromium that a typical stainless steel currently requires.

In the production process, a crystal of chromium chloride is subjected to a high vacuum in which it vaporizes to form chromium and chlorine ions. An ion accelerator propels the ions through a positively charged magnetic field which sweeps up the negatively charged chlorine ions. The positively charged chromium ions slam into the surface of the steel with enough energy to reach depths as great as 100 atomic layers, a few millionths of an inch (0.0000005 mm). When the treated surfaces are put in contact with corrosive liquids such as brine or sulfuric acid, they resist corrosion as well as stainless steel alloy containing 18 percent chromium.

Traditional chrome alloy or chromized steels now used for turbine blades, hydraulic rams, pistons, pump shafts and drop forge dies, could be machined more easily from carbon steels and acquire the protection of chromium implanted on the critical surfaces by ion bombardment. Considerable savings in labor and material could be realized.

Siliconizing involves the impregnation of iron or low-carbon steels with silicon to form a case containing about 14 percent silicon. The process is carried out by heating the work in contact with a silicon bearing material such as silicon carbide or ferro-silicon in an ordinary carburizing furnace at a temperature of 1700 to 1850°F (930 to 1010°C). After the parts have reached the desired temperature, chlorine gas is added as a catalyst to speed the liberation of the silicon from the silicon carbide. The silicon then immediately diffuses into the metal under treatment. Case depth of 0.025" to 0.030" (0.06 to 0.08 mm) is usually produced in low-carbon steel in two hours. Siliconized cases are difficult to machine although the

Metallurgy

Fig. 10-12. Bar with silicon case cut in half. One surface (left) slightly etched, and the other (right) boiled in dilute nitric acid, leaving only a corrosion-resistant shell. (American Society for Metals)

hardness is only Rockwell B80 to 85. The silicon case shown in Fig. 10-12 is resistant to scaling up to 1800°F (about 980°C).

Oil impregnated siliconized cases exhibit very good wear resistance, especially under corrosive conditions. Siliconized cases have been successfully applied in usages involving corrosion, heat or wear as in pump shafts, cylinder liners, valve guides, forgings, thermocouple tubes and fasteners.

Aluminizing or *Calorizing* is a patented process of alloying the surface of carbon or alloy steel with aluminum by diffusion. The process consists essentially of packing the articles in a powder compound containing aluminum and ammonium chloride, then sealing them in a gas-tight revolving retort in which a neutral atmosphere is maintained. The retort is heated for 4 to 6 hours at 1550 to 1700°F (about 845-930°C). After removal from the retort, the article is then heated for 12 to 48 hours at a temperature of 1500 to 1800°F about (815-980°C) to allow further aluminum penetration to depths of 0.025" to 0.045" (0.06-0.11mm).

The usual case content of 25 percent aluminum provides good resistance to heat and corrosion. Calorized parts are reported to remain serviceable for years at temperatures not exceeding 1400°F (760°C) and have been used at temperatures of up to 1700°F (about 930°C) for heat treating pots. Typical aluminized metal parts include heat treat pots for molten salts, cyanide and lead, bolts for high temperature operation, tubes for superheated steam, furnace parts, etc.

Flame hardening. Flame-hardening methods may be used when a hard surface and a soft interior are needed, a combination of properties that may be more difficult and costly to obtain by any other method. Flame hardening is used to surface-harden large and small gray cast-iron machine parts, parts that would not permit heating in a furnace, plus water or oil quenching, for hardening, due to danger of warping and cracking.

Any large steel or cast-iron section requiring a high surface hardness that would be difficult to obtain by the usual methods might be successfully hardened by the flame-hardening method. In this process of surface hardening, heat is applied to the surface of the steel or cast-iron part by an oxyacetylene flame. Only a thin layer of the surface metal is brought up to the hardening temperature, and as the torch moves slowly forward heating the metal, a stream of water follows the torch, quenching and hardening the surface as it becomes heated. The speed of the torch is adjusted so that the heat penetrates only to the desired depth, thereby controlling

Surface Treatments

Fig. 10-13. Flame-hardening machine. Hardening a gear by rotating it between a number of gas burners. After the gear teeth reach the correct hardening temperature, the gear and rotating table are lowered into the quenching tank which is part of the unit. (Massachusetts Steel Treating Corp.)

the depth of hardness. The technique of flame hardening may be carried out in several different ways. Small parts may be individually heated and then quenched. In the case of cylindrical shapes such as shafts, gears, etc., the surface may be heated by slowly rotating the part and exposing the surface to the flame of a torch or a series of gas burners as illustrated in Fig. 10-13.

Upon reaching the hardening temperature and when the heat has penetrated to the desired depth, the part may be quenched for hardening. Both steel and gray cast iron may be treated by the flame-hardening process. In general, a steel to be hardened by this method should have a carbon content of at least 0.40 percent. The best range of carbon is between 0.40 and 0.70 percent carbon. Higher carbon steels can be treated, but care must be exercised to prevent surface cracking. If special alloy steels are used, the low-alloy type seems to respond satisfactorily and works the best. Applications of the flame-hardening process include rail ends, gears, lathe beds, track wheels for conveyors, cams and cam shafts, etc.

Induction heating for hardening. In the induction heating or *Tocco process* as applied to the surface hardening of steel, heating is accomplished through the use of inductor heating coils placed around the surface to be hardened. A high-frequency current at high voltage is transformed into low-voltage current with high amperage and passed through the inductor coils or blocks surrounding but not actually in contact with the surfaces to be hardened. The inductor coils induce a current in the surface of the steel. This induced current rapidly heats the metal to

221

Metallurgy

Fig. 10-14. Set-up for induction hardening of a gear.

Fig. 10-15. Surface hardening of a grinding machine spindle. Upon its reaching the correct hardening temperature, the spindle is removed and quench-hardened.

Surface Treatments

the proper hardening temperature. When the area to be hardened has been subjected to an accurately controlled current of high-frequency for the correct length of time, the electric circuit is opened, and simultaneously the heated surface is quenched by a spray of water from a water jacket built into or around the inductor blocks or heating coil. Figs. 10-14 and 10-15 illustrate two set-ups used. In Fig. 10-14, the work is rotated between the heating coils and quenched by a water jacket surrounding the heating coils and part.

The time cycle for the complete operation is only a few seconds. With split-second heating control and instantaneous pressure quenching, a good surface hardness is obtained with a gradual blending of the hardness into a soft core that is unaffected by heat. This process may be applied to many different carbon and alloy steels, provided the carbon content is high enough to permit quench hardening.

Because special inductors or coils are required for each job and automatic timing has to be worked our for each particular case, the equipment for induction surface hardening is costly. However, because of the short time cycle involved, production from the equipment is high. Maximum returns come from quantity production of the same piece. Close control of the heated zone together with greatly reduced distortion during heat treatment are two distinct advantages of this method over general hardening methods. Figs. 10-16 and 10-17 illustrate the results that may be expected through the use of this process. Fig. 10-17 shows the hardness penetration or case-hardness values that may be expected.

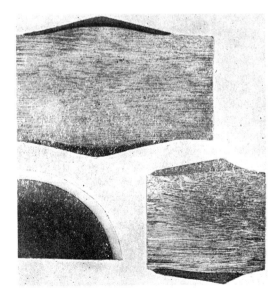

Fig. 10-16. Etched sections of shafting, illustrating the characteristic case-hardening effect obtained by the Tocco induction heating method. (Ohio Crankshaft Co.)

Metallurgy

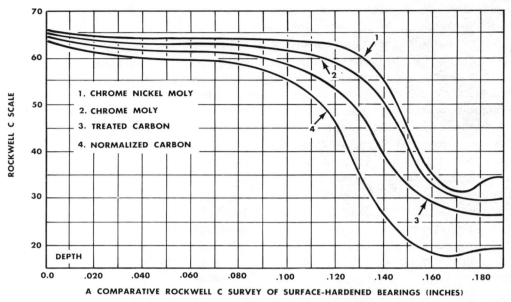

Fig. 10-17. Hardness depth characteristics of four types of steel, induction-hardened by the Tocco process. (Ohio Crankshaft Co.)

COATINGS

Metallic coatings are commonly applied to the base metal by electroplating, hot dipping, spraying or impregnation. Metallic coatings are used to impart some particular characteristic to the surface of the metal, such as improved corrosion resistance, wear resistance, hardness or to enhance the appearance of the metal. The particular metallic coating selected and the process by which it is applied are usually dictated by economics, availability of equipment, and the knowledge and ability to successfully apply the coatings.

Plating

Plating of metal surfaces with another metal is one of the most common methods employed for obtaining satisfactory resistance to corrosion or for bettering appearances. Many different metals may be used in plating, including nickel, silver, chromium, copper, zinc, gold, cadmium, tin, iron, lead, cobalt, platinum, rhodium, and tungsten. Also, many alloys of these metals may be used.

Surfaces of metal objects may be plated

Surface Treatments

by several methods, such as (1) electroplating, which is the depositing of a metal onto the surface of a metal object by electrolysis, (2) by means of dipping the metal object into a molten bath of the metal to form the coating (this method is used extensively for tin plating known as *tin plate*, and for plating of zinc, *zinc galvanizing*), (3) depositing of metal on to a surface by action of a vapor, such as in the *sherardizing process*, where the objects to be coated are placed in a tight drum with zinc powder and heated to 575° to 850°F (about 300 to 455°C). A surface coating of zinc is formed by the action of the zinc vapor which penetrates and forms a zinc surface coating, (4) by means of a spray of atomized molten metal directed onto the surfaces of heated metal objects, known as the *Schoop metallizing process*, (5) a surface plating called *Parkerizing*. (See *Phosphate coatings* below.) (6) a process known as *hard facing* for hardening surfaces of iron and steel, in which a hard metallic layer is welded to a softer metal by the fusion welding process. Hard facing may also be the application of tungsten carbide *particles* to a surface bonded by a brazing metal.

Phosphate coatings. Phosphate coatings, produced by immersion of metal parts in a hot solution of manganese dihydrogen phosphate for 30 to 60 minutes, result in a surface that is gray in color but after oiling or waxing becomes black. Thinner coatings, modifications of this process, are applied in a much shorter time by spraying or dipping in hot phosphate solutions containing catalyzers. In the *Bonderizing* process, which is a modification of the Parkerizing process, a thin coating of protection is obtained by immersion in a hot phosphate solution for a period of 30 seconds or longer. Such a coating is primarily intended as a base for paint and in itself offers only a temporary protection to the surfaces so treated.

A phosphate coating has proved advantageous in resisting excessive wear and seizure of machine parts and to minimize scuffing during the wearing-in period. It has been successfully used on parts such as pistons, piston rings, camshafts, and other engine and machine parts.

Surface oxidation. Oxide films formed on metal surfaces will build up resistance to corrosion and change the appearance of metal objects. Aluminum may be treated to develop a very thick adhering film of surface oxide which enhances its resistance to corrosion by means of electrolytic oxidation. Steel objects with a bright, clean surface may be treated by heating in contact with air to 500° to 600°F (260° to 315°C), and thus a color oxide film is formed (temper color) which adds to sales appeal in some instances and increases the resistance to corrosion. A heavy blue or black oxide film may be formed on steel by heating the steel in a bath of salts of low melting point. These salts are sold under various trade names, and the operation is referred to as *bluing* or *blacking* of steel, being used on gun parts, small tools, spark plug parts, etc.

Metal spray process. Through the use of a gas-fired pistol, metal in the form of wire or powdered metal is fused and then sprayed from the gun and deposited upon the surface that is to be coated. The action is similar to that of a paint gun. See

Metallurgy

Fig. 10-18. Metallizing a splined shaft using a corrosion resistant powdered metal. (Metco, Inc.)

Fig. 10-18. The sprayed metal adheres to the surfaces of the part due to impact of the tiny particles. This adherence is aided by the proper preparation of the surfaces to receive the metallic coating.

Usually a rough surface, such as prepared by sand blasting, is required for good adherence. The coating is not very strong as it is somewhat porous in nature, but it has proved satisfactory for the building up of worn parts of machines, such as shafts, cylinders, rolls, etc., and may be used to build up resistance to corrosion and wear. A number of different metals, including aluminum, zinc, tin, copper, lead, brass, and bronze, may be sprayed onto surfaces by this process. However, the process is limited commercially to the metals having a relatively low melting point. The coating is somewhat thicker than those produced by hot dipping and electroplating methods, ranging from 0.004'' to 0.025'' (0.010 to 0.64 mm).

Hard-facing. In the hard-facing process, air-hardening steels such as high-speed steels, natural hard alloys such as stellite, tungsten carbide, and boron carbide, and special stainless steels may be fused to metals requiring a harder or tougher wear and shock-resistant surface. The hard-facing alloy may be fused to almost any metal part by either the oxyacetylene torch or the electric-arc method of welding.

In applying hard-facing alloys, it is important to prepare the surfaces to be hard faced by grinding, machining, or chipping, so as to clean and remove rust, scale, or other foreign substances. The work to be hard faced is often preheated before applying the hardfacing alloy. This preheating may be carried out in a furnace or by use of an oxyacetylene flame.

If arc-welding methods are used, the atomic hydrogen process is recommended; however, good results can be obtained with straight arc-welding equipment when a short arc and low current are used.

If the oxyacetylene blowtorch is used, it is recommended that the flame be adjusted to a reducing or excess of acetylene flame. The technique employed when using the oxyacetylene blowtorch is to heat the surface to be hard faced to a sweating temperature, heating only a small section at a time. The hard-facing rod is then

Surface Treatments

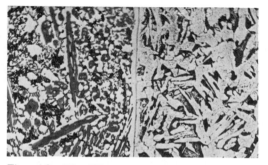

Fig. 10-19. Photomicrograph of the straight line nature of the bond obtained when Haynes Stellite (left) is properly applied to steel (right) with an oxy-acetylene blowpipe.

brought into the flame and allowed to melt and spread over the sweating area. If the operation is properly carried out, the hard-facing rod spreads and flows like solder. Additional hard-facing alloy may be added and the required built-up thickness accomplished.

With ideal welding conditions, the bond between the hard-facing alloy and the parent metal is very strong, in most cases being stronger than the hard-facing alloy alone. Fig. 10-19 illustrates the type of structure found in a hard-faced steel surface using Haynes Stellite. Though there is practically no penetration between Haynes Stellite and steel, the bond is actually stronger than the deposited metal. The deposited metal is pure Haynes Stellite, undiluted with iron from the base metal, and hence possesses maximum wear resistance.

This hard-facing process may be applied to nearly all steels and many alloys. Many of the uses include hard-facing tractor shoes, dies, cutting tools, plowshares, exhaust valve seats, airplane tail skids, pulverizer hammers, dipper bucket teeth, and parts subject to abrasion or impact.

Vapor deposited metallic coatings. Ultra-thin coatings produced by condensation of metallic vapor under high vacuum are finding increasing usage, especially in the development of electronic solid state devices. Articles to be plated may include paper, fabric, wax and glass in addition to metals. The metal to be deposited is vaporized by decomposition of metallic compounds, by sputtering techniques or by evaporation of molten metals. The articles to be plated are carefully cleaned and then placed into a vacuum chamber, Fig. 10-20. The chamber is usually flushed out with an inert gas such as argon and then is evacuated to a very low pressure. The coating is vaporized by one of the methods mentioned and is allowed to condense on the article to be plated. Masks may be used over areas where plating is not desired. Typical articles metallized by this process are the reflectors for sealed beam head lamps, electrical capacitors, the 200" mirror in an astronomy telescope, alloy diaphragms for carbon broadcasting transmitters, high fidelity sound recording waxes, surgical gauzes and integrated electronic circuits. Some coated electronic articles are shown in Fig. 10-21.

Oxide coatings. Certain oxide films on metal surfaces are intended to build up resistance to corrosion or to change the appearance of the metal object. Aluminum and magnesium may be treated by the anodizing process to develop very thick (up to 0.005") (0.01mm) films which

Metallurgy

provide excellent corrosion and abrasion resistance and which can be colored with organic dyes.

Oxide coatings can be produced on iron and steel objects by heating them in alkaline oxidizing solutions in the 283 to 315°F (140 to 157°C) temperature range, rinsing in water and oiling. Fused nitrate baths consisting of conventional quenching and drawing salts will produce black oxide coatings.

Copper may be given a black oxide coating by treatment with strong alkaline solutions containing chlorite or persulphates. Many proprietary solutions are commercially available for decorative and corrosion-resisting oxide coatings on spark plug parts, gun parts, small tools and accessories.

Fig. 10-20. High-vacuum equipment for vapor-deposition of thin films. (Consolidated Vacuum Corp.)

Fig. 10-21. Articles that have been thin-film coated. (Consolidated Vacuum Corp.)

REVIEW

1. For what purposes are surface treatments of metal carried out?
2. Name several kinds of surface treatments for metals.
3. As applied to case carburizing and hardening, what does case mean?
4. During the carburizing process, what is considered the best carbon pick-up? Why?
5. What commercial items are carburized for hardening?
6. What depth of carbon penetration per hour should be expected?
7. What two methods can be used to avoid a brittle case?
8. Describe the low-cost treatment after carburizing. What disadvantages has this treatment?
9. What treatment will refine and fully harden the case of steel?
10. Name the special advantages of special grain-controlled carbon and specialty alloy steels.
11. What type of work is carburized and hardened in cyanide liquid baths?
12. Why must liquid carburizing and hardening baths be hooded and vented?
13. What are the advantages of liquid-bath carburizing and hardening?
14. Describe the nitriding process for steel.
15. How much time is required in the nitriding operation of steel for a shallow case?
16. Name the advantages of nitriding.
17. Define carbonitriding.
18. What is the advantage of ion bombardment in chromizing?
19. What physical improvement is made by siliconizing?
20. What physical properties does aluminizing give to steel?
21. How is the flame hardening carried on?
22. What is the time cycle for an induction hardening operation?
23. Name six methods of plating metals on metals.
24. What products are coated by vapor deposition method?
25. What are oxide coatings used on?

Metallurgy

White hot alloy steel emerging from furnace.

CHAPTER 11  ALLOY OR SPECIAL STEELS

Although by far the largest tonnage of steel manufactured is of the plain carbon type, large quantities of alloy or special steels are manufactured. *Alloy steel* may be defined as plain carbon steel to which metallic elements have been added in large enough percentage to alter the characteristics of the steel. Carbon steel does contain minor amounts of other elements besides the basic iron and carbon, such as silicon and manganese; however, unless these other elements are increased in percentage beyond the usual small percentages found in carbon steel, the steel is not considered an alloy or special steel.[1]

Carbon is the most important element in any steel, whether a carbon or alloy steel. The carbon content determines the hardness that can be obtained by standard hardening methods. The hardness of any steel depends upon carbon content and heat treatments, regardless of the amount or kind of alloying element added. As measured in a standard Rockwell hardness testing machine a hardness of 68 on the C scale is about the maximum hardness that any plain carbon or alloy steel can develop.

Although the maximum hardness may not be materially increased by the addition of any alloying element, properties such as ductility, magnetism, machinability, hardenability, etc., may be drastically changed. In general, it is easier to harden alloy steels than plain carbon steels. Because of this ease of hardening and the greater uniformity accomplished by heat treating, alloy steels are often selected for specific use in preference to plain carbon steels.

Some of the reasons for selecting alloy steels may include the following:

Metallurgy

1. Alloy steels often may be hardened by quenching in oil or even in air to produce their maximum hardness.

2. Because of the less drastic oil or air quench, compared to a water quench required for most carbon steel tooling, alloy steels are less apt to crack or warp.

3. Alloy steels can be hardened to a greater depth and heavy sections of alloy steel can be hardened with less problems of distortion and dimensional instability than with carbon steels.

4. The martensitic crystal structure (the structure of hardened steel) formed in alloy steels is more stable and resistant to tempering effects. This property permits the use of alloy steels in dies and tools subject to high (red-hot) heat in production operations.

5. Some alloy steels develop high resistance to shock in operation due to greater ductility, with the same tensile strength properties exhibited by carbon steels.

6. Some alloy steels exhibit a marked resistance to grain growth and oxidation at high temperatures. Grain growth can result in a weaker, more brittle steel.

7. Alloy steels may exhibit less sensitivity to the effects of cryogenic (sub-normal) or elevated temperatures and perform better under adverse climatic and environmental conditions than carbon steels.

8. Corrosion resistance can be obtained in alloy steels. Stainless steels have marked advantages in both industrial use and in consumer goods.

Behavior and Influence of Special Elements

Upon addition of a special element to steel, the added element may combine with the steel in four ways:

1. Dissolve in both the liquid and solid states of the iron and thus be retained in solid solution in the iron at room temperature.

2. Combine chemically with the carbon in the steel to form a compound carbide, or with the iron to form other compounds such as nitrides.

3. Form an oxide with the oxygen of entrapped air in liquid steel.

4. Remain uncombined as lead does in screw-machine steel.

Nickel, copper and silicon are retained in iron at any temperature to form a solid solution as in #1 above. Elements in solution, in general, increase the strength and hardness of the steel and lower the ductility. Nickel and copper in moderate amounts decrease ductility very slightly, hence the importance of nickel steels.

The elements manganese, chromium, tungsten, vanadium and molybdenum combine with the cementitic constituent in the steel to form a carbide with iron and carbon. Because of their hardness and relative brittleness, the carbide-forming elements, which separate out from solid solution during slow cooling of the steel, seldom develop the desired properties when the carbides exist as free carbides. The value of the carbide-forming steels depends upon the ability to heat-treat the steel so as to dissolve the carbides and retain them in solution upon quenching.

In the third way of combining, elements

such as aluminum, silicon and zirconium may readily oxidize upon being added to steel. The entrapped air bubbles (blowholes) in the ingot, resulting from the pouring of the liquid steel from the ladle, furnishes the oxygen to form the oxides. The oxides of these elements may appear as inclusions in the steel. The effect of these inclusions seems to be to lower plasticity and they appear to contribute to crack formation and ultimate failure.

The fourth behavior or alloy elements is the ability to remain in the steel as free constituents or uncombined. Graphitic carbon, lead and excess copper are elements which behave in this manner. Their principal influence is to break up the continuity of the steel and thus weaken the structure. One beneficial result of discontinuity is better machinability.

ALLOY STEEL STRUCTURES

The steels produced by the addition of alloying elements develop structures and properties that may be divided into four classifications. These four groups are indicated in Fig. 11-1 and Fig. 11-2.

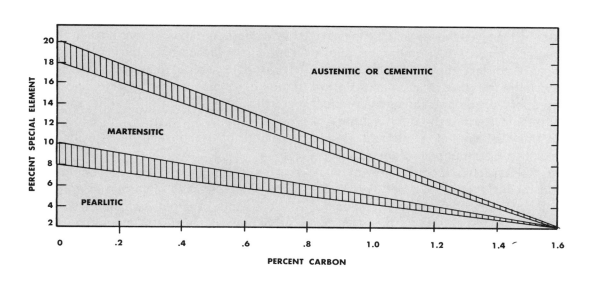

Fig. 11-1. Constitutional diagram of alloy steels. (after Guillet)

Metallurgy

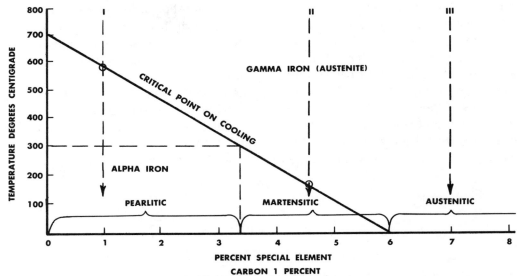

Fig. 11–2. Influence of special element on the position of the critical point. (after Sauvuer)

Group I. Pearlitic Alloy Steel

The diagrams in Figs. 11-1 and 11-2 are only working diagrams which serve as a working theory of the metallic behavior of special elements in iron-carbon alloys. These diagrams should not be used to determine the structural composition of any specific composition of alloy steel if the analysis is to be very accurate. If accuracy is wanted, a student should consult a constitutional diagram involving the particular composition of alloy steel being studied. Constitutional diagrams of many of the common industrial alloy steels are now available. The diagrams in Figs. 11-1 and 11-2 may be considered as introductory diagrams to the study of special alloy steels.

This type of alloy steel contains a relatively small percentage of the alloying element. With a low percentage of carbon, the percentage of the added special element may be as high as 6.0 percent as indicated in Fig. 11-1. The structure and characteristics of pearlitic alloy steel are similar to carbon steels and the microscopic analysis may reveal all pearlite, or a mixture of pearlite and free ferrite, or a mixture of pearlite and free cementite, all depending upon the amount of carbon and the alloying element.

Generally the alloying element lowers the *eutectoid ratio* of carbon to iron and it therefore requires less than 0.85 percent carbon in a special steel to produce 100 percent pearlitic or eutectoid steel. The lower carbon alloys of this group are usually used for structural purposes. They may be hardened, however, by casehardening methods and used for

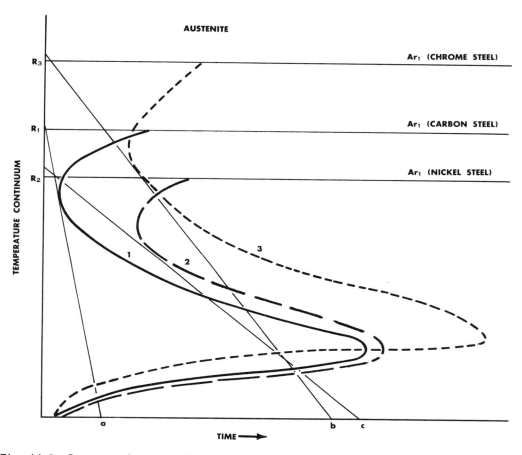

Fig. 11-3. Comparative reaction rate curves for plain carbon steel, nickel steel, and chromium steel. (Bain)

tools. The medium-to-high carbon steels of this group are subjected to heat treatments and used in highly stressed parts of structures and machines or find applications in tools.

Because a great many of these alloy steels have a slower rate of transformation from austenite to pearlite during cooling, they are more easily made martensitic and may often be quenched in oil instead of water or brine. Fig. 11-3 illustrates this effect and indicates the rate of cooling needed in order to produce martensite with a carbon steel, as compared with a chromium and nickel steel. Because they may be hardened fully by oil quenching, these steels are hardened with less danger of warping and cracking. However, due to their slower rate of transformation, these steels will become hardened through their section, leaving no soft core. In a

Metallurgy

tool this increases the danger of breaking in service.

Group II. Martensitic Alloy Steels

As indicated in Fig. 11-1, when the carbon content and amount of alloying element exceed that of the pearlitic type of alloy, a series is formed of compositions that retain a natural martensitic structure upon air cooling. The addition of a greater amount of special element than found in pearlitic steels has resulted in the slowing up of the transformation rate so as to allow undercooling of the austenitic structure to around 200°F (93°C) even when cooled in still air. The alloying element acts as a fast quench in the lowering of the transformation from austenite to pearlite. This effect is indicated in Fig. 11-2, which illustrates the influence of the special element upon the transformation temperature.

If the transformation from austenite takes place above 600°F (315°C), a pearlitic structure is formed. If the transformation upon cooling takes place below 400°F (about 205°C) but above room temperature, a martensitic type of structure is formed. The properties of martenistic alloy steel are not unlike those of fully-hardened and martenistic carbon steel, although they are much more stable above room temperature and resist tempering effects and are therefore used for tools and dies that are subjected to hot work. These steels are usually hardened by air cooling from above their critical temperature and annealed by unusually slow cooling in a furnace.

Group III. Austenitic Alloy Steels

These steels, indicated in Fig. 11-1 and Fig. 11-2, are steels that remain austenitic in structure (gamma iron) upon slow cooling from the temperature of solidification. These steels do not undergo any change in the condition of the iron and therefore exhibit no critical temperature upon cooling. These steels cannot be hardened by heat treatment although they may be cold work-hardened and annealed. Also, if any precipitation occurs with these steels upon slow cooling from a high temperature, they may be reheated and quenched to redissolve the precipitate and keep it in solution. These austenitic steels exhibit great shock strength and low elastic strength and are very ductile. They workharden very rapidly and develop great resistance to wear by abrasion. The chrome-nickel austenitic steels are very resistant to corrosion.

Group IV. Cementitic Alloy Steels

As indicated in Fig. 11-1, some alloying elements, on being added to steel in increasing amounts, fail to convert the steel into an austenitic type. A steel containing 18.0 percent of a special element, and 0.60 percent carbon, upon slow cooling from above its critical temperature, would have a structure of ferrite or martensite with numerous particles of cementite embedded in the ferritic or martensitic matrix. Such a steel has been referred to as a *cementitic* or Group IV steel.

The cementitic type of alloy steel is usually a difficult steel to use. It requires special care to properly anneal the steel so as to make it machinable. Also, these steels are subjected to such hardening treatments as to cause most of the cementite or carbides to be absorbed and retained in a martensitic structure. The excess cementite or carbides add to the wear resistance of the martensitic structure. These steels are largely used in tools, particularly where hardness and resistance to wear are important.

ALLOY STEELS

Nickel Steels

The use of nickel as an alloy in steel dates back many decades to when the U.S. Navy conducted tests on the use of nickel alloy steel for armor plate. Nickel, when added to steel, dissolves to form a solid solution with iron, lowering the critical range to a marked degree and forms a steel that may be made pearlitic, martensitic or austenitic by simply varying the percentage of nickel. The lower carbon pearlitic-nickel steels contain 0.5 percent to 6.0 percent nickel and are particularly suited for structural application because of greater toughness, strength and resistance to corrosion.

Martensitic-nickel steels, because of brittleness and hardness, are little used. They contain from 10 to 22 percent nickel. This range of nickel compositions was selected as the most probable to form martensitic structures upon slow cooling. It is only an approximate range.

The tonnage nickel steels have a microstructure similar to the plain carbon steels; i.e., the structure consists of ferrite, pearlite, and cementite in various amounts, depending upon the ratio of nickel to carbon in the steel. Low-carbon steel may contain up to about 10 percent nickel and remain pearlitic, whereas with higher carbon contents, the nickel is decreased to about 1 percent and the steel remains pearlitic.

In pearlitic nickel steels (nickel from 0.5 to 5.0 percent), nickel has a marked effect in slowing down the rate of transformation from austenite to pearlite during the cooling cycle from above the critical range. This retarding of the transformation rate, as indicated by *curve 2*, Fig. 11-3, allows the use of a slower cooling rate during the hardening operation. These steels can be fully hardened by oil quenching or use of hot salt bath quenches which reduce the warping and danger from cracking during the hardening operation. Also, the slower transformation rate of these nickel steels makes it possible to heat-treat successfully heavy sections and obtain uniform hardening results. The addition of other elements, such as molybdenum and chromium, to the nickel steel increases the effectiveness of the nickel.

Nickel also retards the rate of grain growth at elevated temperatures. This ef-

Metallurgy

fect is valuable when low-carbon nickel steels are subjected to carburizing treatments for purposes of casehardening. Very little grain growth may occur in nickel steels during the carburizing cycle, thus preventing the formation of a coarse-grained core in the finished casehardened part. By maintaining a fine grain in the core during the carburizing cycle, the necessity of a regenerative heat treatment can be avoided, and the use of a single quench and draw can be employed to finish the carburized part.

Nickel in steel also improves the resistance to fatigue failure, increases the resistance to corrosion, and improves the toughness and impact properties of steel. Nickel steels are harder than carbon steels, a characteristic which is of value in parts that are used for wear resistance.

Nickel steels, with nickel from 1.5 to 3 percent are used for structural applications. These steels, as forgings or castings, develop excellent mechanical properties after a simple annealing or normalizing operation. Steels containing approximately 5 percent nickel are famous for their superior behavior in parts that are subjected to severe impact loads. These steels, low in carbon, may be casehardened and make an excellent gear; in the higher carbon contents these steels are hardened and tempered before using.

Although many straight nickel alloy steels are used, the value of nickel as an alloying element may be emphasized by the addition of other alloying elements. Both chromium and molybdenum are used with nickel in some of these special steels. The nickel-chromium, and nickel-chromium-molybdenum alloy steels are capable of developing superior mechanical properties and respond very well to heat treatments. In the higher alloy compositions, 2 percent nickel, 0.8 percent chromium, 0.25 percent molybdenum, these steels have excellent hardening characteristics and develop a uniform hardness in heavy sections. (SAE 4340)

Many corrosion-resistant steels will be found within the range of 10 to 22 percent nickel, but the nickel is usually combined with chromium producing an austenitic steel.

The austenitic nickel steels present a most fascinating study. Research engineers are constantly finding new alloys of this group.

A few of the most common are as follows:

1. 25 to 30 percent nickel is used for corrosion resistance.

2. 20 to 30 percent nickel-iron alloys are non-magnetic by normal cooling and can be made magnetic by cooling to liquid air temperatures.

3. 30 to 40 percent nickel has a very low coefficient of expansion and is called *Invar*.

4. 36 percent nickel and 12 percent chromium alloy is called *Elinvar* and has a nonvariant elastic modulus with temperature change.

5. With over 50 percent nickel, steel develops high magnetic properties.

Chromium Steels

Of all the special elements in steel, none is used for such a wide range of applications as chromium. Chromium dissolves in both gamma iron and alpha iron, but in the presence of carbon it combines with the carbon to form a very hard carbide. To steel, chromium imparts hardness, wear resistance, and useful magnetic properties, permits a deeper penetration of hardness, and increases the resistance to corrosion.

Chromium content in steels varies from a small percentage to approximately 35 percent. Chromium is the principal alloying element in many of the alloy or special steels. Also, a large tonnage of steels is manufactured in which other alloying elements than chromium are added, such as vanadium, molybdenum, tungsten, etc.

The low-carbon or pearlitic-chromium steels are the most important of the chromium steels although stainless steels containing chromium from 12 to 35 percent are becoming more and more important.

When chromium is used in amounts up to 2 percent in medium and high-carbon steels, it is usually for the hardening and toughening effects and for increased wear resistance and increased resistance to fatigue failure. Chromium, when added to steel, raises the critical temperature and slows up the rate of transformation from austenite to pearlite during the hardening process. This slower transformation rate allows the use of a slow quench in hardening and produces deep-hardening effects.

With additions of less than 0.50 percent chromium to the steel, there is a tendency to bring about a refinement of grain and to impart slightly greater hardenability and toughness. Chromium is added to the low-carbon steels for the purpose of effecting greater toughness and impact resistance at sub-zero temperatures, rather than for reasons of increased hardenability. The lower-carbon chromium steels can be casehardened, resulting in greater surface hardness and wear resistance.

The SAE-5120 type of steel is used in the casehardened condition for general purposes. Engineering steels of the SAE-52100 class find many applications where deep and uniform hardening from oil quenching is needed. These steels develop good wear resistance and fatigue resistance with moderately high strength. Steels of this type, the SAE-52100 series, are used for gears, pistons, springs, pins, bearings, rolls, etc.

Chromium is used quite frequently in tool steels to obtain extreme hardness. Applications of such include files, drills, chisels, roll thread dies, and steels commonly used for ball bearings, which contain approximately 1 percent carbon and 1.5 percent chromium. The extreme hardness and resistance to wear associated with the chromium steels seems to be a specific property of the chromium carbides such as Cr_7C_3, Cr_4C, and $(FeCr)3C$ found in these steels.

Increasing the chromium content of steels beyond 2 and up to 14 percent increases the hardenability, the wear resistance, the heat resistance in steels used for hot work, and the corrosion resistance. The high-carbon high-chromium steels, with chromium from 12 to 14 percent and carbon from 1.50 to 2.50 percent, are used when wear resistance is of prime importance, as in knives, shear blades, drawing dies, lathe centers, etc.

Although high percentages of chromium, such as those in the high-carbon high-chromium steels, lead to the proper-

Metallurgy

ty of corrosion resistance, this property requires that the chromium carbides be dissolved at a high temperature and retained in solution by quenching. In order to obtain corrosion resistance, it is necessary to keep the chromium carbides in solution or to lower the carbon content to a value that will reduce the formation of chromium carbides. So-called *stainless* steel is a chromium alloy steel with a relatively low carbon content.

Stainless Steels.

Stainless, or corrosion-resistant, steels include a large number of different alloys. However, modern stainless steel is an alloy made principally from iron and chromium, referred to as *straight chromium steel* and identified as the 400 series of steels, see Table 11-1. Nickel is added to the chromium-iron alloys to form a series of stainless or corrosion-resistant alloys identified by the 300 series of steels, see Table 11-1. Nickel improves the corrosion resistance of the basic iron-chromium alloys and greatly increases their mechanical properties by improving their plasticity, toughness, and welding characteristics. Stainless steel, originally intended for high-grade cutlery and tools, has since developed into a steel of inestimable value for engineering purposes and is of great economic value in present-day ferrous metallurgy because of its resistance to a wide range of corroding materials. Metal waste due to corrosion has become an important engineering problem. It has been stated that no waste, except of human life, is more important. Besides corrosion-resistant and stainless characteristics, other attributes of stainless steels are the property of resisting oxidation and scaling, and the maintenance of strength at elevated temperatures. Types 309 and 310 in Table 11-1 are primarily heat-resistant alloys in which the chromium content is sufficiently high to obtain resistance to scaling up to approximately 2000°F (about 1090°C), with sufficient nickel to maintain this type of steel in its austenitic condition.

The oxide-film theory is offered in an effort to explain the corrosion resistance of chromium alloys. It is believed that an oxide film forms naturally on the surfaces of all metals and protects them to some degree, at least temporarily, from corrosive action. The oxide film that forms under normal atmospheric conditions on the surface of stainless steels of the high chromium-iron type is said to be very stable, tough, and self-healing. If the oxide film is maintained, stainless steels are very effective against any corrosive media. Ordinary compositions of stainless steels do not satisfactorily resist common acids such as hydrochloric, hydrofluoric, sulfuric, or sulfurous, although special compositions that resist such attack have been developed.

Iron-Chromium Alloys

Stainless steels containing from 14 to 18 percent chromium and a maximum of 0.12 percent carbon are magnetic and consist structurally of a solid solution of iron and chromium. They are a ferritic type of stainless steel used for general require-

Alloy or Special Steels

TABLE 11-1. STAINLESS STEEL SPECIFICATIONS.

ALLOY TYPES	CHROMIUM PERCENT	NICKEL PERCENT	OTHER ELEMENT[1] PERCENT	CARBON PERCENT	MANGANESE PERCENT	SILICON PERCENT	STRUCTURE AS-ANNEALED
502	4.0–6.0	0.5 max		0.10 max	0.60 max	0.75 max	Pearlite-Ferrite
501+Mo	4.0–6.0	0.5 max	0.50 Mo	Over 0.10	0.60 max	0.75 max	Pearlite-Ferrite
410	11.5–13.5	0.5 max		0.15 max	0.60 max	0.75 max	Pearlite-Ferrite
414	11.5–13.5	2.0 max		0.15 max	0.60 max	0.75 max	Pearlite-Ferrite
416	12.0–14.0	0.5 max	0.45–0.60 Mo	0.15 max	0.75–1.20	0.75 max	Pearlite-Ferrite
420	12.0–14.0	0.5 max		0.30–0.40	0.50 max	0.75 max	Pearlite
431	14.0–18.0	2.0 max		0.15 max	0.50 max	0.75 max	Pearlite-Ferrite
430	16.0–18.0	0.5 max		0.12 max	0.50 max	0.75 max	Pearlite-Ferrite
430-F	16.0–18.0		S,Se .07 Min	0.12 max	0.75–1.20	0.30–0.75	Pearlite-Ferrite
440	16.0–18.0	0.5 max	0.45–0.60 Mo	0.65–0.70	0.35–0.50	0.75 max	Pearlite
440	16.0–18.0	0.5 max		1.00–1.10	0.35–0.50	0.75 max	Pearlite-Carbide
446	23.0–30.0	1.0 max		0.35 max	1.50 max	0.75 max	Ferrite-Carbide
301	16.0–18.0	7.0–9.0		0.09–0.20	1.25 max		Austenitic
302	17.5–20.0	8.0–10.0		0.08–0.20	1.25 max	0.75 max	Austenitic
304	18.0–20.0	8.0–10.0		0.08 max	2.00 max	0.75 max	Austenitic
303	17.5–20.0	8.0–10.0	S,Se,P[2]	0.20 max	1.25 max	0.75 max	Austenitic
316	16.0–18.0	10.0–14.0	2.00–3.00 Mo	0.10 max	2.00 max	0.75 max	Austenitic
317	18.0–20.0	10.0–14.0	2.00–4.00 Mo	0.10 max	2.00 max	0.75 max	Austenitic
321	17.0–20.0	7.0–10.0	Ti[3]	0.10 max	0.75 max	0.75 max	Austenitic
347	17.0–20.0	8.0–12.0	Cb[4]	0.10 max	2.00 max	0.75 max	Austenitic
309	22.0–26.0	12.0–14.0		0.20 max	1.25 max	0.75 max	Austenitic
310	24.0–26.0	19.0–21.0		0.25 max	1.25–1.75	0.75 max	Austenitic

[1] Sulfur and phosphorus shown in this column when sufficient; not shown if normal
[2] Sulfur and selenium 0.07 min; phosphorus 0.03 to 0.17
[3] Titanium content min = 4 times percent carbon
[4] Columbium content min = 10 times percent carbon

Data from *Stainless Steels, an Elementary Discussion*, pamphlet by Allegheny Ludlum Steel Corp

Metallurgy

ments where resistance to corrosion and heat is needed, but where service conditions are not too severe and where slight discoloration of the surface during service can be tolerated. Steels of this type, that contain less than 0.10 percent carbon with chromium from 14 to 20 percent, are not heat treatable except for an annealing treatment. They cannot be hardened by heat treatment as they do not transform into austenite when heated to an elevated temperature.

These steels are susceptible to grain growth at elevated temperatures, and with a larger grain size they suffer a loss in toughness. The ferritic stainless steels may be cast and forged hot or cold, but they do not machine easily—a characteristic of all stainless steels. They possess a high resistance to corrosion in an ordinary atmosphere, providing they have been highly polished and are free from foreign particles. Improvement in their machinability can be obtained by additions of molybdenum and sulfur, or phosphorus and selenium. Although sulfur improves machinability, it reduces toughness, ductility, and corrosion resistance. The additions of phosphorus and selenium are less detrimental then sulfur so far as the corrosion resistance of these steels is concerned.

Iron-Chromium-Carbon Alloys (Martensitic Stainless Steels)

Everyone is familiar with cutlery stainless steels which give hardness, resistance to abrasion, high strength, and cutting qualities. These steels contain enough carbon so that when they are quenched from the austenitic condition, they develop a martensitic structure that has a hardness value dependent upon the carbon-chromium ratio. Steels of this type are called *martensitic* stainless steels.

These martensitic steels transform to austenite at elevated temperatures and therefore are essentially austenitic at the quench-hardening temperature. The chromium dissolved in the austenite slows up the transformation rate so that steels of this type can be made fully hard, or martensitic, by oil quenching. Some compositions can be hardened by air cooling from the austenitic temperature range. The structure of the hardened steel is similar to that of other steels.

Following a quench-hardening treatment, the fully-hardened steel is usually tempered or drawn within a range up to 1000° F (about 540°C) depending upon the toughness requirements. The annealed condition of this type of steel, after slow cooling from the austenitic temperature range, consists of complex carbides in an alpha-iron structure.

Corrosion resistance of the steel in this condition is poor, and it is therefore necessary to heat the steel into the austenitic range and dissolve the chromium carbides and retain this solution condition by quenching. Because of this requirement with this type of stainless steel, it has been replaced for structural purposes by the low-carbon type of steel that is naturally stainless and does not require a solution heat treatment. However, when a relatively hard stainless steel is required, a higher carbon content is essential, as in types 420 and 440 shown in Table 11-1. The Type 420 steel containing 12 percent chromium and 0.35 percent carbon is known as the regular cutlery grade steel. In the hardened martensitic condition, it develops a hardness of about 55 Rockwell C and attains its maximum corrosion resistance. The Type 440 is a general purpose hardenable stainless steel used for hardened steel balls and pump parts and is particularly adapted to resist corrosion for parts of machinery used in the oil industry.

Chromium-Nickel-Iron Stainless Alloys

The 300 series of stainless steels listed in Table 11-1, contain considerable nickel in addition to chromium. The addition of nickel to these steels is sufficient to render the alloy austenitic at room temperature; these steels are often referred to as *austenitic stainless steels*. Austenitic stainless steels are highly resistant to a large variety of corrosive agents and find a wide range of applications.

The most important of the austenitic stainless steels is the so-called Type *18-8* containing 18 percent chromium and 8 percent nickel and including the modifications represented by Types 301, 303, 316, 321, and 347, Table 11-1. These steels possess extraordinary toughness and ductility, but they are nonmagnetic (gamma iron) and cannot be hardened by heat treatment. They are used in their natural condition as cast, forged, cold worked, and annealed. The only heat treatment given these steels is that of annealing after cold working or annealing for the purpose of stabilization.

Cold working of these austenitic stainless steels may be employed to develop high strengths and increased hardness. The strength of the Type 18-8 may be varied from that of 90,000 psi (620,600 kN/m^2) in the annealed condition to as high as 260,000 psi (1,792,700 kN/m^2) in the cold work-hardened condition, with the hardness varying from 120 to 480 Brinell. The extreme high toughness of austenitic steels results in poor machining qualities which, however, may be somewhat improved by the addition of phosphorus and selenium, or through the addition of sulfur, as in the free-machining austenitic steels (Type 303).

The austenitic chromium-nickel steels, when heated in the temperature range of 800° to 1500°F (about 430° to 815 °C) for a sufficiently long period, suffer from a precipitation of chromium carbides, largely forming at the grain boundaries of the austenitic grains. The precipitated carbides are less resistant to corrosion than the balance of the austenitic structure, and under conditions of marked precipitation of carbides at the grain boundaries, the steel may disintegrate when exposed to a corrosive medium. Fig. 11-4 illustrates a condition of severe intergranular corrosion resulting from exposing a piece of 18-8 austenitic stainless steel to a sulfuric acid and copper sulfate solution after the steel had been subjected to a long heating cycle at 1400°F (760°C). The corrosive medium has disintegrated the steel along the austenitic grain boundaries.

The susceptibility of this type of steel to intergranular corrosion after heating to a red heat is influenced by the combined chromium and nickel contents with respect to the carbon content. Carbide pre-

Metallurgy

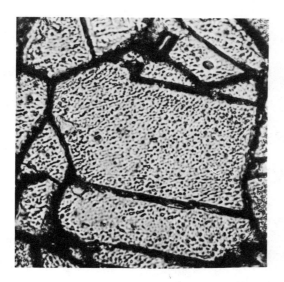

Fig. 11-4. Photomicrograph of stainless steel, Type 301, has been subjected to sulfuric acid and copper sulfate etching solution (Strauss solution). This photomicrograph shows an extreme case of intergranular attack resulting from carbide precipitation, magnified 500 times.

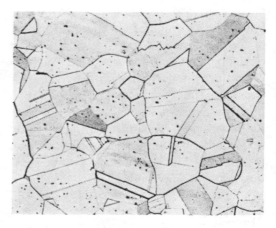

Fig. 11-5. Austenitic stainless steel, Type 301, containing 16 to 18 percent chromium and 7 to 9 percent nickel. Magnified 250 times.

cipitation diminishes rapidly with the reduction in carbon content, but it is difficult to manufacture stainless steels with the carbon content low enough to completely eliminate the possibility of intergranular precipitation. To prevent the precipitation of chromium carbides, elements such as columbium and titanium are added to the steel. Both columbium and titanium are strong carbide formers and combine with the carbon, using it up as special carbides and thus preventing the formation of the detrimental chromium carbides. The carbides of columbium and titanium are not considered harmful to the corrosion resistance of the steel.

Subjecting austenitic stainless steels that have not been treated with either co-lumbium or titanium to a heating cycle, as in welding or forging operations, is dangerous with regard to an intergranular corrosion attack. Stabilization of such steels and the removal of the danger from corrosion may be obtained by a solution heat treatment. The stabilization, or solution heat treatment, usually consists of heating the steel to a temperature range of 1850° to 2100°F (about 1010° to 1150°C), depending upon the composition, for a sufficient length of time (approximately ½ hr.) to get the carbides back into solution, and then cooling rapidly down to below the carbide precipitation range in less than three minutes. The cooling is usually accomplished by a water quench. A microscopic examination of the structure of the steel after this treatment will reveal a homogeneous austenitic structure free from any carbide precipitate, as seen in Fig. 11-5.

Manganese Steels

Manganese is always used in steel, having a beneficial effect in the steel both directly and indirectly, and is always added to steel during its manufacture. Manganese combines with sulfur, iron, and carbon. With no carbon present, it forms a solid solution with both gamma and alpha iron. Manganese has only a mild tendency to combine with the carbon and therefore to form a carbide (Mn_3C), which accounts for one of the principal advantages obtained from alloying manganese with steels: viz., the manganese dissolves in the ferrite matrix, adding materially to its strength. All steels contain manganese from a few hundredths of one percent in many of the low-carbon structural and machine steels to about 2 percent in steels that remain pearlitic in structure.

As the manganese content of the steel is increased, the steel changes from a pearlitic to a martensitic, to an austenitic steel. The pearlitic steels are the most important. A very important steel is one containing up to 1.5 percent manganese. This type is oil hardening and retains its shape; therefore, it is known as *nondeforming steel*. It is used in many types of tools and dies that are to be hardened fully.

Manganese structural steel contains from 1 to 2 percent manganese, and from 0.08 to 0.55 percent carbon.

In tool steels, increasing the manganese content from 0.30 to 1.0 percent changes the steel from water hardening to oil hardening, due to the greater hardenability of the steel with the higher manganese content.

Martensitic manganese steels are not used because of great brittleness and hardness.

Austenitic manganese steels, sometimes called *Hadfield steels*, contain up to 15 percent manganese and 1 to 1.5 percent carbon. In the cast condition, this steel is very weak and brittle because of the presence of free carbides, and it is necessary to subject it to a heat treatment. The cast steel is heated to about 1900°F (about 1040°C) and quenched in water, which increases the strength to 160,000 psi (1,103,200 kN/m^2), with an elongation of 60 to 70 percent in 2''. The properties are those of austenite, with a low elastic limit, but it possesses great wearing power combined with much ductility. The wearing power is due to work hardening under cold deformation. Tests have shown that the initial wear of steel crusher jaws made from this type of steel is greater than the rate of wear after a few tons of stone have been crushed.

Tungsten Steel

Tungsten, like chromium and other alloying elements, dissolves in iron and forms a carbide with the cementite. The carbide formed in steels containing carbon and tungsten is thought to be Fe_4W_2C or Fe_3W_3C. It is apparent that this complex carbide is present in the commercial tungsten steels. The complex carbide formed in tungsten steels is very hard and brittle and imparts hardness to the steel. The tungsten carbide, like many of the other hard carbides found in alloy steels, resists tempering or softening of the mar-

Metallurgy

tensite upon heating. When this carbide is present in sufficient quantities, it imparts high strength and hardness to steels subjected to elevated temperatures, or the property of hot hardness. Tungsten has the following influences on steel:

1. Makes a dense and fine structure in steel.
2. Increases strength and toughness at high temperature.
3. Increases hardenability of steel.
4. Stabilizes martensite, imparting the property of *red hardness,* the ability of a cutting tool to maintain a cutting edge even at a red heat.

Tungsten forms two general types of steel:

1. Carbon type, with the tungsten under 1 percent for general purposes, tungsten from 1 to 2.5 percent for special tools, and 2.5 to 7 percent tungsten for fast finishing tool steels. Parts can be cut at much higher speeds with this steel than with carbon cutting tools, but it cannot be used when subjected to much heating, as can the high-speed steels.
2. Special type, which includes the tungsten-magnet steel, drawing-die steels, steels for pneumatic tools, hot-working steels and high-speed steels.

Molybdenum Steels

Molybdenum is considered one of the most useful alloys to be used in steel. Although the element molybdenum was unknown prior to 1790, the superiority of the iron, which is now known to have contained molybdenum, was recognized about six hundred years ago by Japanese swordmakers.

Molybdenum dissolves in both gamma and alpha iron, but in the presence of carbon it combines to form a carbide. The behavior of molybdenum in steel is about the same as tungsten, but twice as effective. Since large deposits of molybdenum were discovered in this country, the commercial development has been rapid.

Molybdenum, like tungsten, improves the hot strength and hardness of steels. The complex carbide formed by the addition of molybdenum acts to stabilize martensite and resist softening upon heating, similar to the action from additions of tungsten. Molybdenum reduces the tendency toward grain growth at elevated temperatures and slows up the transformation rate from austenite to pearlite. This characteristic of molybdenum steels, increased hardenability, allows the use of air quenching of many of the alloy tool steels instead of oil quenching in the hardening operation.

The sluggishness of the molybdenum steels to any structural change during heat-treating operations leads to longer soaking periods for annealing or hardening and slower cooling rates in order to obtain a well-annealed structure. Molybdenum is often combined with other alloying elements in commercial alloy steels. The addition of other special elements seems to enhance the value of molybdenum in steels. Molybdenum-chromium, molybdenum-nickel, molybdenum-tungsten, and molybdenum-nickel-chromium are common combinations used. The use of molybdenum as a supplementary element in steel is more extensive than its use alone.

Molybdenum structural steels of the

Alloy or Special Steels

SAE-4000 series find many applications intended for heat-treated parts. The molybdenum in these steels strengthens the steels both statically and dynamically, eliminates temper brittleness almost completely, refines the structure, and widens the critical range.

Molybdenum is alloyed with both structural steel and tool steel. The low-carbon molybdenum steels are often subjected to case-hardening heat treatments. Medium-carbon molybdenum steels are heat treated for parts including such applications as gears, roller bearings, and aircraft and automobile parts that are subjected to high stresses. Tool steels for general use and for tools for hot working are often alloyed with molybdenum. Greater hardenability and improvements in physical properties are outstanding reasons for the selection of a tool steel containing molybdenum. As a stabilizer for martensite, molybdenum can be used as a substitute for tungsten in hot working and high-speed steels.

High-Speed Steels

High-speed steels are so named because, when used as tools such as machine cutting tools, Fig. 11-6, they remove metal much faster than ordinary steel. Their chief superiority over other steels lies in their ability to maintain their hardness even at a dull red heat (about 1100° F or 595°C). Due to this property, such tools can operate satisfactorily at speeds which cause the cutting edges to reach a temperature that would ruin the hardness of ordinary tool steels (Table 12-2).

It will be seen from this table that the principal alloying elements are tungsten, molybdenum, chromium, and vanadium. The tungsten type of high-speed steel contains up to 20 percent tungsten, from 2 to 5 percent chromium, usually from 1 to 2 percent vanadium, and sometimes cobalt, which is added for special properties. More recently, molybdenum has been successfully substituted for tungsten in these steels, with molybdenum up to 9 percent and no tungsten, although a more common high-speed steel contains approximately equal amounts of molybdenum and tungsten, 5 percent of each. The most commonly used high-speed steel is referred to as 18–4–1, which means that it contains 18 percent tungsten, 4 percent chromium, and 1 percent vanadium. The carbon content in the usual high-speed

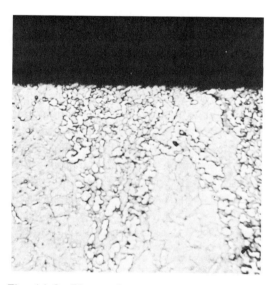

Fig. 11-6. Photomicrograph of the edge of a sample of hardened high-speed steel, showing practically no surface skin. Magnified 500 times.

Metallurgy

steel runs about 0.65 to 0.75 percent although in some grades the carbon content may run as high as 1.30 percent (see Table 12-2).

The molybdenum or molybdenum-tungsten high-speed steels have been largely substituted for the straight tungsten high-speed steels of the 18–4–1 type. Actual shop tests made with heavy and light lathe tools, planer tools, and milling reamers demonstrate that the heavy planer and lathe tools stand up as well as or better than tungsten or the tungsten-cobalt tools. The molybdenum grades of high-speed steels have one drawback to their use: viz., they are much more sensitive to decarburization than the tungsten steels during the hot-working and heat-treating operations. This problem can be overcome by the use of borax which is fused in a thin coating over the steel, forming a protective film during the heating and hot forging and during subsequent forging operations. The borax may also be used during the heat-treating operations, but it causes a cleaning problem and is injurious to the furnace hearth. In the place of borax, a furnace designed with the correct atmosphere will prevent any serious decarburization. Also, it has been found that it is practical to use salt baths for the heat-treating operations since the molten salt protects the surfaces of the molybdenum high-speed steel from decarburization or carburization.

Vanadium Steels

Although vanadium is not a major alloy in many of the special alloy steels, the role it plays as a minor alloying element is very important. Vanadium has a strong affinity for carbon, and when added to steel, it is found partly combined with the ferrite but principally with the carbide or cementite. The presence of vanadium greatly increases the strength, toughness, and hardness of the ferrite matrix of steels. The carbide, or cementite, formed with vanadium is likewise greatly strengthened and hardened and shows little tendency to segregate into large masses.

The most important effect of vanadium in steels is to produce a very small grain size and induce grain control in the steel. The vanadium steels are inherently fine-grained, and this fine grain size is maintained throughout the usual heat-treating temperature range. Steels containing vanadium are also shallow hardening because they have a faster transformation rate from austenite to pearlite during the cooling cycle from the hardening temperature. However, due to the grain-control characteristics imparted by the addition of vanadium, steels may be heated over a wide range of temperatures during the hardening operation without danger of developing a coarse and weak structure. See Fig 11-7.

Another important effect resulting from the presence of vanadium carbides is that these carbides are extremely stable and, when in solution, resist precipitation from the solution. As a result, after vanadium steels have been quench-hardened from the heat-treating temperature, they exhibit a marked tendency to resist loss of hardness during the tempering cycle. This characteristic is of major importance with steels that are used for hot work and contributes to the hot hardness of the special

Alloy or Special Steels

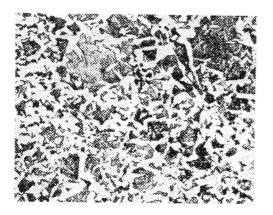 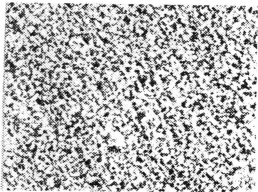

Fig. 11-7. Effect of vanadium addition on grain size of steel slowly cooled from 1550°F (about 845°C): (top) without vanadium, and (bottom) with 0.26 percent vanadium. (American Society for Metals)

high-speed steels. This carbide stability is important in constructional steels and in cast irons used at elevated temperatures.

Vanadium is effective in steels in amounts as low as 0.05 percent, and in many of the common alloy steels the vanadium content runs under 0.30 percent. Alloy steels designed for use at elevated temperatures, such as hot-work steels and high-speed cutting tools, contain up to 4.0 percent vanadium.

Silicon Steels

Silicon as an alloying element dissolves in both gamma iron and alpha iron and does not form a carbide. The solubility of silicon in the gamma iron is approximately 2 percent, whereas it is soluble in the alpha iron up to 18 percent and remains highly soluble even in the presence of carbon. Silicon is present in all steels as a result of the manufacturing process. In low-carbon structural steel and machine steel, the silicon content runs around 0.25 to 0.35 percent. Silicon in this range does not influence the properties of the steel to any extent. With the additions of silicon above 0.50 to more than 2 percent, we find improvement in tensile strength and yield point of steels. A popular spring steel contains approximately 2 percent silicon, 0.7 percent manganese, and 0.5 percent carbon. This steel, after heat treatment, develops excellent strength and toughness and is widely used for coil and leaf springs. Combinations of silicon-chromium and silicon-chromium-manganese are also used for spring steels.

Silicon has been alloyed to steels of the stainless and heat-resisting type, with improvements noted in their ability to resist certain types of corrosion and oxidation at elevated temperatures. The 18–8 (austenitic) type of stainless steel has been improved for some uses by the addition of silicon up to 4 percent. Steels used for

Metallurgy

gas-engine valves in the automotive and airplane industries often contain appreciable amounts of silicon. These steels contain from 1 to 4 percent silicon, 6 to 9 percent chromium, 0.15 percent vanadium, and approximately 0.5 percent carbon. This steel is highly resistant to oxidation and retains high strength and hardness at elevated temperatures.

Alloys of iron and silicon, containing but a very small amount of carbon as an impurity, are used in transformer cores due to their good magnetic properties and high electrical resistance. Such steels contain from 2 to 5 percent silicon and usually are used in sheet form.

Hot-Working Steels

Steels containing from 5 to 15 percent tungsten and several percent of chromium resemble high-speed steels in that they have a tendency to retain their hardness when heated. These steels are used as hot-working tools, such as forging dies or extrusion dies. They need not be as hard as cutting tools, and their carbon content is somewhat lower than that of high-speed steels.

Part or all of the tungsten present in hot-working steels may be replaced by molybdenum, similar to the substitution that may be used in the high-speed steels. Hot-working steels resemble high-speed steels, with this type of steel containing a lower carbon content and somewhat lower percentages of all of the alloying elements. These steels are deep hardening and are of the oil or air-hardening type.

Magnetic Steels

Carbon, chromium, tungsten, and cobalt steels are used for permanent magnets. The superiority, as well as the price, increases in the order given above. These steels are used in the hardened condition. The carbon steels contain from 0.80 to 1.20 percent carbon and are, in fact, carbon tool steels. The chromium magnet steels contain from 0.70 to 1.00 percent carbon and 2 percent or 3 percent chromium. The tungsten steels contain about 0.70 percent carbon and 5 percent tungsten. The best and most expensive cobalt steels contain 35 percent cobalt, together with several percent of both chromium and tungsten. A permanent magnet alloy (*Alnico*) contains approximately 60 percent iron, 20 percent nickel, 8 percent cobalt and 12 percent aluminum. This alloy cannot be forged and is used as a casting hardened by precipitation heat treatments. Alloys of this type have great magnetic strength and permanence.

REVIEW

1. Describe plain carbon steel.
2. What is the approximate maximum hardness developed by plain carbon or alloy steel?

3. Name the reasons for the selection of alloy steels in preference to carbon steels.

4. Name the four ways in which an element added to steel may combine with it.

5. What improvements in steel are contributed by the addition of elements which form a solid solution with the iron?

6. What property makes carbide-forming alloys valuable?

7. What effect does the formation of oxides have upon steel?

8. What improvement results from the addition to steel of elements that remain uncombined?

9. In general, what percentages of special elements and carbon are contained in pearlitic alloy steels?

10. Name the uses of steels in the pearlitic alloy group.

11. How do percentages of carbon and special elements in steels in the martensitic group compare with those in the pearlitic group?

12. Name the uses of martensitic alloy steels.

13. Why have austenitic alloy steels no critical temperature on cooling?

14. What characteristics do austenitic alloy steels have?

15. What use is made of cementitic alloy steels?

16. What qualities make nickel steels suitable for structural applications?

17. What other elements are alloyed with nickel alloy steels to make special steels?

18. Give the composition of three nickel alloy steels and name a particular property for which each is noted.

19. What use is made of low-carbon chromium steels that have been case-hardened?

20. What important property does a high percentage of chromium give to steel?

21. For what is high-carbon chromium steel used?

22. What percentages of carbon and chromium are used in cutlery stainless steels?

23. Why is sulfur added to some stainless steel?

24. What is the strength of the stainless steel known as 18-8 in the annealed and work-hardened conditions? What is the hardness?

25. What use is made of so-called non-deforming steel?

26. Name the special properties of Hadfield steels?

27. What four improvements does tungsten make on the characteristics of steel?

28. What are the two types of tungsten steel?

29. Name one disadvantage of molybdenum high-speed steels.

30. For what are hot-working steels used?

31. How does the addition of vanadium to steel affect its physical properties?

32. Name some uses of silicon steels.

Metallurgy

Electric furnaces make graded steels easily.

CHAPTER 12

CLASSIFICATION OF STEELS

In the early days of steelmaking, great secrecy was maintained by the steelmakers who had developed newer and better structural and tool steels. For many years, tool steels were sold by patented trade names and the steel salesman would not divulge the composition of a tool steel, but instead sold his product on its reputation. Gradually, the need of designers and engineers to be able to evaluate steels as to their mechanical properties led to the classification and standardization of the many different types and compositions of steels used throughout industry.

SAE AND AISI SYSTEMS

Practically all steel on the market today is sold using the specifications and compositions devised by the Society of Automotive Engineers and the American Iron and Steel Institute. These specifications are universally used in industries where steels, and particularly alloy steels, are being used. These specifications are primarily intended for structural applications, as in machines, but they include a limited number of high-carbon and alloy steels that can be used for tools.

In the selection of any of these steels, the user may consult the specifications and select a number of different steels of different compositions that might be applied to the same usage. No definite applications for the different steels are included in the specifications. However, the complete specifications include compositions, heat treatment, and conservative physical properties that might be expected from the steels so that the user may make an intelligent selection.

A numerical index system is used to identify the compositions of the SAE/AISI

Metallurgy

steels which makes it possible to use identifying numerals on shop drawings and blueprints that are partially descriptive of the composition of material covered by such numbers. In order to facilitate the discussion of steels, some familiarity with this nomenclature is desirable. The numerical index is used to identify the chemical compositions of the structural steels. In this system, a four numeral series is used to designate the plain carbon and alloy steels. Five numerals are used to designate certain types of alloy steel.

The basic numerals for the various types of steel in the SAE/AISI system are shown in the following chart. The first digit indicates the type to which the steel belongs; thus 1 indicates a carbon steel; 2, a nickel steel; 3, a nickel-chromium steel, etc. In the case of the simple alloy steels, the second digit generally indicates the approximate percentage of the predomi-

TABLE 12-1. DESIGNATIONS OF TYPES OF STEELS.

SERIES	TYPES AND PERCENTAGE COMPOSITION.
10XX	Nonsulfurized carbon steels
11XX	Resulfurized carbon steels (free machining)
12XX	Rephosphorized & resulfurized carbon steels (free machining)
13XX	Manganese 1.75 (principal alloy)
23XX	Nickel 3.50
25XX	Nickel 5.00
31XX	Nickel 1.25; chromium 0.65 (a binary alloy)
33XX	Nickel 3.50; chromium 1.55 (a binary alloy)
40XX	Molybdenum 0.20 or 0.25
41XX	Chromium 0.50 or 0.95; molybdenum 0.12 or 0.20
43XX	Nickel 1.80; chromium 0.50 or 0.80; molybdenum 0.25
44XX	Molybdenum 0.40
45XX	Molybdenum 0.52
46XX	Nickel 1.80; molybdenum 0.25
47XX	Nickel 1.05; chromium 0.45; molybdenum 0.20 or 0.35
48XX	Nickel 3.50; molybdenum 0.25
50XX	Chromium 0.25; 0.40 or 0.50
50XXX	Carbon 1.00; chromium 0.50
51XX	Chromium 0.80, 0.90, 0.95, or 1.00
51XXX	Carbon 1.00; chromium 1.05
52XXX	Carbon 1.00; chromium 1.45
61XX	Chromium 0.60, 0.80, or 0.95; vanadium 0.12 (0.10 to 0.15)
81XX	Nickel 0.30; chromium 0.40; molybdenum 0.12
86XX	Nickel 0.55; chromium 0.50; molybdenum 0.20
87XX	Nickel 0.55; chromium 0.50; molybdenum 0.25
88XX	Nickel 0.55; chromium 0.50; molybdenum 0.35
92XX	Manganese 0.85; silicon 2.00; chromium 0 or 0.35
93XX	Nickel 3.25; chromium 1.20; molybdenum 0.12
94XX	Nickel 0.45; chromium 0.40; molybdenum 0.12
98XX	Nickel 1.00; chromium 0.80; molybdenum 0.25

nant alloying element. Usually the last two or three digits indicate the average carbon content in *points,* or hundredths of one percent. Thus 2340 indicates a nickel steel of approximately 3 percent nickel (3.25 percent to 3.75 percent) and 0.40 percent carbon (0.38 percent to 0.43 percent).

In some instances, in order to avoid confusion, it has been found necessary to depart from this system of identifying the approximate alloy composition of a steel by varying the second and third digits of the number. An example of such a departure are the steel numbers selected for several of the corrosion and heat-resisting alloys.

The series designation and types are summarized in Table 12-1:

The SAE numbers are the same as the AISI numbers except the latter may be given a letter prefix to indicate the method of manufacture. A few of the AISI steels are not included in the SAE listings.

Small quantities of certain elements are present in alloy steels that are not specified as required. These elements are considered as incidental and may be present to the maximum percentages as follows:

copper 0.35; nickel 0.25; chromium 0.20; molybdenum 0.06

Complete, detailed tables of steel classifications are too extensive to reproduce in this book but are available for reference in many of the industrial handbooks such as *Machinery's Handbook*.

TOOL STEELS

Technically, any steel used as a tool may be termed a tool steel. Practically, however, the term is restricted to steels of special composition that are usually melted in electric furnaces and manufactured for certain types of service. The current commonly used tool steels have been classified by the American Iron and Steel Institute into seven major groups and each commonly accepted group or subgroup has been assigned an alphabetical letter. Methods of quenching, applications, special characteristics and steels for particular industries were considered in this type classification as follows:

GROUP

Water hardening ---------------------------------
Shock resisting ----------------------------------
Cold work ---
Hot work --
High speed --
Special purpose ----------------------------------
Mold steels ---------------------------------------

SYMBOL AND TYPE.

W
S
O - Oil hardening
A - Medium alloy
D - High-carbon—high-chromium
H - H1 to H19 including chromium base
    H20 to H39 including tungsten base
    H40 to H59 including molybdenum base
T - Tungsten base
M - Molybdenum base
L - Low-alloy
F - Carbon tungsten
P -

Metallurgy

The AISI identification and type classification of tool steels is given in Table 12-2.

In the column *Typical Application,* three general classifications are shown, relating to the quenching medium used in the hardening operation. The water hardening steels are the plain carbon steels, or the so-called low-alloy steels, in which the alloying element is not present in large enough percentage to change the heat-treatment characteristics of the steel. The water hardening steels require a water or brine quenching in order to produce maximum hardness. However, it should be recalled that thin sections cool much faster than heavy sections and it is possible to quench a thin section of water hardening steel in oil and obtain satisfactory results.

The oil and air hardening steels are the steels that may be quenched in oil or air from the hardening temperature to attain the maximum hardness. These steels are usually of the higher alloy type, containing such elements as manganese, molybdenum and chromium, all of which retard the transformations during cooling and thus allow a slower quench to obtain the maximum hardness. All of the steels in this group may be hardened by quenching in oil and a few of the higher alloyed compositions may be hardened by quenching in air when the section is such as to allow fast enough cooling. None of these steels may be safely hardened by quenching in water or brine without the danger of hardening cracks.

TABLE 12-2. CLASSIFICATION AND SELECTION OF TOOL STEELS.

AISI TYPE	PERCENTAGE COMPOSITION						TYPICAL APPLICATIONS
	C	W	Mo	Cr	V	Other	
W1	0.60 to 1.40^1	—	—	—	—	—	**WATER HARDENING** Low carbon: blacksmith tools, blanking tools, caulking tools, cold chisels, forging dies, rammers, rivet sets, shear blades, punches, sledges. Medium carbon arbors, beading tools, blanking dies, reamers, bushings, cold heading dies, chisels, coining dies, countersinks, drills, forming dies, jeweler dies, mandrels, punches, shear blades, woodworking tools. High carbon glass cutters, jeweler dies, lathe tools, reamers, taps and dies, twist drills, woodworking tools. Vanadium content of W2 imparts finer grain, greater toughness and shallow hardenability.
W2	0.60 to 1.40^1	—	—	—	0.25	—	
W5	1.10	—	—	0.50	—	—	Heavy stamping and draw dies, tube drawing mandrels, large punches, reamers, razor blades, cold forming rolls and dies, wear plates.

Classification of Steel

TABLE 12-2. CLASSIFICATION AND SELECTION OF TOOL STEELS (cont)

AISI TYPE	PERCENTAGE COMPOSITION						TYPICAL APPLICATIONS
	C	W	Mo	Cr	V	Other	
							SHOCK RESISTING
S1	0.50	2.50	—	1.50	—	—	Bolt header dies, chipping and caulking chisels, pipe cutters, concrete drills, expander rolls, forging dies, forming dies, grippers, mandrels, punches, pneumatic tools, scarfing tools, swaging dies, shear blades, track tools, master hobs.
S2	0.50	—	0.50	—	—	1.00 Si	Hand and pneumatic chisels, drift pins. forming tools, knock-out pins, mandrels, machine parts, nail sets, pipe cutters, rivet sets, screw driver bits, shear blades, spindles, stamps, tool shanks, track tools.
S5	0.55	—	0.40	—	—	0.80 Mn, 2.00 Si	Hand and pneumatic chisels, drift pins, forming tools, knock-out pins, mandrels, nail sets, pipe cutters, rivet sets and busters, screw driver bits, shear blades, spindles, stamps, tool shanks, track tools, lathe and screw machine collets, bending dies, punches, rotary shears.
S7	0.50	—	1.40	3.25	—	—	Shear blades, punches, slitters, chisels, forming dies, hot header dies, blanking dies, rivet sets, gripper dies, engraving dies, plastic molds, die casting dies, master hobs, beading tools, caulking tools, chuck jaws, clutches, pipe cutters, swaging dies.
							OIL HARDENING
O1	0.90	0.50	—	0.50	—	1.00 Mn	Blanking dies, plastic mold dies, drawing dies, trim dies, paper knives, shear blades, taps, reamers, tools, gages, bending and forming dies, bushings, punches.
O2	0.90	—	—	—	—	1.60 Mn	Blanking, stamping, trimming, cold forming dies and punches, cold forming rolls, threading dies and taps, reamers, gages, plugs and master tools, broaches, circular cutters and saws, thread roller dies, bushings, plastic molding dies.
O6[3]	1.45	—	0.25	—	—	1.00 Si	Blanking dies, forming dies, mandrels, punches, cams, brake dies, deep drawing dies, cold forming rollers, bushings, gages, blanking and forming punches, piercing and perforating dies, taps, paper cutting dies, wear plates, tool shanks, jigs, machine spindles, arbors, guides in grinders and straighteners.

Metallurgy

TABLE 12-2. CLASSIFICATION AND SELECTION OF TOOL STEELS (cont)

AISI TYPE	PERCENTAGE COMPOSITION						TYPICAL APPLICATIONS
	C	W	Mo	Cr	V	Other	
							OIL HARDENING (cont)
O7	1.20	1.75	—	0.75	—	—	Mandrels, slitters, skiving knives, taps, reamers, drills, blanking and forming dies, gages, chasers, brass finishing tools, dental burrs, paper knives, roll turning tools, burnishing dies, pipe threading dies, rubber cutting knives, woodworking tools, hand reamers, scrapers, spinning tools, broaches, blanking and cold forming punches.
							COLD WORK (MEDIUM ALLOY, AIR HARDENING)
A2	1.00	—	1.00	5.00	—	—	Thread rolling dies, extrusion dies, trimming dies, blanking dies coining dies, mandrels, shear blades, slitters, spinning rolls, forming rolls, gages, beading dies, burnishing tools, ceramic tools, embossing dies, plastic molds, stamping dies, bushings, punches, liners for brick molds.
A3	1.25	—	1.00	5.00	1.00	—	
A4	1.00	—	1.00	1.00	—	2.00 Mn	Blanking dies, forming dies, trimming dies, punches, shear blades, mandrels, bending dies, forming rolls, broaches, knurling tools, gages, arbors, bushings, slitting cutters, cold treading rollers, drill bushing, master hobs, cloth cutting knives, pilot pins, punches, engraver rolls.
A6	0.70	—	1.25	1.00	—	2.00 Mn	Blanking dies, forming dies, coining dies, trimming dies, punches, shear blades, spindles, master hubs, retaining rings, mandrels, plastic dies.
A7	2.25	1.00[2]	1.00	5.25	4.75	—	Brick mold liners, drawing dies, briquetting dies, liners for shot blasting equipment and sand slingers, burnishing tools, gages, forming dies.
A8	0.55	1.25	1.25	5.00	—	—	Cold slitters, shear blades, hot pressing dies, blanking dies, beading tools, cold forming dies, punches, coining dies, trimming dies, master hobs, rolls, forging die inserts, compression molds, notching dies, slitter knives.
A9	0.50	—	1.40	5.00	1.00	1.50 Ni	Solid cold heading dies, die inserts, heading hammers, coining dies, forming dies and rolls, die casings, gripper dies. Hot work applications: punches, piercing tools, mandrels, extrusion tooling, forging dies, gripper dies, die casings, heading dies, hammers, coining and forming dies.

TABLE 12-2. CLASSIFICATION AND SELECTION OF TOOL STEELS (cont)

AISI TYPE	PERCENTAGE COMPOSITION						TYPICAL APPLICATIONS
	C	W	Mo	Cr	V	Other	
							COLD WORK (MEDIUM ALLOY, AIR HARDENING) (cont)
A10[2]	1.35	—	1.50	—	—	1.80 Mn 1.25 Si 1.80 Ni	Blanking dies, forming dies, gages, trimming shears, punches, forming rolls, wear plates, spindle arbors, master cams and shafts, stripper plates, retaining rings.
							COLD WORK (HIGH-CARBON, HIGH-CHROMIUM)
D2	1.50	—	1.00	12.00	1.00	—	Blanking dies, cold forming dies, drawing dies, lamination dies, thread rolling dies, shear blades, slitter knives, forming rolls, burnishing tools, punches, gages, knurling tools, lathe centers, broaches, cold extrusion dies, mandrels, swaging dies, cutlery.
D3	2.25	—	—	12.00	—	—	Blanking dies, cold forming dies, drawing dies, lamination dies, thread rolling dies, shear blades, slitter knives, forming rolls, seaming rolls, burnishing tools, punches, gages, crimping dies, swaging dies.
D4	2.25	—	1.00	12.00	—	—	Blanking dies, brick molds, burnishing tools, thread rolling dies, hot swaging dies, wiredrawing dies, forming tools and rolls, gages, punches, trimmer dies, dies for deep drawing.
D5	1.50	—	1.00	12.00	—	3.00 Co	Cold forming dies, thread rolling dies, blanking dies, coining dies, trimming dies, draw dies, shear blades, punches, quality cutlery, rolls.
D7	2.35	—	1.00	12.00	4.00	—	Brick mold liners and die plates, briquetting dies, grinding wheel molds, dies for deep drawing, flattening rolls, shot and sandblasting liners, slitter knives, wear plates, wiredrawing dies, Sendzimir mill rolls, ceramic tools and dies, lamination dies.
							HOT WORK (CHROMIUM)
H10	0.40	—	2.50	3.25	0.40	—	Mandrels, extrusion and forging dies, die holders, bolsters and dummy blocks, punches, die inserts, gripper and header dies, hot shears, aluminum die casting dies, inserts for forging dies and upsetters, shell piercing tools,
H11	0.35	—	1.50	5.00	0.40	—	Die casting dies, punches, piercing tools, mandrels, extrusion tooling, forging dies, high strength structural components.

TABLE 12-2. CLASSIFICATION AND SELECTION OF TOOL STEELS (cont)

AISI TYPE	PERCENTAGE COMPOSITION						TYPICAL APPLICATIONS
	C	W	Mo	Cr	V	Other	
							HOT WORK (CHROMIUM) (cont)
H12	0.35	1.50	1.50	5.00	0.40	—	Extrusion dies, dummy blocks, holders, gripper and header dies, forging die inserts, punches, mandrels, sleeves for cold heading dies.
H13	0.35	—	1.50	5.00	1.00	—	Die casting dies and inserts, dummy blocks, cores, ejector pins, plungers, sleeves, slides, extrusion dies, forging dies and inserts.
H14	0.40	5.00	—	5.00	—	—	Backer blocks, die holders, aluminum and brass extrusion dies, press liners, dummy blocks, forging dies and inserts, gripper dies, shell forging points and mandrels, hot punches, pushout rings, dies and inserts for brass forging.
H19	0.40	4.25	—	4.25	2.00	4.25 Co	Extrusion dies and die inserts, dummy blocks, punches, forging dies and die inserts, mandrels, hot punch tools.
							HOT WORK (TUNGSTEN)
H21	0.35	9.00	—	3.50	—	—	Mandrels, hot blanking dies, hot punches, blades for flying shear, hot trimming dies, extrusion and die casting dies for brass, dummy blocks, piercer points, gripper dies, hot nut tools (crowners, cutoffs, side dies, piercers), hot headers.
H22	0.35	11.00	—	2.00	—	—	Mandrels, hot blanking dies, hot punches, blades for flying shear, hot trim dies, extrusion dies, dummy blocks, piercer points, gripper dies.
H23	0.30	12.00	—	12.00	—	—	Extrusion and die casting dies for brass, brass and bronze permanent molds.
H24	0.45	15.00	—	3.00	—	—	Punches and shear blades for brass, hot blanking and drawing dies, trimming dies, dummy blocks, hot press dies, hot punches, gripper dies, hot forming rolls, hot shear blades, swaging dies, hot heading dies, extrusion dies.
H25	0.25	15.00	—	4.00	—	—	Hot forming dies, die casting and forging dies, die inserts, extrusion dies and liners, shear blades, blanking dies, gripper dies, punches, hot swaging dies, nut piercers, piercer points, mandrels, high temperature springs.

Classification of Steel

TABLE 12-2. CLASSIFICATION AND SELECTION OF TOOL STEELS (cont)

AISI TYPE	PERCENTAGE COMPOSITION						TYPICAL APPLICATIONS
	C	W	Mo	Cr	V	Other	

HOT WORK (TUNGSTEN) (cont)

AISI TYPE	C	W	Mo	Cr	V	Other	TYPICAL APPLICATIONS
H26	0.50	18.00	—	4.00	1.00	—	Mandrels, hot blanking dies, hot punches, blades for flying shear, hot trimming dies, extrusion dies, dummy blocks, piercer points, gripper dies, pipe threading dies, nut chisels, forging press inserts, extrusion dies for brass and copper.

HOT WORK (MOLYBDENUM)

AISI TYPE	C	W	Mo	Cr	V	Other	TYPICAL APPLICATIONS
H41	0.65	1.50	8.00	4.00	1.00	—	Cold forming and hot forming dies, dummy blocks, extrusion dies for copper base alloys, hot or cold heading dies, hot or cold punches, gripper dies, thread rolling dies, hot shear blades, trimming dies, spike cutters.
H42	0.60	6.00	5.00	4.00	2.00	—	Cold trimming dies, hot upsetting dies, dummy blocks, header dies, hot extrusion dies, cold header and extrusion dies and die inserts, hot forming and swaging dies, nut piercers, hot punches, mandrels, chipping chisels.
H43	0.55	—	8.00	4.00	2.00	—	Cold forming punches, hot forming dies and punches, swaging dies, hot gripper dies, hot heading dies, hot and cold nut dies, hot spike dies, grippers and headers, hot shear knives, hot working punches, insert punches.

HIGH SPEED (TUNGSTEN)

AISI TYPE	C	W	Mo	Cr	V	Other	TYPICAL APPLICATIONS
T1	0.75[1]	18.00	—	4.00	1.00	—	Drills, taps, reamers, hobs, lathe and planer tools, broaches, crowners, burnishing dies, cold extrusion dies, cold heading die inserts, lamination dies, chasers, cutters, taps, end mills, milling cutters
T2	0.80	18.00	—	4.00	2.00	—	Lathe and planer tools, milling cutters, form tools, broaches, reamers, chasers.
T4	0.75	18.00	—	4.00	1.00	5.00 Co	Lathe and planer tools, drills, boring tools, broaches, roll turning tools, milling cutters, snaper tools, form tools, hobs, single point cutting tools.
T5	0.80	18.00	—	4.00	2.00	8.00 Co	Lathe and planer tools, form tools, cut-off tools, heavy duty tools requiring high red hardness.

Metallurgy

TABLE 12-2. CLASSIFICATION AND SELECTION OF TOOL STEELS (cont)

AISI TYPE	PERCENTAGE COMPOSITION						TYPICAL APPLICATIONS
	C	W	Mo	Cr	V	Other	
							HIGH SPEED (TUNGSTEN) (cont)
T6	0.80	20.00	—	4.50	1.50	12.00 Co	Heavy duty lathe and planer tools, drills, checking tools, cut-off tools, milling cutters, hobs.
T8	0.75	14.00	—	4.00	2.00	5.00 Co	Boring tools, lathe tools, heavy duty planer tools, tool bits, single point cutting tools for stainless steel.
T15	1.50	12.00	—	4.00	5.00	5.00 Co	Form tools, lathe and planer tools, broaches, milling cutters, blanking dies, punches, heavy duty tools requiring good wear resistance.
							HIGH SPEED (MOLYBDENUM)
M1	0.80[1]	1.50	8.00	4.00	1.00	—	Drills, taps, end mills, reamers, milling cutters, hobs, punches, lathe and planer tools, form tools, saws, chasers, broaches, routers, woodworking tools.
M2	085.100[1]	6.00	5.00	4.00	2.00	—	Drills, taps, end mills, reamers, milling cutters, hobs, form tools, saws, lathe and planer tools, chasers, broaches and boring tools.
M3-1	1.05	6.00	5.00	4.00	2.40	—	Drills, taps, end mills, reamers and counterbores, broaches, hobs, form tools, lathe and planer tools, cheeking tools, milling cutters, slitting saws, punches, drawing dies, routers, woodworking tools.
M3-2	1.20	6.00	5.00	4.00	3.00	—	Drills, taps, end mills, reamers and counterbores, broaches, hobs, form tools, lathe and planer tools, cheeking tools, slitting saws, punches, drawing dies, woodworking tools.
M4	1.30	5.50	4.50	4.00	4.00	—	Broaches, reamers, milling cutters, chasers, form tools, lathe and planer tools, cheeking tools, blanking dies and punches for abrasive materials, swaging dies.
M6	0.80	4.00	5.00	4.00	1.50	12.00 Co	Lathe tools, boring tools, planer tools, form tools, milling cutters.
M7	1.00	1.75	8.75	4.00	2.00	—	Drills, taps, end mills, reamers, routers, saws, milling cutters, lathe and planer tools, chasers, borers, woodworking tools, hobs, form tools, punches.
M10	0.85.100	—	8.00	4.00	2.00	—	Drills, taps, reamers, chasers, end mills, lathe and planer tools, woodworking tools, routers, saws, milling cutters, hobs, form tools, punches, broaches.

TABLE 12-2. CLASSIFICATION AND SELECTION OF TOOL STEELS. (Cont.)

AISI TYPE	PERCENTAGE COMPOSITION						TYPICAL APPLICATIONS
	C	W	Mo	Cr	V	Other	
							HIGH SPEED (MOLYBDENUM) (cont)
M30	0.80	2.00	8.00	4.00	1.25	5.00 Co	Lathe tools, form tools, milling cutters, chasers.
M33	0.90	1.50	9.50	4.00	1.15	8.00 Co	Drills, taps, end mills, lathe tools, milling cutters, form tools, chasers
M34	0.90	2.00	8.00	4.00	2.00	8.00 Co	Drills, taps, end mills, lathe tools, milling cutters, form tools, chasers.
M36	0.80	6.00	5.00	4.00	2.00	8.00 Co	Heavy duty lathe and planer tools, boring tools, milling cutters, drills, cut-off tools, tool holder bits.
M41	1.10	6.75	3.75	4.25	2.00	5.00 Co	
M42	1.10	1.50	9.50	3.75	1.15	8.00 Co	Drills, end mills, reamers, form cutters, lathe tools, hobs, broaches, milling cutters, twist drills, end mills, Hardenable to Rockwell C 67 to 70.
M43	1.20	2.75	8.00	3.75	1.60	8.25 Co	
M44	1.15	5.25	6.25	4.25	2.25	12.00 Co	
M46	1.25	2.00	8.25	4.00	3.20	8.25 Co	
M47	1.10	1.50	9.50	3.75	1.25	5.00 Co	
							SPECIAL PURPOSE (LOW ALLOY)
L2	0.50 to 1.10[1]	—	—	1.00	0.20	—	Automotive gears and forgings, arbors, crank pins, chuck jaws and liners, chain feed sprockets, dogs, drift pins, die rings, friction feed disks, gun barrels, gun hoops, jack screws, lead and feed screws, machine shafts, pinions, rivet sets, shear blades, spindles, wrenches, die insert holders.
L3	1.00	—	—	1.50	0.20	—	Bearing races, mandrels, cold rolling rolls, ball bearings, arbors, precision gages, broaches, taps, dies, drills, thread rolling dies, knife edges, cutlery, files, knurls.
L6	0.70[1]	—	0.25	0.75	—	1.50 Ni	Arbors, blanking dies, forming dies, disk saws, drift pins, brake dies, hand stamps, hubs, lead and feed screws, machine parts, punches, pawls, pinions, shear blades, spindles, spring collets, swages, tool shanks, metal slitters, wood cutting saws.
							SPECIAL PURPOSE (CARBON-TUNGSTEN)
F1	1.00	1.25	—	—	—	—	Taps, broaches, reamers.
F2	1.25	3.50	—	—	—	—	Burnishing tools, tube and wiredrawing dies, extruding dies, reamers, taps, paper knives, eyelet dies, inserts for heading or sizing dies, forming dies, blanking dies, piercing punches, cold heading dies, sizing plugs and gages, roll turning tools, deep drawing dies.

Metallurgy

TABLE 12-2. CLASSIFICATION AND SELECTION OF TOOL STEELS. (Cont.)

AISI TYPE	PERCENTAGE COMPOSITION						TYPICAL APPLICATIONS
	C	W	Mo	Cr	V	Other	
MOLD							
P2	0.07	—	0.20	2.00	—	0.50 Ni	Plastic molding dies (hobbed and carburized).
P3	0.10	—	—	0.60	—	1.25 Ni	Plastic molding dies (hobbed and carburized).
P4	0.07	—	0.75	5.00	—	—	Plastic molding dies, die casting dies (hobbed and carburized).
P5	0.10	—	—	2.25	—	—	Plastic molding dies (hobbed and carburized).
P6	0.10	—	—	1.50	—	3.50 Ni	Heavy duty gears, shafts, bearings, plastic molding dies (hobbed and carburized.)
P20	0.35	—	0.40	1.25	—	—	Molds for zinc and plastic articles, holding blocks for die casting dies.
P21	0.20	—	—	—	—	4.00 Ni. 1.20 Al	Thermoplastic injection molds, zinc die casting dies, holding blocks for plastic and die casting dies.

[1]Other carbon contents may be available.   [2]Optional.   [3]Contains free graphite in micro-structure to improve machinability.
Some of these grades can be made with added sulfur to improve machinability.
Source: AISI Tool Steel Products Manual and producers of tool steels.

STAINLESS AND HEAT RESISTING STEELS

The stainless and heat resisting steels possess relatively high resistance to oxidation and to attack by corrosive media at room and elevated temperatures. They are produced to cover wide ranges in mechanical and physical properties. They are melted exclusively by the electric furnace process.

A three-numeral system is used to identify stainless and heat resisting steels by type and according to four general classes. The first digit indicates the class and the last two indicate type. Modification of types are indicated by suffix letters. The meaning of this AISI system is as follows:

Series 2xx... Cr, Ni, Mn steels
Series 3xx... Cr, Ni steels
Series 4xx... Cr steels
Series 5xx... low Cr steels

The chemical composition ranges and limits are given in Table 12-3.

TABLE 12-3. COMPOSITION LIMITS OF STANDARD TYPE STAINLESS AND HEAT RESISTING STEELS.
(Electric furnace steels)

AISI TYPE NUMBER	C	Mn (max)	P (max)	S (max)	Si (max)	Cr	Ni	Mo	Co	Se	Ti	Cb-Ta	Ta	Al	N
201	0.15 max	5.50/7.50	0.060	0.030	1.00	16.00/18.00	3.50/5.50								0.25 max
202	.15 max	7.50/10.00	.060	.030	1.00	17.00/19.00	4.00/6.00								.25 max
301	.15 max	2.00	.045	.030	1.00	16.00/18.00	6.00/8.00								
302	.15 max	2.00	.045	.030	1.00	17.00/19.00	8.00/10.00								
302B	.15 max	2.00	.045	.030	2.00/3.00¹	17.00/19.00	8.00/10.00								
303	.15 max	2.00	.20	.15 min	1.00	17.00/19.00	8.00/10.00	0.60 max²							
303 Se	.15 max	2.00	.20	.060	1.00	17.00/19.00	8.00/10.00			0.15 min					
304	.08 max	2.00	.045	.030	1.00	18.00/20.00	8.00/12.00								
304L	.03 max	2.00	.045	.030	1.00	18.00/20.00	8.00/12.00								
305	.12 max	2.00	.045	.030	1.00	17.00/19.00	10.00/13.00								
308	.08 max	2.00	.045	.030	1.00	19.00/21.00	10.00/12.00								
309	.20 max	2.00	.045	.030	1.00	22.00/24.00	12.00/15.00								
309S	.08 max	2.00	.045	.030	1.00	22.00/24.00	12.00/15.00								
310	.25 max	2.00	.045	.030	1.50	24.00/26.00	19.00/22.00								
310S	.08 max	2.00	.045	.030	1.50	24.00/26.00	19.00/22.00								
314	.25 max	2.00	.045	.030	1.50/3.00	23.00/26.00	19.00/22.00								
316	.08 max	2.00	.045	.030	1.00	16.00/18.00	10.00/14.00	2.00/3.00							
316L	.03 max	2.00	.045	.030	1.00	16.00/18.00	10.00/14.00	2.00/3.00							
317	.08 max	2.00	.045	.030	1.00	18.00/20.00	11.00/15.00	3.00/4.00							
321	.08 max	2.00	.045	.030	1.00	17.00/19.00	9.00/12.00				5 x C min				
347	.08 max	2.00	.045	.030	1.00	17.00/19.00	9.00/13.00					10 x C min			
348	.08 max	2.00	.045	.030	1.00	17.00/19.00	9.00/13.00		0.20 max			10 x C min	0.10 max		
403	.15 max	1.00	.040	.030	0.50	11.50/13.00									
405	.08 max	1.00	.040	.030	1.00	11.50/14.50								0.10/0.30	
410	.15 max	1.00	.040	.030	1.00	11.50/13.50									
414	.15 max	1.00	.040	.030	1.00	11.50/13.50	1.25/2.50								
416	.15 max	1.25	.060	.15 min	1.00	12.00/14.00		0.60 max²							
416 Se	.15 max	1.25	.060	.060	1.00	12.00/14.00				0.15 min					
420	over 0.15	1.00	.040	.030	1.00	12.00/14.00									
430	0.12 max	1.00	.040	.030	1.00	14.00/18.00									
430F	.12 max	1.25	.060	.15 min	1.00	14.00/18.00		.60 max²							
430F Se	.12 max	1.25	.060	.060	1.00	14.00/18.00				.15 min					
431	.20 max	1.00	.040	.030	1.00	15.00/17.00	1.25/2.50								
440A	.60/0.75	1.00	.040	.030	1.00	16.00/18.00		.75 max							
440B	.75/0.95	1.00	.040	.030	1.00	16.00/18.00		.75 max							
440C	.95/1.20	1.00	.040	.030	1.00	16.00/18.00		.75 max							
442	.20 max	1.00	.040	.030	1.00	18.00/23.00									
446	.20 max	1.50	.040	.030	1.00	23.00/27.00									.25 max
501	over 0.10	1.00	.040	.030	1.00	4.00/6.00		.40/0.65							
502	0.10 max	1.00	.040	.030	1.00	4.00/6.00		.40/0.65							

¹Range
²At producer's option; reported only when intentionally added

CLASSIFICATION BY HARDENABILITY

Alloy steels as well as plain carbon steels may be classified according to their ability to harden or to be made martensitic. The ability to harden has been defined as the *hardenability* of the steel. The role played by the hardenability of steels is one of great importance, and any test that can be made to check this characteristic of steel will prove of value in the selection and qualification of any steel for a given application.

It has been determined that the hardenability of steels is dependent upon factors other than chemical composition or alloy content of the steel. Such factors as methods of manufacture, practice in shaping, and variables in treating, all influence the hardenability of steels. Because these

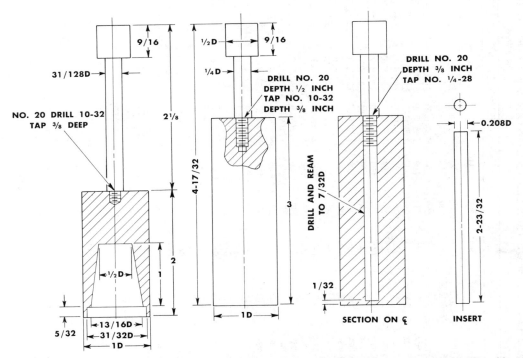

Fig. 12-1. Test specimens for end-quench method of determining hardenability. Usual specimen at center; L-bar for steels of low hardenability at left; drilled bar at right for steels available in small sizes only. (Metal Progress, article by Walter E. Jominy, Research Laboratories of General Motors Research Corp.)

Classification of Steel

variables will influence the hardenability, the usual methods of testing, without carrying out a test for hardenability, may not reveal the complete story about the steel. This being true, several methods have been designed to measure the ability of the steel to harden. The principle behind all of the hardenability tests is to measure the maximum section or thickness of steel that can be made hard or martensitic successfully. The Jominy end-quench test is recommended as a hardenability test for alloy structural and tool steels. This test can be used for both shallow and deep-hardening steels.

The type of test piece used in the Jominy end-quench test is illustrated in Fig. 12-1. The test specimen is heated to the hardening temperature and dropped into a fixture that holds the lower end of the specimen above a spray nozzle. The quenching medium is water, which, by means of a suitable spray nozzle, hits only the bottom of the specimen. By this method, practically all of the heat is extracted from the face of the quenched end of the test piece. After the specimen has been cooled, it is ground with a small flat surface 0.015" deep along the entire length of the test bar, and Rockwell hardness

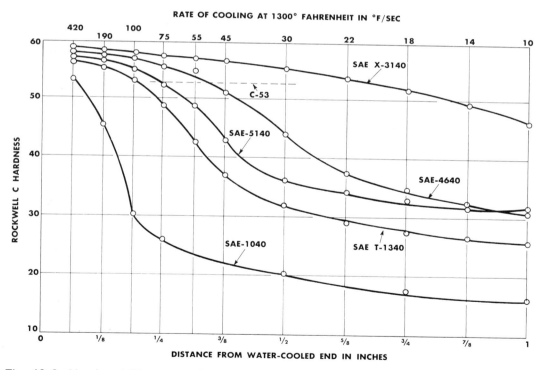

Fig. 12-2. Hardenability curves (end-quench bars) of five steels containing 0.40 percent carbon. (Metal Progress, article by Walter E. Jominy)

Metallurgy

readings are made at ⅛" intervals from the hard end.

The results of the Jominy end-quench test can be plotted as illustrated in Fig 12-2. This test can be used to check the hardenability of steels before they are used, or it can be used to test the hardenability of one steel against another and in this way determine a suitable steel.

REVIEW

1. Explain the SAE numbering system for steels.
2. What are the non-specified elements present in alloy steels?
3. What is the maximum permissable percentages for each?
4. The term *tool steel* is restricted in practice to what steels?
5. What are the seven major groups of tool steels?
6. What are the three general classifications of tool steels?
7. Under what conditions is it possible to quench a water-hardening steel in oil?
8. If oil and air-hardening steels were quenched in water or brine, what would be the danger?
9. What knowledge of what characteristic of steel is of value in the selection and qualification of any steel for a given application?
10. Describe the Jominy end-quench test.

CHAPTER 13 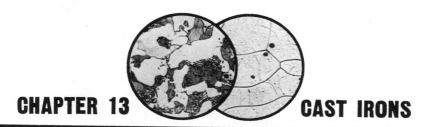 CAST IRONS

History records the production of cast iron chill molds in China as early as the 4th century B.C. Because of its unique properties, cast iron continues to fill a very important role in the modern world.

Cast iron is essentially an alloy of carbon, silicon and iron in which more carbon is present than can be retained in solid solution. The carbon content in cast irons is usually more than 1.7 percent and less than 4.5 percent. The high percentage of carbon renders the cast iron brittle and unworkable by rolling or forging. The basic shape of cast iron products is acquired by *casting*, hence the name *cast iron.*[1]

Silicon, which acts as a graphitizing agent, is usually present in amounts of one-half to two percent, although it may be much higher in certain silicon alloy irons.

Cast iron, like steel, always contains a certain amount of manganese, phosphorus and sulfur because of the refining methods used in its production. Special alloying elements such as copper, molybdenum, nickel and chromium are added, as in steel, for the purpose of altering the chemical and mechanical properties of the iron.

There are many variations in structure and properties available in cast iron. However, for convenience, they may be classified into five main groups:

1. Gray cast iron
2. White cast iron
3. Malleable iron
4. Ductile (nodular) iron
5. Alloy cast irons

The terms *gray* and *white,* when applied to cast iron, refer to the appearance of the fracture of a casting. The gray cast iron fractures with a dark gray fracture, whereas the white cast iron shows a light gray fracture, almost white, as illustrated in Fig. 13-1. Malleable iron gets its name from the ability of this type of cast iron to bend

---

[1]Since expressions of psi (pounds per square inch) in units of tens and hundreds of thousands are not absolute quantities, the conversion units of kN/m^2 (kilo-Newtons per meter squared) have been rounded off appropriately.

Metallurgy

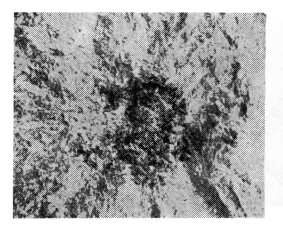 

Fig. 13-1. Fracture appearance of white cast iron, left; gray cast iron, right.

or undergo permanent deformation before it fractures. It is much more ductile and tough than either white or gray cast iron. In each group or classification of cast irons, many variations in composition and in the foundry practice used in the manufacture may result in a great many variations in the structure and physical properties of the cast metal.

GRAY CAST IRON

Although the greatest tonnage of gray iron is produced by cupola melting, other furnaces such as the air furnace, electric furnace, and rocking-type furnace are also used in its manufacture. The cupola is the oldest and cheapest melting method used but is the most difficult to control as to the composition and structure of the final product. In the production of a casting, the molten metal is taken from the furnace and poured into a mold usually made from sand. After the mold has been poured and the metal has cooled sufficiently, the casting is removed from the mold and taken to the cleaning room where adhering sand is removed by one of several processes, and the castings are chipped and ground. Inspection of the castings takes place at several points during cleaning, with a final inspection after the finishing operations. Heat treating, when required, is performed after cleaning.

Metallurgy of gray iron. Carbon in cast iron, as in steel, plays the most important part in metallurgical control of the properties developed in the cast metal. Some of the carbon in gray cast iron is in the combined form similar to the form that carbon assumes in steel, but by far the larger percentage of the total carbon is present in the iron as flakes of graphitic carbon. To control the properties of gray iron, one must control the amount, size, shape, and distribution of the graphitic and combined forms of carbon. The metallurgist tries to control the effect of carbon largely by controlling the composition of the iron, particularly the percentage of carbon and silicon, and by controlling other factors that influence the formation of combined and graphitic carbon. Some of the variables that influence the carbon control include:

1. Melting method used, such as cupola, air furnace, etc. The cupola melting method is not as easily controlled as the air or electric furnaces, and its use may result in undesirable amounts of carbon.

2. Chemical composition. The chemistry of the charge placed in the furnace and the composition of the final product largely determine the properties of this product.

3. Temperatures throughout the entire process are factors that have an important bearing on the final structure and properties of the cast metal.

4. The cooling rate from the temperature at the pouring of the casting to room temperature is a very important factor, as this has a marked influence upon the final structure developed in the casting.

5. Heat treatments that may be applied to the final casting.

Other variables that may have a marked effect upon the characteristics of any casting are factors of casting design and the workmanship relative to the entire process of making a casting. From these established facts, it should be apparent that to make a gray iron casting to meet certain structural and physical requirements is rather a difficult task, and it can be accomplished only with accurate control of all of the variables that are encountered in the process. Control of composition alone will not necessarily result in satisfactory castings; however, if all of the variables including melting and foundry technique can be controlled accurately, then the foundryman can expect to regulate his results by control of composition. The chemical composition, with all other factors remaining constant, largely determines the type of structure and therefore the properties of the final product.

Structure of gray iron. Gray cast iron can be varied in its structural composition within wide limits and develop a wide range of properties and applications. There is no one structure or set of properties for this iron. Microscopic examination of gray iron will reveal the presence of the following structural constituents in varying percentages, depending upon the character of the iron: graphite, pearlite, ferrite, cementite, steadite, and manganese sulfide. The constituents, graphite, and pearlite, are present in varying percentages in all of the grades of gray iron. The presence of ferrite or cementite, in a free condition, depends upon the amount of combined carbon present in the casting. If the amount of combined carbon is below the eutectoid ratio (pearlite), free ferrite will be found in the structure. On the other

Metallurgy

hand, if the combined carbon content is above the amount necessary to form pearlite, then free cementite will be present in the final structure of the casting. The presence of steadite and manganese sulfide depends upon the amount of phosphorus and sulfur present in the casting.

Graphite in gray iron. Graphite is by far the most important constituent of gray iron. The amount, size, shape, and distribution of the graphite flakes found in this iron largely control the final properties of the iron. The carbon present in the form of graphite has a relatively low density and therefore occupies more volume than indicated by the weight percentage determination. A gray iron with 3 percent of its total carbon present as graphite actually has 9.6 percent or more graphite by volume in its structure. Due to its flake-like shape, the graphite breaks up the continuity of the iron and greatly weakens it. If the graphite is present in small, round, well-distributed particles, its weakening effect is much less pronounced, and the iron develops much higher mechanical properties. If the foundry could control the graphitic constituent of gray iron, almost complete control would be exercised over the properties of the metal. The condition of graphitic carbon may be determined by a microscopic examination of the polished surface of the iron.

Large flakes of graphitic carbon produce a soft, weak casting; whereas, fine and short flakes of graphitic carbon result in a stronger casting. This difference in graphitic carbon is illustrated in Figs. 13-2 and 13-3. The nodular shape of graphitic carbon is desired, as illustrated in Fig. 13-4, but this shape of graphite is ob-

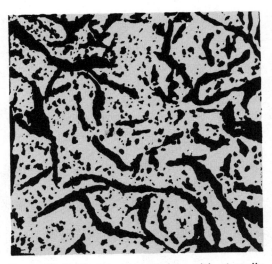

Fig. 13-2. Gray cast iron with tensile strength of 15,000 psi (103,325 kN/m^2) and Brinell hardness of 140. Large flakes of graphitic carbon are responsible for low hardness and strength (polished, not etched). Magnified 100 times.

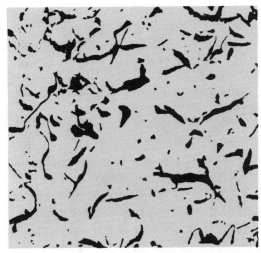

Fig. 13-3. Gray cast iron with tensile strength of 35,000 psi (241,092 kN/m^2) and Brinell hardness of 195. Note much smaller flakes of graphitic carbon than in Fig. 13-2 (same conditions).

Cast Irons

Fig. 13-4. Malleable cast iron, shows nodular shape of graphitic carbon (polished, not etched); tensile strength, 52,000 psi (360,197 kN/m^2); elongation 16 percent in 2''; Brinell hardness, 118. Magnified 100 times.

tained in annealing of white cast iron, thus producing malleable cast iron.

The graphitic phase of carbon in gray cast iron may be influenced by many factors. Included among them are:

1. Ratio of carbon to silicon. Silicon acts as a graphitizer, and the more silicon present the more and larger will be the flakes of graphite.

2. Additions of steel to the melt. Steel additions reduce the total carbon and dilute the graphitic carbon present in the melt. This reduces the opportunity for the carbon to graphitize and develop many and large flakes.

3. Temperature used in melting. Very high melting temperatures seem to dissolve the graphite present in the melt and thus reduce the tendency to graphitize.

4. Fast cooling. Rapid solidification of the casting will reduce the action of graphitization. The cooling rate is largely controlled by the size of the cast section. This is one of the large factors affecting the character of a casting. If the foundry is asked to cast a section that is wedge shaped, or that has a very obvious change in section from thin to thick, this causes more difficulty in controlling the graphitic carbon constituent than any other single factor. The thin sections may not graphitize at all and therefore will be hard and difficult to machine, whereas the thick section of the same casting will graphitize nearly completely into large flakes of graphite and cause a very weak, soft condition.

5. Special alloys added to melt. The principal effect of alloys used in the foundry is the control they exert over the formation of graphitic carbon. Nickel added helps to graphitize the carbon in a thin section, whereas chromium reduces the graphitization in a heavy section.

Pearlite in gray iron. The constituent *pearlite* in gray iron is the same as the pearlite present in steel. Carbon present in the combined form (Fe_3C) will combine with the ferrite constituent upon cooling through the critical range of the iron, thus developing the pearlitic constituent. Due to the presence of phosphorus, silicon, and manganese, a combined carbon content ranging from 0.50 to 0.89 percent may develop a structural matrix in iron which is largely pearlitic. It is possible to have all of the carbon present in the gray iron as graphite, in which case a structure of ferrite and graphite results, with no pearlite, and the iron is soft and weak. As the amount of combined carbon in-

Metallurgy

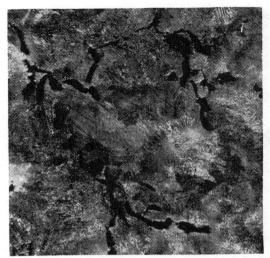

Fig. 13-5. Pearlitic gray iron, etched to bring out the fine pearlitic structure which forms the matrix of this iron. Magnified 250 times.

creases, the iron changes from all ferrite and graphite to pearlite and graphite. A casting consisting of pearlite and graphite is known as a *pearlitic gray iron* and is considered the best possible structure for strength, yet it retains good machinability. Fig. 13-5 illustrates the type of structure referred to as a pearlitic gray iron, consisting of graphite flakes with a matrix of pearlite.

The pearlite constituent influences the properties of gray iron as follows:

1. The strength of gray iron increases as the amount of pearlite increases.
2. The finer the constituents in pearlite, (i.e., ferrite and cementite layers), the stronger the iron.
3. Finer grain size gives increased strength.
4. The Brinell hardness increases with the increase in the fineness and amount of pearlite in going from a ferritic to a pearlitic condition.
5. Increase in the fineness and the amount of the pearlite and decrease in grain size result in some increase in toughness, noticeable in machining operations, but produce a correspondingly smoother finish.

If more combined carbon is present than can form with the pearlite during slow cooling through the critical range of the iron, a free cementite is present which adds materially to the hardness.

Free cementite in gray iron. The presence of free cementite in gray iron will increase the difficulty of machining and lower the strength of the iron. Cementite is very hard and brittle and its presence in the free form as cementite particles may be accidental or may be for the purpose of increasing the hardness and wear resistance of the iron. Ordinarily, free cementite

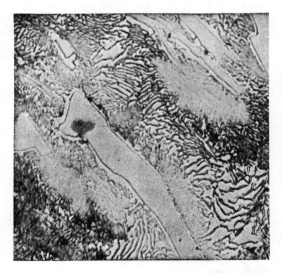

Fig. 13-6 White cast iron, magnified 500 times. (Same as Fig. 13-5) Pearlite, dark; cementite, white.

in large amounts is found only in hard, white iron or chilled irons. Fig. 13-6 illustrates the large percentage of free cementite found in a white iron casting.

Ferrite in gray iron. Ferrite is a naturally soft, ductile constituent possessing good tensile properties. There is no free ferrite present in gray iron unless the combined carbon content is below that which produces a pearlitic matrix. The presence of free ferrite does not impart ductility to the gray iron because the continuity of the ferrite matrix is broken by graphite flakes. A highly ferritic gray iron is not ductile as it means less pearlite and more graphitic carbon, resulting in a gray iron of less strength and softer than those containing a pearlitic matrix.

Steadite in gray iron. *Steadite* is a eutectic constituent formed from the presence of phosphorus, which, as an iron phosphide, combines with the iron to form a low-melting-point constituent which melts at about 1750°–1800°F (955°–980°C). Steadite is hard and brittle and is considered undesirable in irons when good mechanical properties are required. Steadite contains about 10 percent phosphorus, and a 1 percent phosphorus iron forms about 10 percent steadite. Phosphorus increases the fluidity of the molten iron, which is desirable in some types of castings, such as ornamental iron castings. However, while phosphorus increases the fluidity, it also decreases the strength, and for this reason it is better to obtain fluidity by higher tapping and pouring temperatures when strength is desired.

Minor constituents of gray iron. Minor constituents of gray iron include manganese sulfide (MnS), oxides, and gases. Manganese sulfide is present as relatively small inclusions scattered throughout the iron and has no appreciable effect on the strength of the iron. Oxides and gases weaken the iron and promote lack of density and soundness. Oxidation of cast iron results in a marked growth in the volume of the iron, which causes trouble when the iron is subjected to elevated temperatures.

Influence of rate of cooling upon structure of gray iron. The rate at which the cast metal cools from the beginning of freezing to below the critical range determines to a large degree the type of structure developed in the final casting. Fast cooling through the cooling zone and down to a relatively low temperature will result in the formation of little or no graphitic carbon. Such an iron will be hard and brittle. Very slow cooling will allow for marked graphitization with little or no combined carbon and result in a soft and weak iron.

There is no easy way to control the relative rates of cooling of castings cast in sand molds, as the rate of cooling is largely governed by the volume and surface area of the individual castings. Light sections with large surface area will cool much faster than heavy sections with small surface area. The effect of the cooling rate upon the structure developed in casting can be illustrated by a section in the shape of a wedge. See Fig. 13-7.

The thin point of the wedge will cool relatively fast and will consist structurally of a white iron, cementite, and pearlite. If the cooling rate is fast through the critical range of the iron, a martensitic matrix will

Metallurgy

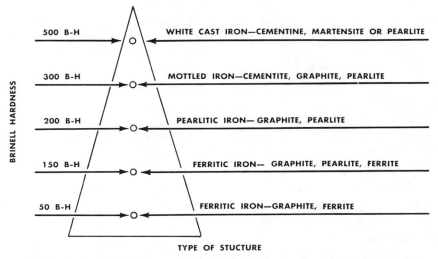

Fig. 13-7. General influence of section on structure and Brinell hardness.

be formed. The heavy end of the wedge, which cools relatively slowly, will contain a structure of ferrite and graphite with no combined carbon. This end will be very soft and weak.

From this illustration, it should be appreciated that any marked change in section in any given casting design may cause trouble for the foundry when a uniform structure throughout the entire casting is specified. The foundryman can avoid a marked difference in the structure and properties of a casting with thin and thick sections by employing various foundry techniques, largely those of composition control. It is possible to obtain a uniform structure in castings of varying section thickness; however, it is much better to design the casting with as uniform a section as possible.

Carbon-silicon ratio in gray iron. One of the most important factors affecting the structure and properties of a casting is the amount of carbon and silicon used and the ratio of the carbon to the silicon. This factor of foundry control is illustrated in Fig. 13-8, which points out the ratio of carbon to silicon required for irons that fall in the gray iron, mottled iron, and white iron classes. The ratio of carbon to silicon in Fig. 13-8 has to be modified, depending upon the section size of the casting and the presence and amount of other elements or alloys contained in the iron. However, silicon is by far the strongest graphitizer of all of the elements that may be added to cast iron; consequently it is the predominant element in determining the relative proportions of combined and graphitic carbon present in the final cast-

Cast Irons

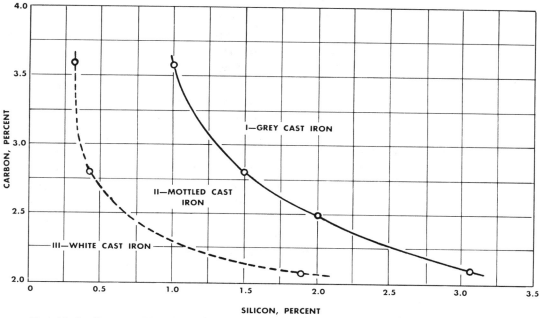

Fig. 13-8. Composition limits for gray, mottled, and white cast irons (Tanimura).

ing. Thus, the foundryman controls the structure and properties of his casting largely by adjusting the silicon content relative to the amount of carbon and the section size of the casting.

In general, the foundryman uses less silicon for a high-carbon iron and more silicon for a low-carbon iron. High-strength gray irons are of the pearlitic type, with the graphitic carbon in fine, well-distributed flakes. These high-strength gray irons usually have a low carbon content, around 2.50 to 2.65 percent, with the silicon running around 1.50 to 1.70 percent. The structure of such an iron is illustrated in Fig. 13-5 and 13-9. The following indicates the range of chemicals that might be available in gray cast irons:

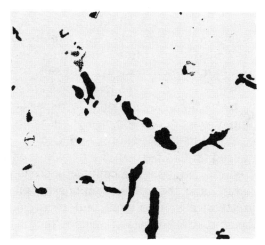

Fig. 13-9. Polished gray iron (Same as in Fig. 13-5 but not etched), showing the type of graphite flakes present in high-strength iron. Magnified 250 times.

Metallurgy

	PERCENT
Carbon	2.50 to 3.50
Silicon	0.60 to 2.40
Manganese	0.50 to 1.20
Phosphorus	0.05 to 1.00
Sulfur	0.04 to 0.25

plus special alloying elements, including nickel, chromium, molybdenum, and copper.

Effect of nickel on gray iron. Nickel is added to gray cast iron in amounts from 0.25 to 5.0 percent. Nickel will assist in the graphitizing of the carbon, but it is only about one-half as effective as silicon in this respect. However, about 3.0 percent silicon seems to be the limit in addition of this constituent in preventing the formation of free cementite or hard iron, whereas the effect of nickel extends beyond that percentage so that greater additions of this constituent act as a great aid in rendering iron gray that otherwise would be hard and contain free cementite.

Additions of nickel in small amounts will reduce the effects of chilling or rapid cooling and soften thin cast sections by eliminating hard, free cementite, thus promoting machinability. Nickel promotes density and freedom from porosity by permitting the use of a lower silicon content without causing chill or hard spots to appear in light sections. As nickel is increased up to 4 percent the hardness and strength of heavy sections increases slightly, and the lighter sections become progressively more gray and more machinable. Unlike silicon, nickel progressively and uniformly hardens the matrix of the iron by changing the structure of coarse pearlite to a finer and harder pearlite and finally to martensite. Nickel also helps in refining the grain and promotes dispersion of the graphite in a finely divided state, thus improving the strength and toughness of the iron.

Effect of chromium on gray iron. Chromium is a carbide-forming element, and chromium carbide is much more stable than the iron carbide (cementite) found in plain cast iron. By stabilizing the carbide (cementite), chromium acts opposite to the graphite-forming elements, silicon and nickel. Chromium increases the chill effect, i.e., retards graphite formation and increases the hardness and strength. This effect of chromium is illustrated in Fig. 13-10. Chromium additions promote the formation of more finely laminated and harder pearlite, thus increasing the strength and hardness of the iron matrix. Experience has shown that combinations of nickel and chromium develop better results than if they are used alone as alloying elements. Nickel, which is a graphitizer, and chromium, which is a carbide stabilizer, seem to supplement each other.

Effect of molybdenum on gray iron. Molybdenum as an alloying element in gray iron has but little effect in stabilizing the carbide (cementite) but is used to increase the strength of iron, due to its effect upon the decomposition of austenite to pearlite. Molybdenum decreases the rate of change from austenite to pearlite during the cooling cycle in cast iron as it does in steel. This results in a harder and stronger iron due to the formation of much finer pearlite, approaching a martensitic-type structure. In gray irons, molybdenum improves the strength, hardness, impact strength, fatigue strength, and strength at elevated temperatures.

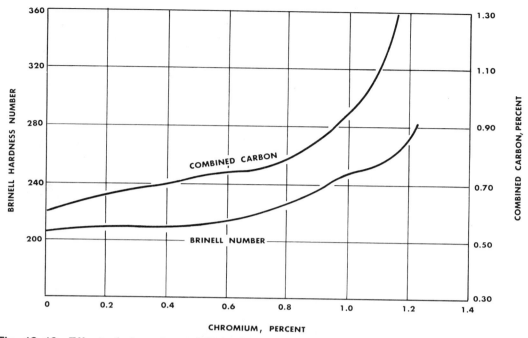

Fig. 13-10. Effect of chromium on Brinell hardness and combined carbons of arbitration bars (2.5 percent Silicon; 3.3 percent Carbon) (Trantin). (Cast Metals Handbook, 1935 Edition, American Foundrymen's Association)

Effect of copper on gray iron. Copper, like silicon, when alloyed in cast iron, functions primarily as a graphitizing agent. However, its ability to act as a graphitizer is only about one-tenth that of silicon. Copper may be added to gray iron in amounts up to 3 percent. A large percentage of the copper is thought to dissolve, forming a solid solution with ferrite. The remaining copper, in excess of its limits of solid solubility, is dispersed as microscopic or submicroscopic particles in the iron. The addition of copper refines the pearlite and increases the tensile strength and the Brinell hardness. It improves the wear resistance of the iron, its resistance to acid and atmospheric corrosion, and when combined with chromium, improves its strength and resistance to oxidation at elevated temperatures.

Properties of Gray Cast Iron

Gray cast iron is widely used in the machine tool field and in general engineering and industrial applications because of its ability to be cast in intricate shapes, its low cost, and its wide range of useful properties. Included among its outstanding properties are the following:

1. Excellent machinability.
2. Excellent wear resistance.
3. Tensile strength in the as-cast condition, varying from 20,000 psi (137,900 kN/m^2) for the soft, weak irons to 70,000 psi (482,700 kN/m^2) for the high-strength cast irons.
4. Compressive strength is usually three to four times the tensile strength, ranging from 70,000 to 200,000 psi (482,700 to 1,379,000 kN/m^2).
5. Gray cast iron has no well-defined elastic limit and may be loaded up to 80 percent or more of its maximum strength.
6. It is the only ferrous alloy that may be varied in stiffness. Its modulus of elasticity, at a load of 25 percent of ultimate strength, ranges from 12,000,000 to 21,000,000 psi (82,740,000 to 144,795,000 kN/m^2).
7. The hardness of cast iron can be controlled within certain limits and may extend from 100 to 350 Brinell hardness.
8. Included in its properties are its valuable damping capacity for vibrations and its good resistance to corrosion.

The tensile strength of gray iron is important from an engineering design viewpoint. Tensile strengths may be varied from 20,000 psi (137,900 kN/m^2) for soft, weak irons, to over 70,000 psi (482,700 kN/m^2) for the high-strength gray irons and irons in the heat-treated condition. Modern practice classifies gray cast irons according to their minimum tensile strengths.

Compressive strength. The compressive strength of gray iron is one of its most valuable properties. Compressive strengths developed are from three to five times the tensile strength and will run from 70,000 psi (482,700 kN/m^2) to over 200,000 psi (1,379,000 kN/m^2) for some of the higher strength irons.

Hardness. The hardness of gray iron may be varied within wide limits and may run from 100 Brinell to over 700 Brinell for iron that has received heat treatments, such as surface hardening. Gray irons with a Brinell hardness value of 180 to 200 constitute the major group and are more commonly used than any other class.

Transverse strength. A method for testing cast iron that is readily made in the foundry without elaborate equipment and with a test bar in the unmachined condition is known as the *transverse bend test.* A test bar of standard dimensions is cast and broken on supports spaced 12, 18, or 24 inches apart. The load required to rupture the bar is measured, and the value obtained can be reasonably correlated with the tensile strength of the iron. Also, the maximum deflection taken at the center of the bar indicates the ductility and toughness of the iron.

Fatigue strength. The fatigue strength, or endurance limit, for gray iron follows the values for tensile strength. The ratio of fatigue strength to tensile strength runs about 0.40 to 0.57. For the common gray irons, this would be about 9000 to 26,000 (62,100 to 179,300 kN/m^2) psi. It has been demonstrated during endurance tests that soft gray irons were less sensitive to notches and grooves or other surface discontinuities than is expected for most materials.

Gray iron is also apparently less sensitive to notch brittleness than most materials. This has been explained by pointing out that gray irons contain many internal notches caused by the presence of graphite flakes; therefore, the addition of external notches does not make so much difference. Also, gray irons are less rigid with lower stiffness values, i.e., lower modulus of elasticity, than other iron alloys; this reduces the effect of stress concentrations so that the effect of a notch as a stress raiser in producing fatigue failure is less apparent. Usually the modulus of elasticity, at 25 percent of the ultimate strength, ranges from 12,000,000 to 18,000,000 psi (82,740,000 to 124,110,000 kN/m^2) whereas steel has a modulus of elasticity of 30,000,000 psi (2,068,500 kN/m^2).

The lower modulus of elasticity or stiffness developed in gray irons results in an iron having greater damping effects to vibrations set up in service. This characteristic of gray iron makes it desirable for parts of machine tools where vibrations cause trouble in accurate machining operations, producing inaccuracies and rough finishes. The stiffer the iron, the lower the damping capacity. With a material of high strength and high modulus of elasticity, the vibrations may be allowed to build up to a serious intensity, whereas a high damping material with a fair strength will damp them out.

Wear resistance. Gray cast iron is widely used for machinery parts to resist wear. The graphitic carbon in gray irons acts as a lubricant and prevents galling when metal to metal contact occurs. Increasing the hardness of the iron improves its resistance to wear. A minimum hardness of 200 Brinell is usually specified for parts that are manufactured for wear resistance. However, Brinell hardness is not an accurate criterion of wear resistance for the type of microstructure, and alloys used in the iron seem to have far more influence upon this important property. Increasing the hardness by heat treatments is often practiced for parts such as ways of lathes, cylinder liners, cams, gears, sprockets, etc., and this practice does increase the wear resistance of the iron. In the hardening process, however, only the pearlitic constituent is changed to a harder structure. The graphite flakes remain as they were in the as-cast condition.

Heat resistance. Very little change in the strength of gray iron is observed at elevated temperatures up to 800°F (about 425°C). Above this temperature, in the average types of gray iron, there is a sharp drop in the strength and hardness, and in most cases a permanent loss in strength occurs. Alloyed gray irons, with a fine grain size and small graphite flake size, suffer less from loss of strength at elevated temperatures. Gray irons subjected to elevated temperatures up to 700°F (about 370°C) show good resistance to wear and galling, which makes them useful for applications in automobile, steam, and Diesel engines. However, gray irons grow or develop a permanent increase in volume when subjected to repeated cycles of heating and cooling in excess of 700°F (about 370°C).

Growth of cast iron is largely due to graphitization of the combined carbon in the iron and to corrosion and oxidation in the iron. It is thought that oxidizing gases penetrate the iron by way of the graphite flakes and attack the constituents of the iron, especially the silicon. This growth is

Metallurgy

serious as it reduces the strength and causes a very weak, brittle condition.

Heat Treatments of Gray Iron

Gray cast iron may be subjected to various heat treatments for the purpose of increasing its usefulness and improving its mechanical properties. The heat treatments usually applied to gray iron include (1) stress-relief annealing, (2) annealing, (3) hardening and tempering, and (4) nitriding.

Stress-relief annealing. Gray cast iron for machine parts is often subjected to a stress-relief annealing treatment for the purpose of removing any of the residual internal stresses present in the castings as they are received from the foundry. Also, the treatment tends to soften any hard spots and hard corners that occur in the castings, resulting from chills. The stress-relief annealing treatment consists of heating the castings to within a temperature range of 750° to 1250°F (about 400° to 675°C) for a period of several hours, followed by slow cooling to below 700°F (370°C). After reaching this low temperature, the doors of the furnace may be opened and the castings cooled to room temperature in air.

Fig. 13-11 illustrates a furnace loaded with gray iron castings for machine tool parts that are being treated in this manner. This treatment will remove the internal stresses and allow the castings to be finish-machined to accurate dimensions without the fear of warping. Also, this treatment will improve the machinability of castings having hard corners and hard

Fig. 13-11. Gray iron castings for machine tool parts being given a stress-relief annealing to remove internal stresses and to improve machinability. (Massachusetts Steel Treating Corp.)

spots by breaking down the combined carbon.

Annealing. The purpose of annealing gray cast iron is to increase the softness of the iron for economy in machining. Castings that are hard and difficult to machine because of the amount of combined carbon in their structure can be annealed so as to graphitize the combined carbon and restore easy machinability. The process consists in controlling the degree of transformation of the combined carbon to graphitic or free carbon. The annealing treatment selected depends upon the structure and composition of the iron being treated and the degree of softening desired.

The usual practice is to heat the castings in a furnace to around 1500° to 1600°F (815° to 870°C), hold the castings at this temperature for an hour or more, depending upon the maximum section of the casting. Follow by a slow cool to around 700°F (370°C), at which temperature the doors of the furnace may be opened and the castings removed. The maximum temperature used in this treatment depends upon the composition of the casting. Annealing temperatures from 1200° to 1800°F (650° to 980°C) have been used. The time of undergoing heat influences the degree of graphitization and is governed by the required softness.

Hardening and tempering. Gray irons of the pearlitic type can be hardened and tempered just as steels can. The pearlite in the structure of the iron can be changed to austenite upon heating to above the critical temperature of the iron, and upon rapid cooling (quenching), transformed to martensite. The casting thus treated can be tempered to relieve the quenching stresses, increase the toughness of the iron, and reduce the maximum hardness.

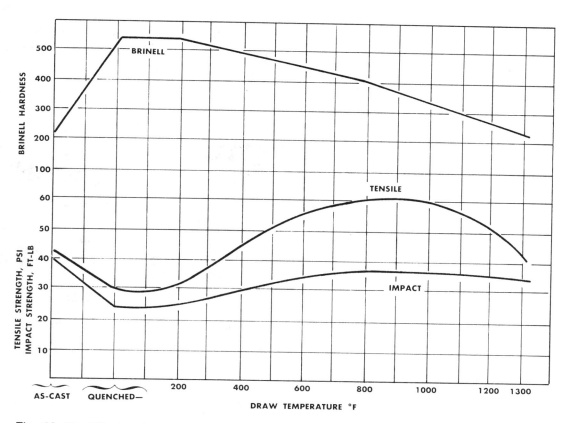

Fig. 13-12. Effects of tempering temperatures on the properties of gray cast iron. (Cast Metals Handbook, 3rd Edition, American Foundrymen's Association)

Metallurgy

The purpose of hardening and tempering gray iron is to strengthen and to increase the wear and abrasion resistance of the iron. Ordinary pearlitic gray irons are heated to 1500°F (815°C) and oil quenched although water is used as a quenching medium to a limited extent. Water quenching will cause more warping and danger from hardening cracks. The hardening of ordinary gray irons by this treatment will increase the Brinell hardness range from around 180-200 to 400-500. Following the hardening operation, the casting may be tempered from 400° to 800°F (about 205° to 430°C), depending upon the desired properties. Fig. 13-12 illustrates the change in tensile strength and hardness that might be expected from hardening a pearlitic iron and tempering within a temperature range of 200° to 1200°F (93° to 650°C).

Gray iron castings can be locally hardened by either the flame-hardening or induction-hardening method. Local heating of massive castings can be carried out safely by either process. If this is followed by local quench-hardening, the heat treater is able to harden surfaces without heating the entire casting to the quenching temperature. This insures a very hard surface without the danger of cracking and marked distortion that would occur if an attempt should be made to heat and quench the entire casting. Fig. 13-13 shows a fractured section of gray iron that has had one surface hardened by local heating and quenching, using the flame-hardening method. Fig. 13-14 illustrates the type of pearlitic structure in this iron before hardening, and Fig. 13-15 shows the martensitic structure found in the

Fig. 13-13. Fractured section of flame-hardened gray cast iron. Light edge of top shows the depth of hardness. Surface hardness, Rockwell C55.

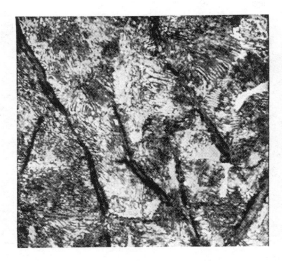

Fig. 13-14. Photomicrograph of gray cast iron, polished and etched. This iron is the same as that shown in Fig. 13-2. Structure consists of graphite flakes, matrix of pearlite, with a small percentage of ferrite (white). Magnified 500 times.

Cast Irons

Fig. 13-15. Photomicrograph of the edge of flame-hardened gray cast iron. Structure is martensitic with graphite flakes. Magnified 500 times.

hardened region of the casting after treatment. It will be seen from Fig. 13-15 that the hardening operation did not affect the graphitic constituent of the casting; it only changed the pearlite to martensite.

Nitriding of Gray Iron

Gray cast irons suitable for surface hardening by the nitriding process are essentially alloy irons containing aluminum and chromium. A typical analysis contains:

	PERCENT
Total carbon	2.62
Graphitic carbon	1.63
Combined carbon	0.99
Silicon	2.44
Manganese	0.60
Chromium	1.58
Aluminum	1.37

Castings of this composition are usually given a hardening and stabilizing heat treatment prior to the nitriding operation. This consists of an oil quench from 1550° to 1600°F (840° to 870°C), followed by a draw at 1100°–1200°F (about 595° to 650°C) for two hours. The nitriding is carried out in the usual way, heating to 950°F (about 510°C) in an atmosphere of approximately 30 percent dissociated ammonia for a period up to 90 hours. A case depth of 0.015″ may be expected with this treatment. A surface hardness of 800 to 900 Vickers diamond hardness test using 10-kg load is obtained after nitriding a finish-machined casting of this type. To obtain local nitriding, a thin coating of tin paint may be used as a stop-off material similar to the treatment used for steels.

WHITE AND CHILLED CAST IRONS

Iron castings that are classified as white or chilled iron have practically all of their carbon in the combined condition as cementite. These castings are relatively hard and brittle due to the high carbon content which forms a large percentage of hard and brittle cementite. Cast irons that contain low carbon, 2.0 to 2.5 percent and low silicon (see Fig. 13-8), when cast in sand molds and cooled slowly, solidify and cool to room temperature without graphitizing any of their carbon. These irons are naturally white irons.

A white cast iron that has been slowly cooled in the mold has a structure of pearlite and free cementite similar to that of a high-carbon steel, except that the white cast iron contains much more free cementite. The microstructure of this type of iron is illustrated in Figs. 13-6 and 13-16. The greatest tonnage of this type of iron is used for the manufacture of malleable iron. The addition of alloys such as nickel, molybdenum, and chromium to white iron results in the formation of a much harder iron. The type of structure formed by the alloy additions is usually martensite and free cementite, with the alloys acting as hardening agents and thus preventing the formation of pearlite from austenite during the cooling cycle. The structure of an alloyed white iron is illustrated in Figs. 13-17 and 13-18. Irons of this type are used for parts of machinery and industrial equipment where extreme hardness and excellent wear resistance

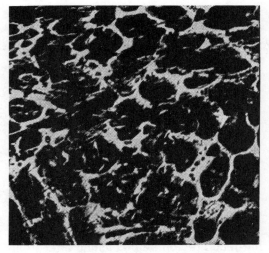

Fig. 13-16. White cast iron, polished and etched. White constituent is free cementite, and the dark constituent is pearlite. Magnified 100 times.

are required, such as in crusher jaws and hammers, wearing plates, cams, and balls and liners for ball mills.

White cast iron can also be made by rapid cooling or chilling of an iron which, if cooled slowly, would be graphitic and gray. If cast iron is cooled relatively fast in the mold, the carbon does not have an opportunity to graphitize and remains combined. Also, rapid cooling prevents the formation of a coarse, soft pearlite and adds to the hardness of the casting. If the cooling is fast enough, martensite may be formed instead of pearlite.

Cast Irons

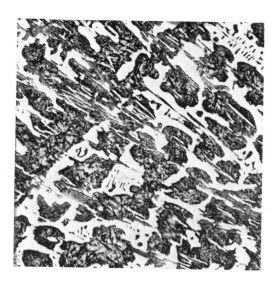

Fig. 13-17. Photomicrograph of the structure of white cast iron alloyed with molybdenum. Structure is martensitic with cementite. Rockwell C-58. Magnified 100 times.

Fig. 13-18. Same as Fig. 13-17. Magnified 1000 times. Structure martensitic with cementite.

Local hardening effects may be produced by a local chill in the molds, which results in a hard, white iron surface on a casting that might ordinarily be soft and gray if cooled slowly. Fig. 13-19 illustrates this effect. Factors that influence the depth of the chill include the ratio of carbon to silicon (high-silicon content decreases the depth of chill), thickness of the casting, thickness and temperature of

Metallurgy

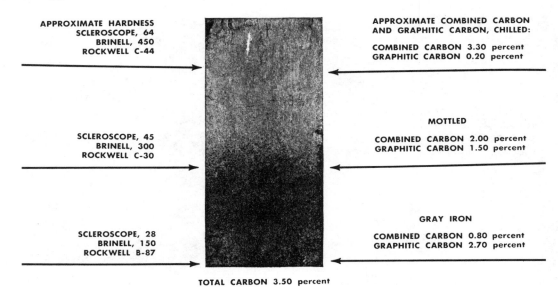

Fig. 13-19. Fracture of a chilled casting, showing the approximate combined carbon content, graphitic carbon content, and hardnness of the white, mottled, and gray iron in the as-cast condition. (Cast Metals Handbook 3rd Edition, American Foundrymen's Association)

the metal placed in the mold to act as a chill, the time that the cast metal is in contact with the metal chiller, and the use of alloy additions.

Chilled iron castings are extensively used for railroad car wheels and for many diversified applications where resistance to wear and compressive strength are major requirements. Applications for chilled iron include rolls for crushing grain and ore, rolling mills for shaping metals, farm implements, and equipment, and cement-grinding machinery.

MALLEABLE CAST IRON

Malleable cast iron is made by a process involving the annealing of hard, brittle white iron which, as the name implies, results in an iron that is much more ductile (malleable) than either white or gray cast iron. Malleable cast iron is not malleable in the sense that it is as forgeable as steel or wrought iron, but it does exhibit greater toughness and ductility as compared to other forms of cast iron. Also, malleable iron is softer than gray cast iron and exhibits easier machining characteristics. Because of these characteristics, malleable iron can be used in applications where greater toughness and resistance to shock are required, such as in farm implements, plows, tractors, harrows, rakes, etc., and finds many applications in automobile parts, hardware, small tools, and pipe fittings in spite of its greater cost as compared to gray cast iron.

Malleable iron is made from white iron castings by a high-temperature, long-time annealing treatment. The original white iron castings are made of a low-carbon, low-silicon type of iron—an iron that will solidify in a mold and cool without the formation of graphitic carbon. The iron is usually melted in a reverberatory type of furnace, commonly known as an air furnace. Occasional heats of white iron are melted in the electric furnace or the open-hearth furnace. Some heats are melted in the cupola, but it is difficult to melt irons of this composition due to the high temperatures required for them.

Annealing or Malleablizing

A white iron casting is first made by casting a controlled composition of metal into a sand mold. Upon cooling, this casting is hard and brittle because of its structure of combined carbon (Fe_3C). In the annealing process, the castings of white iron are placed in cast iron pots, or rings, and surrounded by a packing material which should be sufficiently refractory in nature so as not to fuse to the castings at the annealing temperatures. Sand is commonly used as a packing material, although squeezer slag, crushed air-furnace or blast-furnace slag (used alone or mixed with mill scale), or other forms of iron oxide may also be used. The purpose of the packing material is to protect and support the castings from warping during the annealing cycle.

The object of the annealing cycle is to change the combined carbon or cementite (Fe_3C) of the white iron to a graphitic carbon (temper carbon) found in malleable iron. The decomposition of the combined carbon to graphitic carbon is as follows:

$$Fe_3C \rightarrow 3Fe + C \text{ (graphite)}$$

Cementite (Fe_3C) is unstable at a red heat and decomposes to graphite and ferrite upon heating and slow cooling. The packed castings are placed in a furnace of the box or car type,. and a slow fire is started. The temperature is increased at such a rate as may require two days to reach an annealing temperature of 1550° to 1600°F (840° to 870°C). After reaching this temperature range, the castings are held there from 48 to 60 hours. The castings are then cooled slowly at a rate of not more than 8° to 10°F per hour (about 5°C/hr) until the temperature has dropped to around 1300°F (700°C). The castings may be held at this temperature, 1250° to 1300°F (675° to 700°C), for a period up to 24 hours, or the doors of the furnace may be opened after the castings have been slow-cooled to 1250°F (675°C) and the pots removed and allowed to air-cool. The castings are shaken out as soon as their temperature permits handling.

The annealing cycle should result in all of the combined carbon in the original white iron being completely decomposed to a graphitic or temper carbon condition, and the final structure of the iron should

Metallurgy

consist of ferrite and graphite. The graphitization of the combined carbon starts as soon as the castings reach a red heat. The initial heating to 1600°F (870°C) for 48 hours causes the free or excess cementite of the white iron to graphitize, and the slow cooling cycle to 1250° to 1300°F (675° to 700°C) allows the graphitization of the cementite that is precipitated from solution during this period. When the temperature of the iron falls below the critical temperature on cooling or the Ar_1 point (1250°–1300°F [675° to 700°C]), the balance of the cementite dissolved by the iron is precipitated with the formation of pearlite. A soaking period or very slow cool while this change is taking place will alow the graphitization of the cementite portion of the pearlite. This will complete the graphitization of all the possible cementite contained in the original white iron.

The time required for the complete annealing cycle varies from five to seven days. A so-called accelerated annealing cycle that requires from one-third to one-half the time required by the large annealing ovens in general use has been used successfully. The shorter annealing cycle is accomplished by the use of better designed furnaces of the gas or electric-fired type and, because of the rapid temperature changes and the accurate control possible with these newer furnaces, requires much less total time. Furnaces used in the shorter annealing cycle are the smaller batch-type, continuous car-type, or kiln-type furnaces. Also, if the annealing can be carried out without the use of a packing material, considerable reduction in the annealing time is gained. This is because of more rapid and even heating and cooling of the metal.

Black-Heart Malleable Iron

Because of the ductile nature of malleable iron, and due to the presence of graphitic carbon, the structure appearance upon fracturing will show dark or black with a light decarburized surface; thus the fracture will appear with a white edge and black core. This is known as *black-heart malleable iron,* the structure appearance of which is illustrated in Fig. 13-20. The

Fig. 13-20. Black-heart malleable cast iron showing picture-frame fracture formed by decarburized surface. Magnified 2½ times. (Roys)

Cast Irons

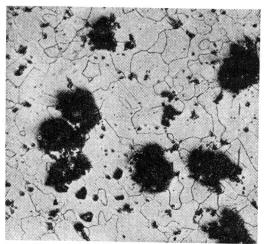

Fig. 13-21. Photomicrograph of malleable iron etched to show ferrite grains. The dark spots are graphitic (temper) carbon. Magnified 100 times.

white picture-frame edge is due to a decarburized surface. The graphite present in a fully annealed malleable iron differs from the flake-like graphite found in gray cast irons in that it is formed into a nodular shape called *temper carbon*. The principal constituents of malleable iron are ferrite and nodular graphite, see Fig. 13-21. The ferrite matrix of fully annealed malleable iron contains the silicon, manganese, and phosphorus in a solid solution condition.

The decarburized or white surface found in malleable iron results from a decarburization or burn-out of the carbon during the annealing cycle. Decarburization can be avoided by the use of a carbonaceous packing material or by annealing in a controlled atmosphere. If annealing is carried out using an iron oxide such as mill scale or iron ore for packing material, it is possible to completely decarburize or burn out all of the carbon in the original iron. This will result in a malleable iron that fractures with a light fracture appearance and is known as *whiteheart malleable iron*. Such a type of iron is seldom manufactured as it is difficult to machine, and its mechanical properties are inferior to those of the black-heart malleable iron.

Properties of Malleable Iron

The chemical composition of malleable iron is controlled within the limits specified for the various grades manufactured. The average chemical composition is as follows:

PERCENT
Carbon................................1.00 to 2.00
Silicon.................................0.60 to 1.10
Manganese.......................less than 0.30
Phosphorus......................less than 0.20
Sulfur..................................0.06 to 0.15

This composition is only approximate and is changed to suit the requirements of the final product. All of the carbon should be in the graphitic form, with no combined carbon present. The amount of phosphorus and sulfur is not objectionable in that the phosphorus does not produce any marked cold-brittleness and, as the iron is not hot-worked, hot-shortness or brittleness caused by the sulfur has no appreciable effect. The average properties of malleable iron are as follows:

Tensile strength.......................................
...................54,000 psi (372,300 kN/m^2)
Yield point...36,000 psi (248,200 kN/m^2)
Elongation, 2'' (5.0 cm)... 15 percent min

291

Metallurgy

Brinell hardness 115
Izod impact strength ....
.... 9.3 ft-lb (135.7 N/m)
Fatigue endurance limit ....
.... 25,000 psi (172,400 kN/m^2)

Modifications in composition and heat treatment may alter these properties, and malleable irons that exhibit much higher mechanical properties are made. One of the modern developments in metallurgy has been the manufacture of high-strength malleable irons known as pearlitic malleable irons.

Alloy Malleable Irons

Some producers of straight malleable iron also manufacture a malleable iron to which they add a small amount of copper and molybdenum. These alloy malleable iron castings have numerous applications since they have a yield point that approximates 45,000 psi (310,300 kN/m^2) and an ultimate strength that often exceeds 60,000 psi (413,700 kN/m^2) accompanied by an elongation in some instances as high as 20 percent or more in 2 inches. These irons are reported as exhibiting excellent machining properties even when the hardness exceeds 200 Brinell, and a very fine surface finish is obtained. Castings for use in valve and pump parts render very good service, showing excellent resistance to wear and corrosion.

The addition of copper to malleable iron apparently accelerates graphitization during annealing treatments and also strengthens the ferrite, while at the same time greatly improving the iron by increasing its susceptibility to heat treatments following the usual annealing cycle. Copper additions from 0.70 to over 2.0 percent will make possible an improvement in the physical properties by a precipitation heat treatment that consists of heating the annealed iron to 1290° to 1330°F (about 695° to 720°C), quenching and then drawing at about 940°F (about 505°C) for three to five hours. The quenching temperature is not high enough to redissolve the graphitic carbon but dissolves much of the copper which is precipitated in a finely divided form by reheating to 940°F (505°C).

At this temperature copper is relatively insoluble and precipitation occurs at a fast rate. The precipitated copper increases the hardness and strength without much loss in ductility. Precipitation hardening, due to the copper precipitation, also occurs in the regular annealing cycle during the malleabilizing treatment. By controlling the original composition of the casting, marked improvement in physical properties is obtained as compared with the straight malleable iron compositions. The structure of malleable iron alloys with copper, or copper-molybdenum malleable irons, consists of nodular graphitic carbon with a matrix of ferrite, in which we find a precipitate of copper.

Pearlitic Malleable Cast Irons

Straight malleable irons have all of their carbon in the graphitic or temper carbon condition, whereas the so-called pearlite malleable iron retains some of its carbon in the combined condition as cementite, similar to steel and gray cast iron. The structure of the pearlitic malleable iron

Cast Irons

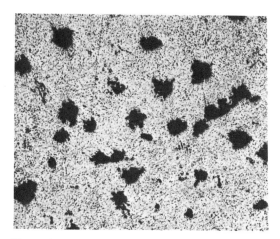

Fig. 13-22. Pearlitic malleable cast iron. Dark nodular temper carbon (graphite). Matrix is fine mixture of ferrite and cementite, polished and etched. Magnified 100 times.

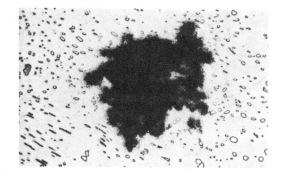

Fig. 13-23. Same as Fig. 13-22. Magnified 750 times. Dark nodular temper carbon (graphite). Matrix consists of cementite particles in ferrite field.

differs from that of the straight malleable iron in that the matrix consists of a pearlite-like structure of ferrite and cementite. Figs. 13-22 and 13-23 illustrate the structure found in this type of iron. It consists of graphite nodules, or temper carbon, with a matrix of spheroidized cementite in ferrite. Pearlitic malleable iron can be made by several different treatments, such as (1) modifying the composition of the original white iron from that used in the manufacture of straight malleable iron, (2) making use of a short annealing cycle, or else (3) subjecting a straight malleable iron casting to a special heat treatment. A brief discussion of these treatments follows.

Modifying composition. By modifying the composition of the white iron, it is possible to retard the action of graphitization during the regular annealing cycle so that some combined carbon is retained in the final product. Careful control of the silicon-carbon ratio and the additions of manganese, molybdenum, and copper are common practice in securing a retention of combined carbon in the annealed iron. A lower silicon and carbon content retards the action of graphitization. Also, copper, manganese, and molybdenum have about one-fourth the graphitizing power of silicon, and addition of these elements retards graphitization during the annealing cycle.

Short-cycle annealing. A common practice in making a pearlitic malleable iron is to employ a shorter annealing cycle than that used when complete graphitization is wanted, as in the making of a straight malleable iron. Manganese additions can be made to the molten metal in the ladle prior to casting as an aid in retarding complete graphitization. A typical analysis for making a pearlitic malleable iron is as follows:

Metallurgy

	PERCENT
Total carbon	2.40
Silicon	0.92
Manganese	0.32
Manganese added to ladle	0.63

This iron is cast to form a white iron similar to that made for straight malleable iron. The addition of manganese to the ladle helps to retard graphitization during the casting cycle and during the annealing cycle that follows. A practice followed for this white iron requires about 30 hours at 1700°F (930°C), followed by cooling to below 1300°F (700°C) in the annealing furnace, and subsequent reheating to 1300°F (700°C) for about 30 hours, followed by cooling to normal temperature.

The total time consumed by this annealing cycle is about 75 hours, as compared with five to seven days for the annealing of a straight malleable iron. The shorter annealing time prevents complete graphitization and the annealing time at the lower temperature, 1300°F (700°C), puts the combined carbon in a spheroidized condition. As a result of this treatment, a cast iron is produced containing a matrix of spheroidized cementite and ferrite in which the nodules of temper carbon are present.

Special heat treatments. It is possible to obtain a pearlitic type of structure in a straight malleable iron by simply reheating a completely graphitized iron to a temperature high enough to give the ferritic matrix an opportunity to dissolve some of the nodular temper carbon, i.e., by heating above the critical temprature of the iron. By controlling the cooling rate from the solution temperature, the dissolved carbon may be precipitated as a coarse spheroidized cementite or as a fine laminated pearlite. Accurate control of the complete cycle is necessary in order to obtain the desired amount of combined carbon and type of structure.

Properties of Pearlitic Malleable Iron

The properties of pearlitic iron depend upon the character of the matrix, i.e., the amount of combined carbon as cementite and the size, shape, and distribution of the cementite particles. In general, pearlitic malleable iron has a higher yield strength and ultimate strength and lower elongation than normal malleable iron. It machines less readily and has a higher Brinell hardness.

Applications include use where strength, rigidity and wear resistance are important factors, such as in gears, sprockets, air tools, brake drums, cams, crankshafts, and wearing pads.

DUCTILE IRONS

Ductile iron is a high strength iron that can be bent. It is also called *nodular iron* or *spherulitic graphite cast iron* (SG iron) because of its microstructure. Ductile iron

Cast Irons

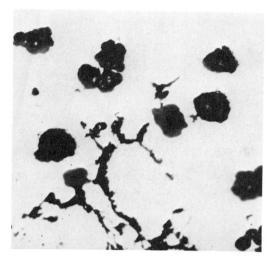

Fig. 13-24. Ductile cast iron partially nodularized. Graphite flakes and nodules in a ferrite matrix, magnified 250 times. Nital etch.

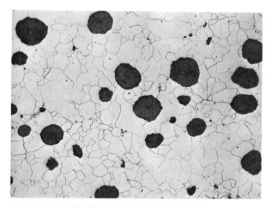

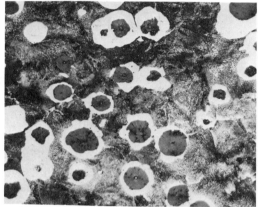

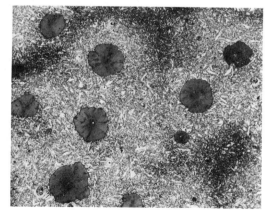

Fig. 13-25. Ferritic ductile cast iron (top). Pearlitic ductile cast iron (middle). Martensitic ductile cast iron (bottom). (International Nickel Co.)

consists of graphite nodules (microscopic spheres) dispersed evenly throughout the cast iron. Fig 13-24 shows a structure which has been partially nodularized and has both graphite flakes and graphite nodules. The graphite, an allotropic form of carbon, in the form of nodules has a minimum influence on the mechanical properties of ductile iron, but in the amount present, it is significant since the ease with which the iron can be melted and cast into complex shapes is dependent upon a relatively high carbon content. Typical structures of ductile irons are shown in Fig. 13-25 and are ferritic, pearlitic and martensitic. This range gives a wide choice of properties for use in industry.

Ductile iron is made by the addition of a

Metallurgy

Fig. 13-26. Typical applications of ductile iron castings. Roller gear for farm implement (top). Hollow automotive shaft (middle). Diesel engine connecting rod (bottom).

small amount of magnesium to the ladle of cast iron. The addition of the magnesium causes a vigorous mixing reaction, resulting in a homogeneous spheroidal structure in the ductile iron casting. Ductile irons combine the processing advantages of cast iron with the engineering advantages of steel. Low melting point, good fluidity and excellent castability are advantages as a casting material while good machinability, wear resistance, high strength, toughness, ductility, hot workability, weldability and hardenability resemble properties of steel. Some typical applications of ductile iron in industry are shown in Fig. 13-26.

WROUGHT IRON

Wrought iron is produced in lots of 300 to 1500 pounds (136 to 680 kg) in what is known as a puddling furnace. This is a furnace somewhat similar to the open hearth, with combustion taking place at one end only. The furnace is lined with iron oxide in the form of mill scale or ore. Cold pig iron is charged into the furnace and as this melts its carbon is eliminated by the iron oxide lining. When practically all of the

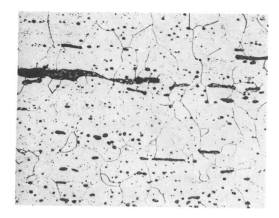

Fig. 13-27. Photomicrograph of wrought iron, taken at 100 magnifications, shows the fibrous slag embedded in a matrix of nearly pure iron.

Fig. 13-28. Wrought iron bars fractured to show the fibrous, hickory-like structure which is characterisitc of the material. (A.M. Byers Co.)

carbon and other impurities have been eliminated, the metal has a higher melting point and begins to form into a pasty mass. This pasty mass of metal is formed into a ball, which is then removed from the furnace. The bulk of the slag is then squeezed from this mass by means of a mechanical squeezer and the iron is then rolled into bars. The material is now a mixture of high purity iron and some slag (about 2 or 3 percent). Figs. 13-27 and 13-28 show fibrous structure of wrought iron. Bars are then cut into lengths and piled, with alternate layers at 90° angles. These stacks of bars are wired, heated to welding heat, and rolled to desired shapes. The purpose of this stacking and rolling is to get a fine and uniform distribution of the slag.

Another process for manufacturing synthetic wrought iron has been developed which yields a material which cannot be distinguished chemically or metallographically from puddled iron, but can be welded somewhat more easily. In this process, pig iron is melted in a cupola and blown to soft steel in a bessemer. The steel is then poured into liquid slag, special precautions being taken to insure thorough mixing. The excess slag is then poured off and squeezed out. After this squeezing, the spongy material is rolled into bars for fabrication into pipe and similar products.

Despite the fact that many different processes have been used in the manufacture of wrought iron, the characteristics and basic principles of metallurgy used in producing it have remained unchanged. The iron silicate or slag was at first considered an undesirable impurity; however, we know now that the slag is responsible for many of the desirable properties of wrought iron, particularly its resistance to fatigue and corrosion.

The principal value of wrought iron lies

Metallurgy

in its ability to resist corrosion and fatigue failure. Its corrosion resistance, directly attributable to the slag fibers, is also a result of the purity of the iron base metal and its freedom from segregated impurities. Because of its softness and ductility, wrought iron finds common use in the manufacture of bolts, pipe, staybolts, tubing, nails, etc.

ALLOY CAST IRONS

A number of high-alloy cast irons have been developed for increased resistance to corrosion and temperature and for other special purposes. These irons may be grouped as follows:
1. Corrosion resistant alloy cast iron
2. Heat resistant alloy cast irons
3. High strength gray irons
4. Special purpose alloy cast iron

Corrosion resistance. These irons are compounded to provide improved resistance to the corrosive action of acids and caustic alkalis. There are three general classifications of these irons, based on the use of silicon, nickel, chromium and copper to enhance corrosion resistance.

High-silicon irons known as *Duriron* or *Durichlor* find wide use in the handling of corrosive acids. By alloying up to 17 percent silicon, a protective film is formed to resist boiling nitric acid and sulfuric acids. However these alloys are inferior to plain gray cast irons in resistance to alkalis. They also have poor mechanical properties, particularly low shock-resistance. They are also hard to machine. The use of this high-silicon alloy is mainly as drain pipes in chemical plants and laboratories.

High-chromium irons with 20 to 35 percent chromium resemble high silicon-irons in the resistance to acids, salt solutions, caustic soda solutions and for general atmospheric exposure. They are used in annealing pots, melting pots for lead and aluminum, conveyor links and other industrial parts subjected to high temperature corrosion. They are also quite shock resistant and are machinable.

High-nickel irons, known commercially as *Ni-Resist* are used in chemical industries as pump housings for pumps handling acids and in the handling of caustics in manufacturing plants. This metal has superior corrosion resistance over 18-8 stainless steels, although not as good as high-silicon irons. While high-nickel irons have a relatively low tensile strength they are tough, machinable and have good foundry properties.

Heat resistance. White cast irons are naturally resistant to growth at continuous high temperatures as they do not contain free graphite that would be oxidized. Silicon and chromium are added to increase resistance to heavy scaling from high temperature. Their alloys form a light oxide surface coating that is impervious to oxidizing atmospheres. High-temperature strength is increased by the addition of molybdenum and nickel. Aluminum is ad-

Cast Irons

ded to reduce both scaling and growth at high temperatures.

Special Purpose Alloy Cast Iron

Cast irons may require special properties, such as controlled expansion or peculiar electrical and magnetic properties. High-nickel alloys of the austenitic type have a coefficient of expansion about 1/5 that of steel. They also have excellent impact characteristics at cryogenic temperatures. These irons also have high electrical resistance and non-magnetic properties. Aluminum additions of 18 to 25 percent in cast iron produce a metal with high electrical resistance while magnetic permeability can be controlled by additions of chromium.

FOUNDRY PRACTICE

The casting of metals to obtain a desired shape is perhaps the oldest metal process known to man. Casting involves pouring molten metal into a prepared mold cavity of the desired shape and allowing the metal to solidify.

Fig. 13-29 shows a schematic of a mold cavity for a casting. The process begins with the making of a wood, metal or plastic pattern of the object to be cast. Metal patterns are used when production runs are high due to the fact that a metal pattern resists the abrasive wear of the moulding sand. Wooden patterns are made first as a master pattern and the permanent pattern in metal is cast from it.

Wooden patterns are used to make molds for casting runs of limited production. Plastic patterns are used for one-of-a-kind castings as the pattern is left in the sand, and melts and gassifies when the hot metal is poured into the mold. This is known as the *lost wax process,* not because the process was lost in the dim past but because the plastic (wax) gets lost when the hot metal contacts it.

Patterns are made slightly oversize to allow for the shrinkage of the cast metal in cooling. The shrinkage of cast iron amounts to about ⅛"; of steel, about ¼". Special shrink rules are used by the pattern makers. These are oversize to the particular shrink desired.

The mold cavity, the clear space in the section shown in Fig. 13-29, is made by the pattern. The pattern is positioned in the molding flask and prepared sand is packed around it. When the pattern is removed from the sand its image is the cavity in the sand. When the castings themselves require cavities in the metal, these cavities are made by cores which are shaped to size in special sand and baked hard. The cores are positioned within the mold in core prints as shown and left in place while the hot metal is poured.

If the mold is used without drying, it is called a *green sand mold.* Or, it may be

Metallurgy

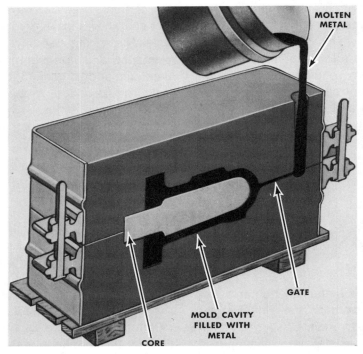

Fig. 13-29. Diagram of a mold cavity for casting.

baked until it has the properties of a soft brick and it is referred to as a *dry sand mold*. Molds for casting iron are washed inside with a graphite paint or dusted with lamp-black in order to form a smooth casting with a minimum of sand sticking to its surface. Molds must be ventilated by making small holes through the sand to allow the gases to escape during pouring.

The molten metal is poured into the mold at the pouring basin, Fig. 13-29, and flows down the sprue to fill the cavity.

When iron is cast into a metal mold, commonly called a *permanent mold,* the surface of the metal is chilled very rapidly and becomes very hard. It will also be white in color because of the absence of graphite. For this reason, such items as crusher rolls are cast in chill molds to produce hard white iron. In the case of castings which are to have certain hard areas, the sand in the mold adjacent to these areas is replaced with metal which causes the metal to harden in these chill spots.

When the casting solidifies, it is pulled from the sand, the cores knocked out, cleaned and annealed. It is then ready for machining.

Cast Irons

Melting Practices

Metals to be cast are usually melted in a cupola. Electric arc and induction furnaces are often used for non-ferrous alloys and high quality stainless steels, tool steels and specialty metals. In the case of the production of extremely high quality metals, the vacuum electric induction furnace is used. This furnace has a controlled atmosphere which eliminates any dissolved gas or oxidization problems.

Two problems that arise in the melting of metals by conventional means are: (1) the absorption of gases into the liquid metal, and (2) the oxidization of the metals in the atmosphere. These problems are alleviated, in the case of iron by the addition of aluminum which deoxidizes the metal. Temperature-control also helps to minimize these problems.

Cast iron is made from pig iron and scrap iron and steel. Alloy irons are usually made from carefully selected scrap steel containing the required alloying elements. Alloys are usually added to the molten iron in the form of ferro-alloys. The fuel used in a cupola is coke and the operation must be carefully controlled to prevent an increase in the carbon content of the melt.

Inoculants. An *inoculant* is an addition made to a melt to alter the grain size or structure of the cast metal. It is usually done late in the melting operation. A very important inoculation treatment for cast iron is the addition of magnesium to produce ductile iron. The addition of as little as 0.04 percent magnesium alters the graphite flakes to produce the microscopic spheres which are found in ductile iron.

REVIEW

1. What is the percentage range of carbon in cast iron?

2. What is the function of silicon in cast iron?

3. Name four common cast iron alloying elements.

4. Name the five main groups of cast irons.

5. Why is *gray iron* so named?

6. What five variables influence the carbon control in the production of cast iron?

7. What is the most important constituent of gray cast iron?

8. What two constituents are found in gray iron but not in steel?

9. What influence has the pearlitic constituent upon the properties of gray iron?

10. Whay type of a structure in a gray iron is considered strongest and yet is easily machined?

11. What influence has the presence of free cementite in a gray iron?

12. Why is phosphorus undesirable in a casting? Under what conditions may it be used?

13. What effect has the rate of cooling upon the structure of a gray iron casting?

Metallurgy

14. For what reasons are alloying elements added to gray iron?

15. How are gray iron castings classified according to the specifications of the American Society for Testing and Materials?

16. Why are gray iron castings less susceptible to notch brittleness than most materials?

17. What treatment is given to gray iron castings to relieve stresses?

18. What outstanding properties has gray iron?

19. How do the hardening and tempering of gray iron castings compare to similar treatments given steel?

20. What type of gray iron responds best to such heat treatments as hardening and tempering?

21. What are the advantages of flame-hardening of gray iron castings?

22. What is *white cast iron*?

23. What is *chilled cast iron*?

24. List the uses or applications of both white and chilled cast irons.

25. Describe the method of making *malleable cast iron*. Name several uses of malleable cast iron in industry.

26. Describe the temper carbon condition found in malleable iron.

27. Describe the average properties of malleable iron.

28. What alloys may be added to malleable irons? Why are they added?

29. What approximate strength and ductility are developed in *pearlitic malleable cast irons*?

30. Describe the common practice used in the annealing cycle for the pearlitic malleable irons.

31. Describe the structure developed in pearlitic malleable iron.

32. What special heat treatments may be employed in order to obtain a pearlitic malleable iron?

33. Describe the microstructure of *ductile iron*.

34. How is ductile iron made?

35. What are some typical applications of ductile iron in industry?

36. Of what importance is slag in *wrought iron*?

37. What is the principle commercial value of wrought iron?

38. Name the four groups of *alloy cast irons* and the industrial uses for each.

39. Describe a pattern as used in casting metals.

40. What three materials are patterns made from?

41. Why are patterns oversize?

42. Define *sprue* and *mold cavity*.

43. Casting metals are usually melted in what common device? Describe the melting process.

44. What is melted to produce cast iron?

45. In what form are most alloys added to molten cast iron?

46. What is an *inoculant*?

CHAPTER 14  WELDING METALLURGY

The joining of two pieces of steel by hammer welding is an ancient process dating back thousands of years, the squeezing and hammering of the pasty iron and slag mixture in the production of primitive wrought iron by hammer welding.

The well-known Damascus swords were produced over a thousand years ago by hammer-welding strips of high-carbon steel to low-carbon steel and then forging the composite metal into a finely laminated metal strip. Medieval armor and weapons were also produced by metal smiths skilled in the art of hammer-welding.

Today, welding constitutes one of the most important procedures of the metal working industry in the fabrication and assembly of metal parts. Through the process of welding, simple metal parts are joined together to form complicated structures.

A weld is defined as, *"a localized coalescense* (melting together) *of metals"*; a *"crystallization into union"*. This joining is usually accomplished by the application of heat, with or without fusion (melting), with or without the addition of filler metal, and with or without the applied pressure.

Fig. 14-1 illustrates the major welding processes in use today. Basically the welding together of two metal parts may be accomplished either by *fusion* (the melting and solidification of metal) or by *pressure* as in forge welding and spot welding. Fusion welds are by far the most common, and many such welds ordinarily do not require the application of pressure. In fusion some of the metal in each part is melted and flows together to form a single, solid piece when cooled.

Welding is a metallurgical process. *Weldability* is defined as the ease with

Metallurgy

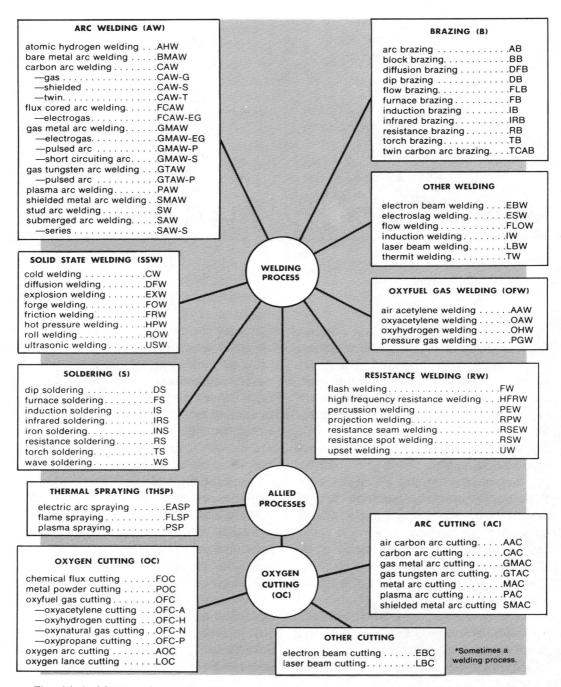

Fig. 14-1. Master chart of welding and allied processes. (American Welding Society)

Welding Metallurgy

which the required degree of joinability and performance can be obtained with a given process and procedure. This involves three conditions: (1) the metallurgical compatibility of the process chosen with the job to be done, (2) the production of a mechanically sound joint, and (3) the serviceability of the joint under special conditions.

Metallurgical compatibility means that the base metal will join with the weld metal without forming unacceptable alloys, undesired grain growth, or changes in chemical composition.

Mechanical soundness relates to the presence of cracks in the base metal and in the weld metal as well as porosity, inclusions, shrinkage and incomplete penetration.

Serviceability means that in addition to meeting the first two conditions the weld must meet physical requirements such as strength, ductility, notch toughness, and corrosion resistance.

Heat is the most essential requirement for the joining of two or more pieces of metal together by welding. In the two most common welding methods (fusion and pressure), the heat energy used to liquefy the metal is secured from an electric arc, a gas flame or electrical resistance. Other types of welding obtain the necessary heat from fuel-fired forges or furnaces, in the case of hammer-welding, or mechanical energy as in friction welding.

In fusion welding the base metals are always melted and, in some cases, extra *filler* metal is added. There are a number of types of fusion welding; these include groups of processes as: arc welding, gas welding, electron beam welding and laser beam welding.

It has been said that fusion welding involves all the principles of metallurgy. In fusion welding, for example, the metal is melted, refined by fluxes, alloyed, and resolidified. All of the defects common to castings may be found in welds, including blow holes, large columnar dendrites, segregation of constituents, and so on.

GAS WELDING (OXY-FUEL) PROCESSES

The process of gas welding involves the burning of a combustible gas in an oxidizing atmosphere. A medium temperature mixture used in welding low-melting point metals is air and acetylene gas. The air and the gas are under moderate pressure when mixed in a hand-held welding torch. The flame is directed to the work surface and obtains fusion of the parts by melting the metals in contact.

A higher heat mixture is obtained by using oxygen and hydrogen gases. These gases are also mixed and burned in a handheld welding torch and used for medium temperature welding operations such as welding aluminum and magnesium, and in brazing and braze-welding.

Extremely high heat is obtained by using a mixture of oxygen and acetylene gas. The flame, which reaches a tempera-

Metallurgy

ture of over 6000°F (3300°C) melts all commercial metals so completely that metals to be joined actually flow together to form a complete bond without the application of any mechanical pressure or hammering. In most instances, some extra metal in the form of a wire rod is added to the molten metal in order to build up the seam for greater strength. On very thin metals the edges are merely flanged and melted together.

The gas welding method utilizes a torch or blowpipe which mixes the fuel gas and the oxygen in the correct proportions and permits the mixture to flow to the end of a tip where it is burned, Fig. 14-2. The appearance of the flame coming from the tip indicates the type of combustion taking place. Variation of the flame characteristics is accomplished by altering the proportions of the gases. In Fig. 14-3 the three types of flames are shown. The neutral flame results from an approximate one-to-one mixture of the gases and is chemically neutral in respect to carbonizing or oxidizing the metal being welded. The neutral flame is the one most used in gas welding operations.

When an excess of oxygen is forced into the gas mixture, the resulting flame is an *oxidizing* flame. This is sometimes used in brazing. With an excess of acetylene gas in the mixture, the flame is *reducing* (carbonizing), its excess carbon will enter into the molten metal. The result will be a pitted and brittle weld.

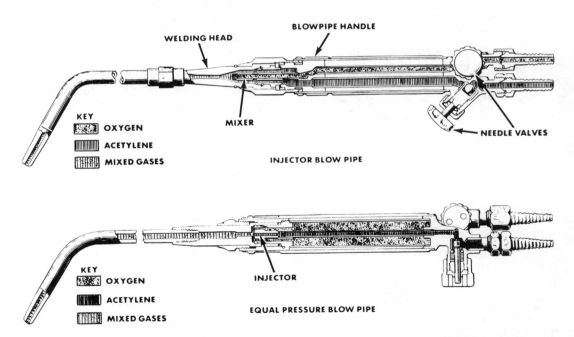

Fig. 14-2. The injector (above) and the equal pressure (below) gas welding torches are the main types in use today. (Linde Co.)

Welding Metallurgy

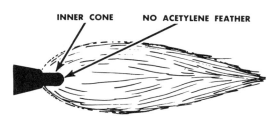

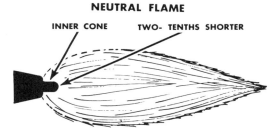

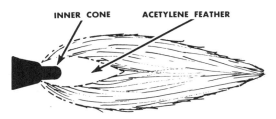

Fig. 14-3. Three types of flame: neutral, oxidizing, and reducing or carbonizing.

ARC AND GAS SHIELDED-ARC PROCESSES

Mentioned earlier in the chapter, carbon arc welding, using either one or two carbon rods, utilizes shielding methods to improve the quality of the weld. Inert gas and/or flux is added by means of the coated filler metal rod. Inert gas and/or flux can also be added by the use of flux-cored rod.

Carbon-Arc Welding

There are a number of types of arc welding. One that is widely used and requires a minimum amount of equipment is carbon arc welding. An electric arc is struck between two electrodes of carbon or between one carbon electrode and the

Metallurgy

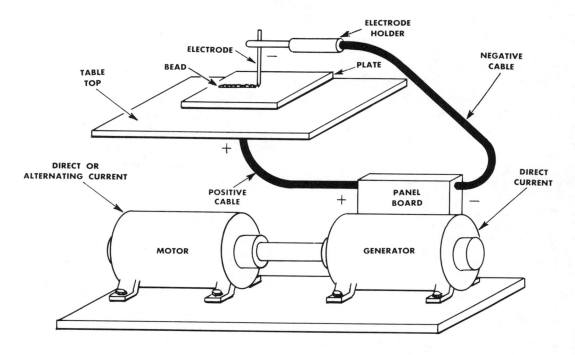

Fig. 14-4. Electric arc welding.

metals which are to be welded. Any additional metal required may be supplied from a filler rod made of the metal being welded which is melted by the intense heat of the electric arc. Fig. 14-4 shows a typical arc welding set-up in which a single electrode is being used. When a metal rod is used as an electrode, it also acts as a filler rod.

The electric arc between the carbon electrodes is characterized by an extremely high and constant temperature. This temperature is approximately 11,000°F (6100°C) as compared to approximately 6,000°F (3300°C) for an oxyacetylene gas flame. Since the arc is struck between the two electrodes, they must be made of permanent material to continuously support the arc, but not melt in the welding operation. When the arc is between an electrode and the metal to be welded, the electrode is either consumable or relatively non-consumable. Rods of carbon or tungsten are used in non-consumable operations, while a wide range of steel alloys with lower melting points than carbon and tungsten are used in consumable (filler) operations.

Gas Shielded Metal-Arc Welding

Early electric arc welding utilized bare metal electrodes. Contamination of the

Welding Metallurgy

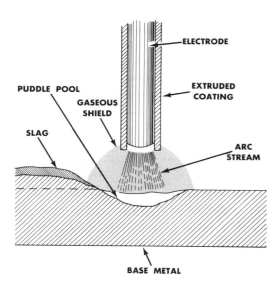

Fig. 14-5. Weld puddle and flux covering of electric arc.

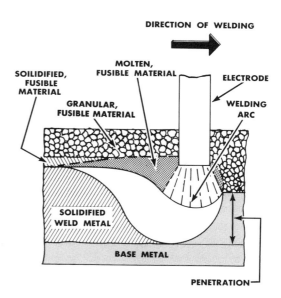

Fig. 14-6. Schematic of arc welder with non-combustible tungsten rod.

weld was a frequent occurrence which produced brittle welds with low tensile strength. This contamination is prevented by the use of shielded electrodes. These special electrodes have a heavy coating of various substances which either melt or release an inert gas. Fig. 14-5 shows a cross-section of a coated electrode in process of welding. Melted coatings, known as *fluxes,* also shield the molten metal from the atmosphere but act as scavengers to reduce oxides, form a slag to float off the impurities and blanket the hot metal to slow the cooling rate.

The specialized gas metal arc welding processes listed in Fig. 14-1 are electro-gas, pulsed arc and short circuiting arc. These are used in specific manufacturing processes that require the enhancement of the thermal qualities of gas welding by the use of super-imposed electric arcs.

Tig welding. The gas tungsten arc welding process is commonly referred to as TIG welding. TIG stands for *Tungsten Inert Gas* and the process is a type of shielded arc where a cloud of inert gas such as argon, helium or nitrogen is placed around the arc. This cloud of inert gas excludes the contaminates in the room air, especially oxygen, and protects the quality of the weld. The TIG welder has a non-combustible rod of tungsten in a special holder that permits the inert gas to be directed directly into the weld area. A refinement of the special pulsed arc gives certain production and quality advantages when incorporated into the TIG method.

Metallurgy

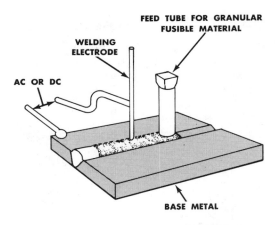

Fig. 14-7. Submerged arc process of protecting the electrode.

Submerged Arc Welding

A second method of protecting the metal while welding is the submerged arc process shown in Figs. 14-6 and 14-7. In this process the welding is shielded by a blanket of granular, fusible material placed on the work. This method is used in automatic, high-production welding and the electrode furnishes the filler metal. The electrode is in the form of heavy wire and is coiled in long lengths and fed into the arc automatically. The shield material is fed from an overhead hopper directly ahead of the electrode, and the unused material is picked up by a vacuum hose and returned to the hopper for re-use.

Plasma Arc Welding

Plasma welding is a process which utilizes a central core of extreme temperature surrounded by a sheath of cool gas.

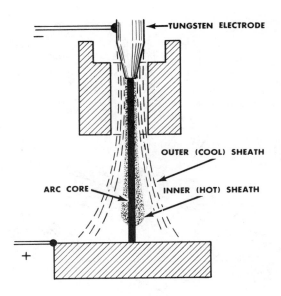

Fig. 14-8. Plasma welding uses a central core of extreme heat surrounded by a sheath of cool gas. (Thermal Dynamics Corp.)

The required heat for fusion is generated by an electric arc which has been highly intensified by the injection of a gas into the arc stream. The superheated columnar arc is concentrated into a narrow stream and when directed onto metal makes possible butt-welds to one-half inch or more in thickness in a single pass without filler rods or edge preparation. See Fig. 14-8.

In some respects, plasma welding may be considered as an extension of the conventional gas tungsten arc welding. The main difference is that the plasma arc column is constricted and it is this constriction that produces the much higher current density. The arc gas, upon striking the metal, cuts or keyholes entirely

Welding Metallurgy

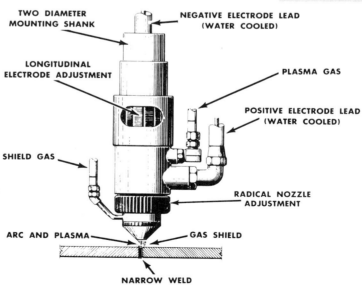

Fig. 14-9. Torch for plasma welding. (Thermal Dynamics Corp.)

Metallurgy

through the piece, producing the small hole which is carried along the weld seam. During this cutting action, the melted metal in front of the arc flows around the arc column, then is drawn together immediately behind the hole by surface tension forces and reforms in the weld bead.

The specially designed torch, Fig. 14-9, for plasma welding can be hand-held or machine-mounted for automated applications. The process can be used to weld stainless steels, carbon steels, Monel metal, Inconel, titanium, aluminum, copper and brass alloys. Although for many fusion welds no filler rod is needed, a continuous filler wire feed mechanism can be added for various fillet types of weld joints.

Mig Welding

Metal inert gas welding is similar to TIG welding. Regular weld metal is used for the electrode instead of tungsten. It melts in the arc and adds filler metal to the weld. The MIG process is widely used in automatic welding equipment.

Other Welding Processes

Due to different production requirements, many different and exotic methods of welding exist. They are listed under the heading *Other Cutting* in Fig. 14-1. The following methods are most used in certain industries.

Electron Beam Welding

Electron beam welding is essentially a fusion process. Fusion is achieved by focusing a high power-density beam of electrons on the area to be joined. Upon striking the metal, the kinetic energy of the high velocity electrons changes to thermal energy, causing the metal to heat and fuse. The electrons are emitted from a tungsten filament heated to approximately 2000°F (1100°C). Since the filament would quickly oxidize at this temperature the welding must be done in a vacuum chamber to avoid oxidation.

Electron beam welding can be used to join materials ranging from thin foil to

Fig. 14-10. Electron beam welding a heat-treated gear cluster. (Sciaki Bros.)

about 2" (5.1 cm) thick. It is particularly adaptable to the welding of refractory metals such as tungsten or molybdenum, and metals which oxidize readily, such as titanium and beryllium. It is also used in joining dissimilar metals, aluminum, stan-

Welding Metallurgy

dard steels, and ceramics. Fig. 14-10 shows a schematic diagram of an electron beam welding machine.

Laser Welding

Laser welding is like welding with a white-hot needle. Fusion is achieved by directing a highly concentrated beam to a spot about the diameter of a human hair. The highly concentrated beam generates a power intensity of one billion or more watts per square centimeter at its point of focus. Because of its excellent control of heat input, the laser can fuse metal next to glass and even weld near varnished coated wires without damaging the insulating properties of the varnish.

Since the heat input to the workpiece is extremely small in comparison to other welding processes, the size of the heat affected zone and the thermal damage to the adjacent parts of the weld are minimized. Thus it is possible to weld heat-treated alloys without affecting their heat-treated condition. A weldment can be held in the hand immediately after welding.

The laser can be used to join dissimilar metals and other difficult-to-weld metals such as copper, nickel, tungsten, aluminum, stainless steel, titanium and columbium. Furthermore, the laser beam can pass through transparent substances without affecting them. This makes it possible to weld metals that are sealed in glass or plastic. Because of the fact that the heat source is a light beam, the effects of atmospheric contamination of the weld joint is not a problem. The current application of laser welding is largely in aero-

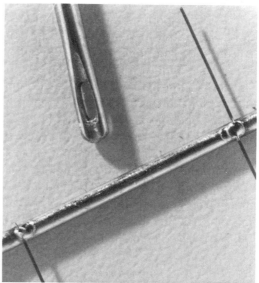

Fig. 14-11. (top) Laser beam vaporizing a 0.020" dia hole in tungsten at 10,700°F (about 5925°C) in 1/1000 sec. (bottom) Laser beam welds a 0.0003" tungsten and a 0.020" nickel wire in microminiature electronic manufacturing. (Hughes Aircraft Co.)

Metallurgy

space and the electronic industries where extreme control in weldments is required. Its major limitation is the shallow penetration. Present day equipment restricts it to metals not over 0.020 inches (0.50 mm) in thickness. Fig. 14-11 shows applications of the laser beam in microelectronic, miniature welding.

SOLID STATE WELDING

Pressure welding is a metallurgical joining process in which the metals are brought into close contact by mechanical pressure. There are a number of special types of pressure welding processes, some of which are related to special production methods and metals. Cold welding, diffusion welding and explosive welding are some of these special pressure methods. Most production pressure welding operations, however, involve the generation of heat. Heat can be generated by heating the metals in a flame as in the ancient forge method used by smiths over the centuries.

Friction Welding

A welding process using mechanical energy as a source of heat is inertia or friction welding. Friction welding is shown in Fig. 14-12 and requires axial pressures in the range of 1500 psi (10,300 kN/m^2)[1] and a speed of 1500 rpm to weld a one-inch diameter steel bar. A stainless steel bar will require 12,000 psi (82,740 kN/m^2) at 3000 rpm. Fig. 14-13 shows typical examples of parts welded by inertia welding.

The bonding temperatures in pressure welding may be above the temperature of recrystallization as in the various types of electrical resistance welding. Friction welding employs a bonded temperature ½ to ⅔ of the melting temperatures of the metals being joined. However, in the ultrasonic welding method, the temperatures may be as low as 35 percent of the melting temperature.

Ultrasonic Welding

Fig. 14-14 shows a schematic diagram of ultrasonic equipment. A transducer vibrates the welding tip horizontally while pressure is applied vertically by an anvil. The vibratory rate is 20 to 40 kilocycles per second and the motion disturbs the oxide film of the metal plates being welded. This places the metal surfaces in intimate contact and causes the weld to take place as the friction between the surfaces heats the metal.

---

[1] Since expressions of psi (pounds per square inch) in units of tens and hundreds of thousands are not absolute quantities, the conversion units of kN/m^2 (kilo-Newtons per meter squared) have been rounded off appropriately.

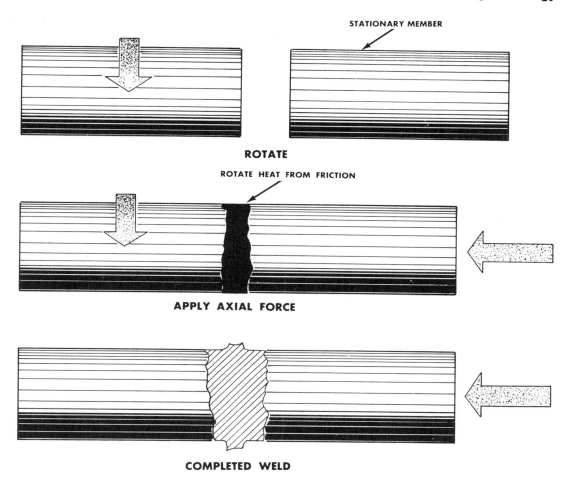

Fig. 14-12. Inertia welding. (top) Schematic of principle involved. (bottom) Photomicrograph of aluminum bronze welded to carbon steel, magnified 75 times. Interface temperatures were below the melting points of either face, avoiding shrinkage, gas porosity, and voids.

Metallurgy

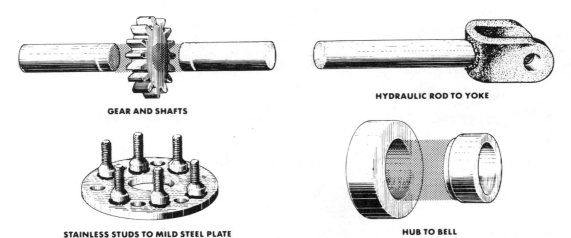

Fig. 14-13. Typical example of parts welded by inertia welding. (Caterpillar Tractor Co.)

Fig. 14-14. Ultrasonic welding. Schematic drawing of set-up. Transducer vibrates welding tip at 20 to 40 kilocycles. Moving tip disturbs oxide film, placing metals in close contact and causing them to weld.

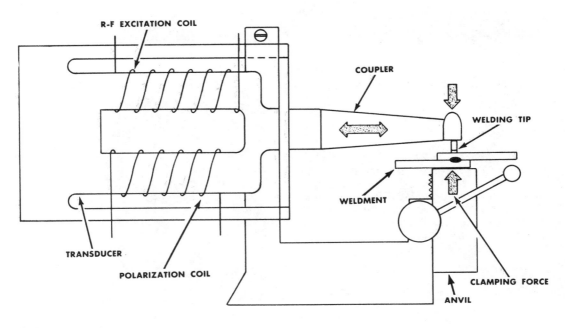

316

RESISTANCE WELDING

Another type of pressure welding that should be treated as a separate field is resistance welding. In this type of welding, the operation is carried out in specially designed welding machines which utilize electrical energy to produce the heat of fusion. In resistance welding the parts are hand-held or positioned in fixtures that are part of the machine.

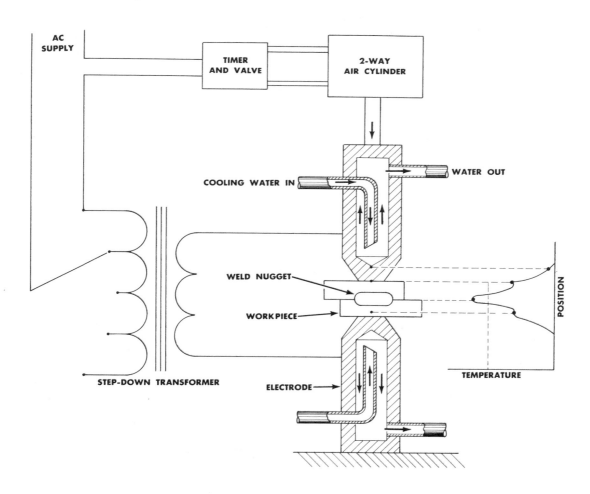

Fig. 14-15. Resistance welding, schematic of equipment. (Sciaki Bros.)

Metallurgy

The heat of fusion is obtained in the weld area from the resistance of the metal being welded to the passage of a low voltage-high amperage electric current. This current is generated by special transformers that are an integral part of the welding machine. Resistance welding devices include: spot welders, butt or upset welders, seam welders, projection welders, stud welders, and flash welders. Fig. 14-15 is a schematic drawing of the spot welder equipment. The two pieces to be spot-welded together are placed between the welding points (electrodes). The electrodes are water-cooled to protect them from the high heat involved. Pressure on the electrodes varies from 4000 to 8000 psi (27,580 to 55,160 kN/m^2). The electrical current is from a low-voltage, high amperage transformer whose secondary terminals are connected to the electrodes with heavy, flexible copper alloy strips.

Operation of spot welder consists of four cycles: (1) pressure applied by the cylinder, (2) momentary passage of electrical current through the closed electrodes, (3) cooling hold, and (4) release. The electrical current is controlled by an electric timer controlled by a pressure switch in the hydraulic circuit. This current passing through the high resistance of the metal pieces between the electrodes, causes a small quantity of the metal to melt at the point of pressure at the juncture of the metals.

The timing of the current flow is critical as it must be long enough to melt a small amount of metal but not too long to melt

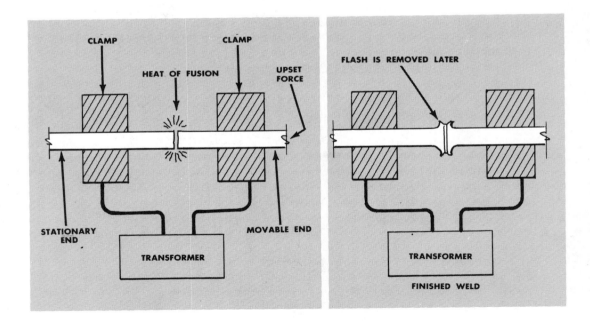

Fig. 14-16. Butt or upset welding.

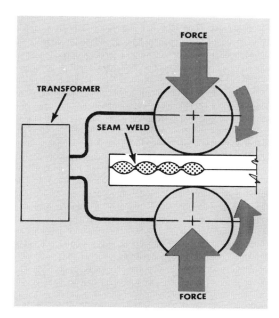

Fig. 14-17. Seam welding.

completely through the metal. When the current flow ceases, the molten metal cools and solidifies to form a *button.* The pressure-hold portion of the weld cycle is timed to insure complete solidification of the button before pressure is released and another weld is made. The water-cooling in the electrodes removes heat from the copper alloy points to prevent softening, peening and wear.

Butt welders, Fig. 14-16, also called *upset welders* are used to join bars end-to-end. Extensions to drills are also welded in this type of machine. The basic four cycles of operation are used, except that in the third cycle a cam-driven mechanism rams the hot plastic ends of the two bars together, thus upsetting or forging the weld. The excess metal around the upset called the *flash* is later removed.

Seam welders, Fig. 14-17 are specialized spot welders using rotating wheel electrodes which pinch the two pieces of metals and weld a continuous seam of overlapping spot welds.

In *projection welding,* one of the pieces to be welded, such as the tapping plate in Fig. 14-18, has projections punched up in a prior operation in a punch press. The current flow is directed through these projections when the tapping plate is forced against the other piece of metal. The electrode plates are flat, hard bronze, water-cooled as were the spot-welder points. When the electric current melts the projections, the pressure on the electrode plates force the metal pieces together and a weld is formed. Fig. 14-19 shows the before and after conditions in a projection welding operation. The dark areas repre-

Metallurgy

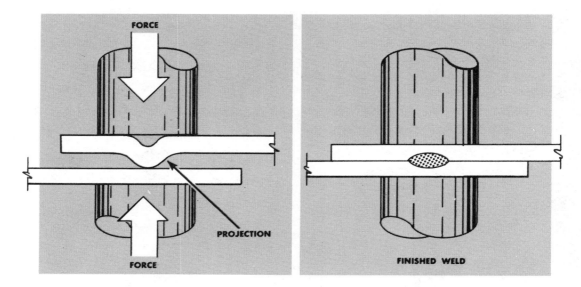

Fig. 14-18. Projection welding.

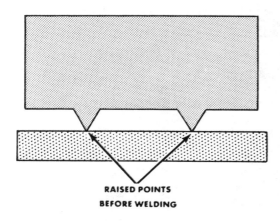

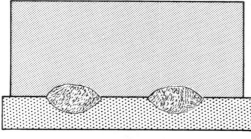

Fig. 14-19. In projection welding, the weld area has been preformed with raised points.

sent molten metal which, when they solidify, firmly hold the weld together. These areas are referred to as *buttons* and a good weld, when pulled apart, will result in a button pulling a hole in one of the metal sheets. This is a common test of production welding in the manufacture of automobile bodies and is referred to as "pulling a button".

Flash Welding

Flash welders, as shown in Fig. 14-20, are a special type of butt or upset welder used for welding thin sheets of steel, end-to-end. The thin sheets, such as 18 or 20 gage (1.21 to 0.91 mm) steel, are positioned with their edges not quite touching and then clamped in place. One clamp, made of conducting metal such as bronze, is stationery while the other clamp is on a moveable slide. As the moveable clamp moves toward the stationary clamp the current is turned on and as the irregular edges of the two pieces of metal touch, the resistance is high and the metal melts rapidly, flashing off in sparks. The movement continues by cam action until a predetermined burn has heated both edges of the metal pieces uniformly. Then the cam upsets the pieces, forcing them to butt together as the current is turned off. The plastic metal is butt-welded and the clamps are released.

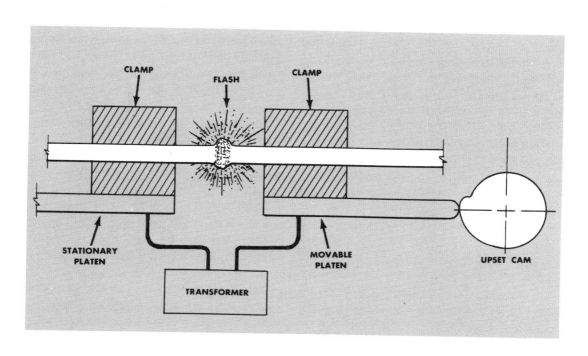

Fig. 14-20. Flash welding.

Metallurgy

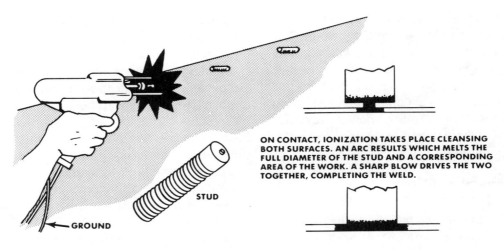

Fig. 14-21. Stud welding, Graham method. (Republic Steel Corp.)

Stud Welding

A specialized form of projection welding known as stud welding is used to fasten a stud vertically to a surface by a resistance welding method. As shown in Fig. 14-21, the stud has a projection formed on its face. The diameter and length of the projection vary with the diameter of the stud and the material being welded. The welding gun is air operated. A collet holds the stud, which is attached to the end of the piston rod. Constant air pressure holds the stud away from the metal until the time of the weld. The piston then drives the stud against the flat surface and a high amperage-low voltage electrical discharge takes place. The flow of current creates an arc which melts the entire area of the stud-end and the corresponding area of the flat surface. A weld is completed with little heat penetration, no distortion and practically no fillet build-up.

BRAZING

Ferrous and non-ferrous metals may be joined by lower temperature methods than those occurring in fusion welding. Brazing is an ancient joining method that takes its name from the original copper alloys used in past ages, brass and bronze. Heat for brazing may be obtained in many ways. The brazing methods listed in Fig.

Welding Metallurgy

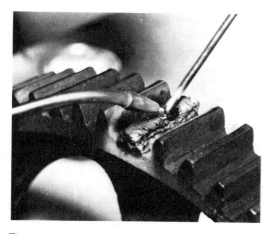

Fig. 14-22. Building up a missing gear tooth with a bronze weld. (Linde Co.)

14-1 are basically an enumeration of the particular methods of applying the necessary heat. Brazing temperatures are related to the melting point of the metal being joined and using a non-ferrous filler metal having a melting point below that of the base metal. Brazing temperatures from gas torches are usually over 800°F (430°C). The filler metal is distributed between the closely fitted surfaces of the joint by capillary action. Silver soldering is a typical example of such a brazing operation. Most silver solders are copper-zinc brazing alloys with the addition of silver in percentages from 9 to 80 percent. A typical silver solder used in industry contains 65 percent silver, 20 percent copper and 15 percent zinc and has a melting point of 1280°F (690°C).

Bronze welding is often referred to as brazing but it varies from brazing in that the base metal is not melted but simply brought up to what is known as a *tinning temperature* (dull red color) and a bead deposited over the seam with a bronze filler rod. Bronze welding is particularly adaptable for repairs to cast iron heating boilers, malleable iron parts and the joining of various dissimilar metals (cast iron and steel, etc.). See Fig. 14-22.

An essential factor for bronze welding is a clean metal surface. Adhesion of the molten bronze to the base metal will take place only if the surface is chemically clear. Even after a surface has been thoroughly cleaned certain oxides may still be present on the metal surfaces. These oxides can only be removed by means of a good flux. One of the most common fluxes in use is a white, colorless crystalline mineral called *borax,* a hydrous sodium borate.

The flux is applied by dipping the heated rod into the powdered flux. The flux adheres to the surface of the rod and can be transferred to the weld. Some rods come prefluxed. The torch flame must be adjusted to be slightly oxidizing and applied to the base piece until the metal begins to turn red. Then a little of the rod is melted onto the hot surface and allowed to spread along the entire crack or seam. The flow of this thin film of bronze is known as the *tinning* operation. Once the base metal is tinned sufficiently the filler rod is melted to fill the crack or complete the bead.

There are many advantages to bronze welding. Since the base metal does not have to be heated to a molten condition, there is less possibility of damage to the base metal. Equally important is the elimination of stored-up stresses, so common to fusion welding. One precaution which

Metallurgy

must be considered in bronze welding is not to weld a metal that will be subjected to a high temperature in later use since bronze loses its strength when heated to 500°F (260°C) or above. In addition, bronze welding should not be used on steel parts that must withstand unusually high stresses.

SOLDERING

Soldering, a joining process using lower melting alloys, is similar to brazing but is done at temperatures below 800°F (430°C). Solder is an alloy of two or more metals that has a melting point lower than the metals being joined and with an affinity for, or be capable of uniting with the metals to be joined. Soft solders are used for joining tin plate and other metal sheets. A 90-tin, 10-lead solder melts at about 410°F (210°C) and a 70-tin, 30-lead alloy at 384°F (196°C). Plumbers solder consists of 2 parts lead and 1 part tin. Common solder, called half-and-half, consists of equal parts of lead and tin and melts at 360°F (182°C). Tin added to lead lowers the melting point of the lead until at 356°F (180°C) or 68 percent tin, the melting point rises with an increase in tin content until the melting point of pure tin is reached, 450°F

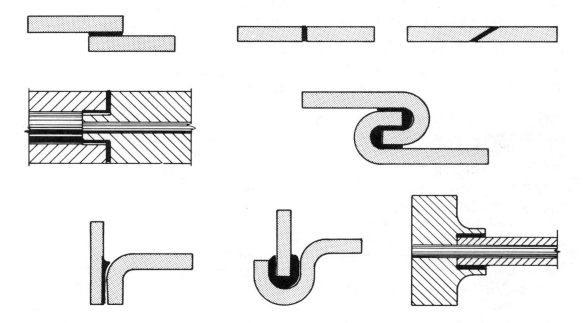

Fig. 14-23. Configuration of brazed and soldered joints.

(232°C). Hard solders may be any solder with a melting point above that of the tin-lead solders; more specifically they are the brazing solders, silver solders or aluminum solders applied with a brazing torch.

The soldering methods listed in Fig. 14-1 list the various means of applying the necessary heat. Each method is related to a particular requirement of the production process or to the metals used. Soldering of mechanical assemblies is usually done using a low-temperature gas torch. Soldering of electrical components is usually done with an electrical resistance *soldering iron* and the solder used should have at least 40 percent tin as the electrical conductivity of lead is only about half that of tin. Wiping solders are used in the automobile industry for the external finishing of joints or metal blemishes to give a smooth surface for painting. These solders, also known as Tinman's solder, contain one part of tin and two of lead.

High production soldering and brazing are done in furnaces which are generally conveyorized. Induction heating and resistance heating are used in some cases to provide high speed and localized heat in automated processes. A molten bath of the filler metal is used in the dip-soldering process. The bath provides both the required heating medium to heat the parent metal and also the filler metal in a molten condition. Modern copper plumbing utilizes a unique joining method in that each fitting has a solder hole into which wire solder is melted after the joint is assembled and heated. The solder fills the joint clearance which is held to 0.0001" to 0.005" (0.025 to 0.127mm). Fig. 14-23 shows a few of the many configurations of brazed and soldered joints.

EFFECTS OF WELDING

Fusion welding is a complex metallurgical process resulting in an end-product, the weld, which may not be acceptable for a number of reasons. As the metal is melted, unwanted oxides and gases may contaminate the molten metal. The high temperature of the weld affects the metal adjacent to the weld as it is heated and cooled thus undergoing a heat treatment. This may harden or soften the metal. These effects are almost as true of high-heat friction methods as of fusion methods.

The solidification of the weld metal is illustrated in Fig. 14-24. The grain growth begins at the edges of the solid metal as the heat of the molten weld metal flows out into the solid metal. Grain growth is columnar dendritic along the length of the weld. Cracking may occur during the final stages of solidification if excessive shrinkage occurs. Cracking can be minimized by careful pre-heating and controlled cooling of the weld area.

The metal on either side of the weld, being heated close to its melting point, is

Metallurgy

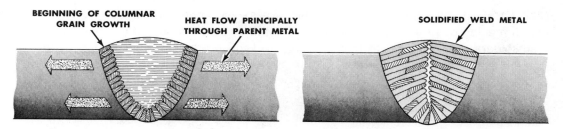

Fig. 14-24. Solidification in single-pass arc weld: (left) Nucleation and columnar grain growth; (right) Solidified weld metal.

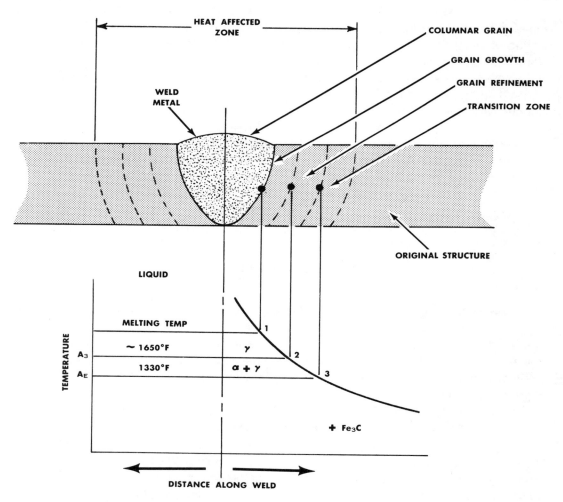

Fig. 14-25. Heat affected zone around an arc weld in low-carbon steel.

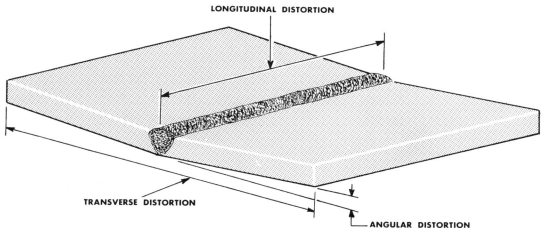

Fig. 14-26. Distortion encountered in butt-welded plates.

subject to grain growth. Coarse grain metal is less desirable than fine-grained structures as it is usually harder and more brittle. Fig. 14-25 shows the heat affected zone around an arc weld made in low-carbon steel.

A shop truism states that *"a good weld is stronger than the metal being welded".* The reason for this observation is that the limiting feature in most welds is not the weld filler metal but rather the parent metal. The analysis of the parent metal, its thickness and its reaction to the heat of welding, determines the efficiency of the final weld.

When fusion welding is employed to form a rigid structure such as a machine tool housing, reaction stresses occur due to the resistance to thermal expansion and intraction by the structure.

Types of distortion that are to be resisted by the total structure are illustrated in Fig. 14-26. These stresses must be removed from the weldment before machining to prevent distortion of the finished housing. Welding stresses are usually reduced by a process called *stress annealing.* In this process, a steel weldment is placed in an oven or furnace and heated to a temperature of 1200°F (about 650°C) held at this heat for an hour or more to become uniformly heated throughout and then slowly cooled to room temperature.

REVIEW

1. What is the original method of welding?
2. Define a weld.
3. What is the most common welding process used?
4. What three conditions must be met in welding?
5. What is the most essential requirement in welding?
6. What are the two main groups of welding processes?
7. What metallurgical processes are involved in welding?
8. What problems can arise in fusion welding?
9. What two types of electrodes are used in arc welding?
10. What is the temperature of an electric arc?
11. Why are shielded arcs used?
12. Describe a TIG welder.
13. What are specific advantages in the plasma welding of butt joints?
14. Describe the MIG process.
15. Why is electron beam welding done in a vacuum?
16. Name some operational advantages of laser welding.
17. Define resistance welding.
18. Name six types of resistance welders.
19. Friction welding develops temperatures much less than the melting temperature of the base metals. Why is this advantageous?
20. Name the four operational cycles of a spot welder.
21. Why are spot welding electrodes water-cooled?
22. What is the difference between brazing and bronze welding?
23. What is the most common brazing flux?
24. What is the composition of a typical silver solder?
25. What is the composition of Plumber's solder?
26. What dual purpose does a dip solder bath provide?
27. Why is "a good weld stronger than the metal being welded"?
28. How are structural distortions from welding corrected?

CHAPTER 15

POWDER METALLURGY

Powder metallurgy is the process whereby metallic shapes are manufactured from metallic powders. The process involves the production of metallic powders and the subsequent welding of these powders into a solid of the required shape. Although at the present time this process is making no large tonnage of metallic shapes, compared with the tonnage produced by the melting, casting, and forging methods, the tonnage of shapes being produced from metal powders is increasing every year.

The powder metallurgy process is limited to the manufacture of shapes of a few pounds in weight, but some 1600 years ago the famous iron pillar in Delhi, India, weighing 6½ tons (5900 kg), was made from iron particles of sponge iron similar to the iron used in this modern process. Ultimately, the powder metallurgy process may become as important to the metalworking industry as the die-casting method. The powder metallurgy process has definite limitations and many difficulties to overcome, which distinctly limit its application at the present time.

The modern metallurgist cannot claim credit for the discovery of the powder-metallurgy process. Metallic powders and solids from metallic powders have been made for many years. Powder metallurgy was used in Europe at the end of the 18th century for working platinum into useful forms. Platinum was infusible at that time. However, it is known that the early Incas in Ecuador manufactured shapes from platinum powders by a similar process long before Columbus made his famous voyage. Wollaston in 1829 developed a technique that proved very successful in the manufacture of a malleable platinum from

Metallurgy

platinum powder. This allowed forging of the resultant solid like any other metal.

The first modern application of the powder metallurgy process was the making of filaments for incandescent lamps. The first metal filament was made from metallic osmium powder, produced by mixing osmium or its oxide with a reducing material that also served as a binding material. The mixture was extruded or pressed to form a filament, which was heated to reduce the oxide and then sintered into a solid form. Similar techniques were developed using tungsten, vanadium, zirconium, tantalum, and other metals. The first successful metal filament was made from tantalum, but with the discovery by Coolidge that tungsten sintered from tungsten powder could be worked within a certain temperature range and then retain its ductility at room temperature, tungsten became the most important filament material.

METAL POWDERS

Metal powders are now available from many of the common metals and some alloys. Metals and alloys such as aluminum, antimony, brass, bronze, cadmium, cobalt, columbium, copper, gold, iron, lead, manganese, molybdenum, nickel, palladium, platinum, silicon, silver, tantalum, tin, titanium, tungsten, vanadium, zinc, carbides, and boron, have been successfully produced in powder form.

The importance of metal powders and of shapes produced from the powders is great when you consider the role played by this process in the illumination field in making filaments for lamps, or the tiny contact points made by this process for use in relays, etc. Also, applications of this process in the manufacture of the oilless bearings and in the manufacture of metallic shapes used in machines and tools make this process very interesting to the average individual connected with the metal-working industry and to the layman who comes in contact with metals in every aspect of his everyday life.

Methods of Making Metal Powders

Metal powders may be considered as raw materials for the fabrication of metal shapes and not as end-products in themselves. The powders are usually made with some idea as to the requirements of the applications in which they will be used. Most of the metal powders are tailor-made to suit the requirements of each application. Due to the many requirements demanded of the metal powders, varied means of manufacture are necessary.

The metal powders may be distinguished from one another and classified by a study of the following characteristics: (a) particle size, (b) particle shape, (c) surface profile, (d) solid, porous, or spongy nature, (e) internal grain size, (f) lattice distortion within each particle, and (g) the impurities present and their location, and whether or not they are in solid solution or exist as large inclusions or as surface and grain-boundary films.

The shape of metallic particles is such as to require study to determine whether or not the shape is angular, dendritic, or fernlike, ragged and irregular, or smooth and rounded. It is obvious that such a study of all these factors relating to metal powder is relatively arduous and requires skill and special techniques. The manufacturer of metallic powders tries to control all of these characteristics and improve techniques so as to manufacture a uniform product in each batch. The methods used in the manufacture of metallic powders include grinding, machining, stamping, shotting, granulation, atomizing, vapor condensation, dissolving constituent, calcium hydride, chemical precipitation, carbonyl process, fused electrolysis, oxide reduction, and electrolytic deposition.

The *mechanical methods*, such as machining and milling, used in the manufacture of metal powders make a relatively coarse powder. The cost is usually high, and the powders made by these methods are usually treated to remove the cold work-hardening received in the process.

In the *shotting process* a rather coarse particle is made by pouring molten metal through a screen or fine orifices and allowing the shot to fall into water. Sometimes this process is used as a breakdown step in the production of a finer powder.

The *granulation process* depends upon the formation of an oxide film on individual particles when a bath of metal is stirred in contact with air. The molten metal is stirred vigorously while cooling, and, as it passes through a mushy state, the metal is granulated by the oxide films which form on the surfaces of the particles, preventing them from coalescing. This method produces a relatively coarse powder with a high percentage of oxide.

In the *atomization process*, molten metal is forced through a small orifice and broken up by a stream of compressed air, steam, or an inert gas. Fine powders may be made by this process, but it requires special nozzles and careful control of temperature, pressure, and the temperature of the atomizing gas. Although oxidation can be prevented by use of an inert gas, little oxidation occurs in this process if air or steam is used. Apparently, the particles oxidize and form a thin protective coating upon the surfaces, thus preventing excessive oxidation from occurring. This process may be used for such metals as tin, zinc, cadmium, aluminum, and other metals having a relatively low melting point. This process allows the production of metal powders with controlled fineness and with a uniform particle size.

Metal powders may also be made by the *vapor condensation process*. In this process, the metal is heated to the vapor state, which, upon cooling, is condensed to a solid powder state. Slight oxide films prevent coalescence of the vapor into a massive liquid state upon cooling. Zinc dust is made in large quantities by this process.

In the *carbonyl process*, metal powders are made by the formation of a carbonyl gas. Carbonyls are produced by passing a carbon monoxide gas over a sponge metal at suitable temperatures and pressures. The carbonyl formed may be a gas or a liquid at room temperature. Upon heating, these carbonyls decompose to form a metal and carbon monoxide gas. The metal formed by this reaction is a very fine dust with the particle size not over a few microns in diameter. Iron and nickel powders may be made by this process.

In the *oxide reduction process* for making a metal powder, the metal is usually in an oxide form, such as iron oxide, or ores of iron such as magnetite and hematite. The oxide is ground to the desired fineness and then reduced to the metallic state by passing a reducing gas over the heated oxide powder. Gases such as hydrogen and carbon monoxide have been successfully used for the reduction process. The powder formed by this process has particles of a spongelike nature, and it is ideal for cold pressing due to its softness and plasticity.

Sponge iron is also made by heating iron ores in contact with charcoal at relatively low temperatures, similar to the process that was used in the production of the irons by early man. The most common metal powders produced by the oxide reduction process include powders of iron, tungsten, copper, nickel, cobalt and molybdenum.

In the *electrolytic deposition process*, metal powders are made by depositing the metal, as in an electroplating process. By proper choice of electrolyte, regulation of the process as to temperature, current density, circulation of the electrolyte, and proper removal of the deposited metal at the cathode in the bath, a metal powder may be produced that is very pure and free from oxides. The powder produced by this method is dendritic or fernlike in particle shape and is a powder of low apparent density. The electrolytic powder is quite resistant to oxidation, and, upon storage, retains this characteristic until it has been pressed and sintered.

Powders made by the *chemical precipitation process* are tin, silver, selenium, and tellurium. In this process, the metal powder to be made is precipitated from a solution by another metal. Tin powder can be obtained from a stannous chloride solution, while zinc and silver can be obtained from a nitrate solution by the addition of either copper or iron.

Production of alloy powders. Alloy metal powder is usually made by mixing metal powders to form an alloy and then heating the mixed powders in a reducing atmosphere to some temperature below the melting point of either powder. Upon heating, diffusion of one metal into the other occurs. The degree of diffusion and uniformity of the alloying depends upon the metals employed and the treatment they receive. Complete diffusion rarely takes place, and therefore only partial alloying occurs. Fig. 15-1 illustrates the microstructure of a 90 percent copper-10 percent tin powder mixture that was pressed at 40 tons/in^2 (551,600 kN/m^2) and sintered at 1950°F (1065°C). Very little diffusion occurred between the copper (dark) and the tin (light). To obtain a more homogeneous structure, pre-alloyed powder should be used. Diffusion is aided by the heating that is carried out following briquetting. If the sintering temperature is relatively high and the time cycle long, diffusion is given an opportunity to greatly increase the alloying effect.

If mixed metal powders are to be alloyed before pressing or briquetting, cak-

Powder Metallurgy

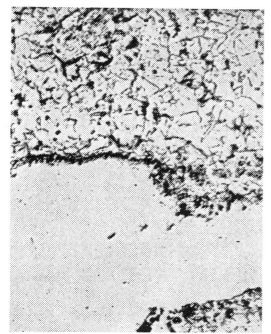

Fig. 15-1. Pressed and sintered alloy powder of 90 percent Cu and 10 percent Sn, pressed at 40 tsi (551,600 kN/m^2) and sintered at 1950°F (1066°C). Shows little or no diffusion between the copper and tin. Copper is dark and tin is light. Magnified 250 times.

ing of the mixed powders may take place during the diffusion cycle, especially if the temperature is in excess of the melting point of one of the constituent metals. The caked metal may be subsequently ground to a powder, but this may produce some work-hardening and require another annealing heating so as to produce a plastic alloy that will press easily. Alloyed powders are harder than the pure metals and consequently are harder to press or compact into a briquette.

Characteristics of Metal Powder

The characteristics of any metal powder should receive close study because to develop intelligent specifications for any given metal powder one must understand what makes the powder behave as it does. At the present time, there are no hard and fast specifications for metal powders, as most of the specifications that are being used by the different manufacturers have been developed by purely empirical means. Some of the factors that are important to the behavior of any powder will be reviewed briefly.

Chemical composition. The chemical composition of metal powders is not as important as their physical and mechanical properties. However, composition and impurities will greatly influence the characteristics of any powder. The most important factor is the amount of oxygen that may be found in the powder as oxides. Oxides may be present in the powder from the method of refining and may form on the surfaces of powder that is exposed to air, as in storage or handling. Oxides may greatly weaken the final structure or product.

Thin oxide films that form during storage or handling may have little detrimental effect because the powder may be cold weldable and the thin films can be reduced by the atmosphere used during the sintering operation. However, it may be found that any oxide film will reduce the ability of the powder to cold-press into a satisfactory shape; accordingly cracks

Metallurgy

and ruptures will occur in the cold-pressed shape. In iron powders, the presence of carbon (as cementite, Fe_3C), silica, sulfur, and manganese may greatly influence the plasticity of the iron powders, making them difficult to cold press. Carbon in the form of graphite may be desirable from the viewpoint of lubrication during cold pressing, and its presence during sintering may result in changing the iron to a product similar to steel. Graphitic carbon is often added to iron powders for the purpose of making a harder iron similar to steel.

Physical structure of the powders. The structure of the powder greatly influences such characteristics as plasticity and ability to be cold-pressed or made into briquettes. It influences pressures required in pressing, flow characteristics, and the strength of the final product. The study of the powder will reveal whether or not the powder is angular in shape, solid, porous and spongy, crystalline, dendritic or fern-like, or ragged and irregular in shape. The type of structure that is desired in any given powder is usually determined from experience by the metallurgist.

Flow characteristics. Powders will fill the die cavities by gravitational flow. A powder is required to flow uniformly, freely, and rapidly into the die cavity. If the flow is poor, it becomes necessary to slow down the process of cold pressing in order to get sufficient powder into the die. The flow rate of any powder is largely influenced by the size of the powder particles, distribution of size, particle shape, and freedom from even minute amounts of absorbed moisture.

Apparent density. The weight of a known volume of unpacked powder is called its *apparent density*. It is usually expressed as grams per cubic centimeter (g/cc). The apparent density is a very important factor affecting the compression ratio that is required in order to press the powder to a given density. If the apparent density of a given powder is one-third that of the required density, then three times the volume of powder is required to produce a given volume; i.e., the loose powder in a die cavity should be compressed from 3 inches (7.6 cm) in height to 1 inch (2.54 cm), or the stroke of the punches used in the dies should close up 2 inches (5 cm). It is obvious that low apparent density or bulkiness of the powder will require a longer compression stroke and relatively deep dies to produce a cold-pressed piece of a given size and density.

Particle size. Particle size is one of the most important characteristics of any powder. The most commonly used method of measuring the size of any metal powder particle is to pass the powder through screens having a definite number of meshes to the linear inch. In the screening method, the size of the particle is measured by a square-mesh screen of standard weave which will just retain the particle. However, most powders are composed of nonspherical particles, and therefore particle size is not a concise method of measurement. Most frequently the powders as used are made up of various sizes of particles. Their size is commonly reported by screening out the coarse, then the medium, and then the fine particles and reporting thus: 66 percent—100 to 200 mesh; 17 percent— 200 to 325 mesh; and 17 percent—325 mesh.

This means that 66 percent of the powder by weight will pass the 100-mesh screen but not the 200-mesh screen, etc. Other methods used for sizing powders include microscopic measurement and sedimentation, both being used to determine the size of particles when they are smaller than the finest mesh size of screens.

Molding of Powders

In the process of fabricating parts from powdered metals, the most important step is the one involving the welding together of the metallic powder to form a solid which will yield the proper shape and the properties required of the finished part. We may think of this step as one comparable to that used in the process of welding two pieces of wrought iron or low-carbon steel, this being accomplished by pressure and heat properly applied; this is called *pressure welding*.

We are all familiar with the forge or hammer-welding process practiced for thousands of years. We are also familiar with the seizure that occurs when two surfaces rub together without proper lubrication or through failure of an oil film. All this is a clue as to the action that takes place when we weld metal powders together to form a solid or so-called briquette. Although a good weld cannot be made between metals at room temperature by pressure alone, when the metal particles are relatively fine and plastic, a welding may occur that is satisfactory from the viewpoint of handling, although little or no strength will be developed. This is particularly true when metal powders such as iron, copper, etc. are used; however, gold powder can be cold-welded by pressure, producing a reasonably strong bond or weld. Under pressure at room temperature, metal powders that are plastic and relatively free from oxide films may be compacted to form a solid of the desired shape, having a strength (green strength) that allows the part to be handled. This result has been called *cold welding* or briquetting.

The welding under pressure of the metal particles in order to form a solid briquette of the shape desired requires the use of pressures varying from 5 to 100 tons/in^2 (68,950 to 1,379,000 kN/cm^2). Relatively light loads are used for the molding or briquetting of the softer and more plastic metals and for producing porous metal parts such as those employed in porous bearings. Pressure up to 100 tons/in^2 (1,379,000 kN/cm^2) is necessary when maximum density is needed and when pressing relatively hard and fine metal powders.

Commercial pressing is done in a variety of presses which may be of the single mechanical punch-press type or the modern double-action type of machine that allows pressing from two directions by moving the upper and lower punches, synchronized by means of cams. These machines also incorporate movable core rods, which make it possible to mold parts having long cores and assist in obtaining proper die fills and help in the ejecting operations of the pressed part. Fig. 15-2 illustrates a mechanical press used for briquetting metal powders.

The molding of small parts at great speeds and at relatively low pressures can

Metallurgy

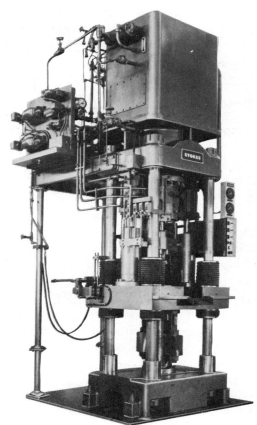

Fig. 15-2. Hydraulic compacting press, 300 ton capacity, for large metal compacts. (Stokes Powdered Metal Press Dept., Pennsalt Equipment Div.)

be best accomplished in the mechanical press. However, large parts and parts to be molded at relatively high pressures are best molded in hydraulic presses. The rate of production of parts by the powdered metal process will vary depending upon the size of the part and the type of press used. Mechanical presses can produce parts at the rate of 300 to 30,000 per hour. To obtain a production of 30,000 parts per hour requires rotary presses in which more than one set of dies is incorporated into the design of the machine. The single punch press with a double action can be used for production of parts up to 2000 pieces per hour.

A press to prove satisfactory should meet certain definite requirements, among which are the following: (1) sufficient pressure should be available without excessive deflection of press members; (2) the press must have sufficient depth of fill to make a piece of required height dependent upon the ratio of loose powder to the compressed volume, this being referred to as the *compression ratio*; (3) a press should be designed with an upper or lower punch for each pressing level required in the finished part, although this may be taken care of by a die designed with a shoulder, or a spring-mounted die which eliminates an extra punch in the press; (4) a press should be designed to produce the number of parts required. Large presses operate on a production of 6 to 8 pieces per minute, while small single-action presses can produce 60 or more parts per minute. Small rotary presses may have a production capacity of 500 per minute. Figs. 15-3 and 15-4 illustrate the changing from powder to briquetted shape in mechanical presses.

Punch and die equipment are of prime importance in any powder metallurgy application. Unit pressures are high, and the best of steels and workmanship must be used to obtain satisfactory operation and reasonable die life. Materials used for dies and punches vary from hard carbide alloys to ordinary carbon steels. The punches are usually made from an alloy punch steel that can be hardened by oil quench-

Powder Metallurgy

Fig. 15-3. The change from powder to briquetted part in two to six seconds is a fitting example of manufacturing speed attainable by powder metallurgy. (Chrysler Corporation)

Fig. 15-4. Close-up of a briquetting press showing a part being ejected from the die. (Chrysler Corporation)

ing. Dies are often made from air-hardening alloy steels of the high-carbon-high-chromium type. The usual method used in die design is to make an experimental die, and the knowledge gained from making a part in this die is used in the final die design.

Factors affecting die design include: powder mix, depth of die fill, pressures employed in briquetting controlled by the density requirements of the part, rate of

Metallurgy

production, abrasive nature of metal powders, need of any lubricant, method of ejection of part from die, volume change during sintering such as growth or shrinkage, and the necessity of repressing after sintering in order to produce parts within the size tolerance allowed. By a careful study of these factors, the designer is able to make proper drawings for a production die.

Lubricants are used in the molding or briquetting of powders for the purpose of protecting the die from excessive wear and to aid in the flow and ease of pressing the powders into a briquette. Lubricants used include graphite, stearic acid and zinc, aluminum, and lithium stearate. The lubricants are usually added to the metal powder and thoroughly mixed before pressing. Aluminum stearate is used for the aluminum alloy powders, and lithium stearate or zinc stearate for the iron powders.

Graphite also may be used as a lubricant with iron powders. If enough graphite is added to the iron powder, a carburization of the powder occurs during the sintering operation, resulting in the formation of a higher carbon alloy. The total carbon content of the sintered alloy can be controlled by the amount of graphite added to the original powder.

Sintering

Heating of the cold-welded metal powder is called the *sintering operation*. The function of heat applied to the cold-welded powder is similar to the function of heat during a pressure-welding operation of steel, in that it allows more freedom for the atoms in the crystals; it gives them an opportunity to recrystallize and remedy the cold deformation or distortion within the cold-pressed part. The heating of any cold-worked or deformed metal will result in recrystallization and grain growth of the crystals, or grains, within the metal. This action is the same one that allows us to anneal any cold work-hardened metal and also allows us to pressure-weld metals. Therefore, a cold-welded or briquetted powder will recrystallize upon heating, and upon further heating, the new crystals will grow, thus the crystal grains become larger and fewer. If this action takes place throughout a cold-pressed powder, it is possible to completely wipe out any evidence of old grains or particle boundaries and have a 100 percent sound weld.

Sintering temperature and time. The sintering temperatures employed for the welding together of cold-pressed powders vary with the compressive loads used, the type of powders, and the strength required of the finished part. Aluminum and alloys of aluminum can be sintered at temperatures from 700° to 950°F (371 to 510°C) for periods up to 24 hours. Copper and copper alloys, such as brass and bronze, can be sintered at temperatures ranging from 1300°F (704°C) to temperatures that may melt one of the constituent metals. Bronze powders of 90 percent copper and 10 percent tin can be sintered at approximately 1600°F (870°C) or lower for periods up to 30 minutes. Compacts of iron powders are usually sintered at temperatures ranging from about 1900°F to 2200°F (about 1040 to 1200°C) for approximately 30 minutes.

When a mixture of different powders is

to be sintered after pressing and the individual metal powders in the compact have markedly different melting points, the sintering temperature used may be above the melting point of one of the component powders. The metal with a low melting point will become liquid; however, so long as the essential part or major metal powder is not molten, this practice may be employed. When the solid phase or powder is soluble in the liquid metal, a marked diffusion of the solid metal through the liquid phase may occur, which will develop a good union between the particles and result in a high density.

Most cold-pressed metal powders shrink during the sintering operation. In general, factors influencing shrinkage include particle size, pressure used in cold welding, sintering temperature, and time employed during the sintering operation. Powders that are hard to compress will cold-shrink less during sintering. It is possible to control the amount of shrinkage that occurs. By careful selection of the metal powder and determination of the correct pressure for cold-forming, it is possible to sinter so as to get practically no volume change. The amount of shrinkage or volume change should be determined so as to allow for this change in the design of the dies used in the process of fabricating a given shape.

Furnaces and atmospheres. The most common type of furnace employed for the sintering of pressed powders is the continuous type. This type of furnace usually contains three zones. The first zone warms the pressed parts, and the protective atmosphere used in the furnaces purges the work of any air or oxygen that may be carried into the furnace by the work or trays. This zone may be cooled by a water jacket surrounding the work. The second zone heats the work to the proper sintering temperature. The third zone has a water jacket that allows for rapid cooling of the work, and the same protective atmosphere surrounds the work during the cooling cycle.

Fig. 15-5 illustrates a furnace of this type. The work may be placed on trays or suitable fixtures and pushed through the furnace zones at the correct rate either by hand methods or by mechanical pushers. A mesh-belt type of conveyor furnace can be used whenever the temperatures of sintering allow it. Such furnaces are used for sintering, operating successfully up to 2000°F (1100°C), although temperatures in excess of 1600°F (870°C) may cause damage to the conveyor belt.

Protective atmospheres are essential to the successful sintering of pressed metal powders. The object of such an atmosphere is to protect the pressed powders from oxidation which would prevent the successful welding together of the particles of metal powder. Also, if a reducing protective atmosphere is employed, any oxidation that may be present on the metal powder particles will be removed and thus aid in the process of welding. A common atmosphere used for the protection and reduction of oxides is hydrogen. Water vapor should be removed from the hydrogen gas by activated alumina dryers or refrigerators *before* it enters the furnace.

A cheap high-hydrogen gas may be generated by the dissociation of ammonia, forming a mixture of gases for a protective atmosphere containing 75 percent

Metallurgy

Fig. 15-5. A sintering furnace with three zones; purging, sintering, and cooling. (Westinghouse Electric and Manufacturing Corp.)

hydrogen and 25 percent nitrogen. The most common type of protective gas used is manufactured from partially burned hydrocarbon gases such as coke oven gas, natural gas, or propane. The gas is mixed with air in a suitable gas converter or generator and then cooled and let into the furnace. By control of the gas-air ratio, a variation in the composition of the manufactured gas is obtained so that a gas can be generated that will prove suitable for a given composition of metal powder. The importance of the atmosphere cannot be too highly stressed. Pure iron will oxidize in an atmosphere suitable for copper sintering; therefore, the selection of the atmosphere depends upon the composition of the pressed metal part.

For sintering copper and many of its alloys, a gas on the reducing side is usually satisfactory, provided no free oxygen is present. For sintering iron at 1950° to 2100°F (1065 to 1150°C), a gas containing approximately 18 percent CO, 1 percent CO_2, 2 percent CH_4, and 32 percent H_2, dried to a dewpoint of 40°F (4°C) has proved satisfactory. A ratio of 5¾:1 between air and propane gas will produce an atmosphere of this composition.

Hot Pressing

If metal powders could be pressed at temperatures high enough to sinter them to a solid at the same time they are briquetted, much lower pressures and temperatures might be used. The problems involved in hot pressing include the following: oxidation of the powders, excessive wear on dies, selection of alloys for

Powder Metallurgy

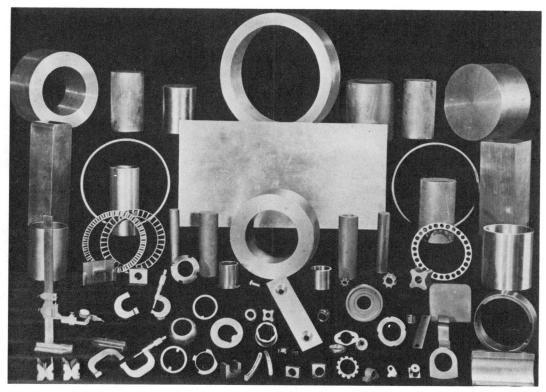

Fig. 15-6. Bearings, tools, and machine parts made by the powder metallurgy process. Some are porous, some dense, some self-lubricating, some dry, but nearly all units were made directly from dies without the necessity of machining. Some of these parts are made 200 times faster and have features impossible to obtain by other methods. (Chrysler Corporation)

punches and dies that will resist the temperatures employed, method of heating and control of temperature, pressures that are to be used, and annealing during or after the pressing cycle. Excellent properties have been obtained from hot pressing of iron and brass powders, using much lower pressures than those needed for the cold-pressing and sintering process, but only in the laboratory.

Applications

Fig. 15-6 illustrates some of the many shapes that have been successfully made by the powder metallurgy process. The manufacture of porous metal bearings from iron, brass, bronze, and aluminum alloys has proved to be a major application of this process. Bearings made by this

Metallurgy

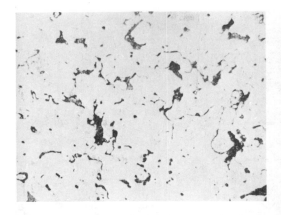

Fig. 15-7. Structure of a porous metal bearing made from iron powder. The dark areas represent voids in the metal bearing. Photomicrograph magnified 250 times.

Fig. 15-8. Self-lubricating aluminum oilite bearings. Advantages of this product are, its 77 percent weight-saving over bronze and the elimination of the electrolysis. Most uses of the new bearings are in aircraft. (Chrysler Corp.)

process are pressed and sintered at temperatures that will produce a part which is more or less porous and spongy. Fig. 15-7 illustrates the microstructure of a porous iron bearing showing the voids (dark) present in the finished part. The voids are more or less connected so that if the part is soaked in oil, the oil is drawn into the porous bearing and held there until pressed out by pressure from a load applied in service. Fig. 15-8 illustrates porous bearings made from aluminum powder, resulting in a lightweight bearing.

Some machine parts, particularly gears, have been successfully made by the powder metallurgy process using powders of iron, iron-carbon (steel), brass, bronze, etc. Figs. 15-9 and 15-10 illustrate the type of microstructure obtained by pressing a reduced iron powder at 40 tons/in^2 (551,600 kN/m^2) and sintering at 2100°F (1150°C) for 30 minutes. The presence of

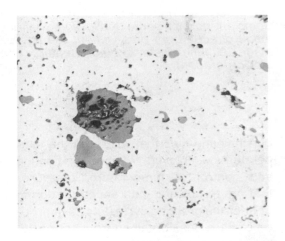

Fig. 15-9. Photomicrograph of a pressed and sintered reduced iron powder. Dark areas are oxide inclusions; polished section, magnified 250 times.

Powder Metallurgy

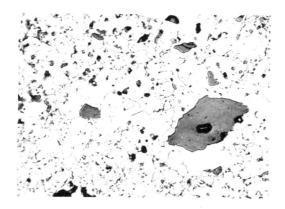

Fig. 15-10. Same as Fig. 15-9, but polished and etched to bring out the iron or ferrite grains. Magnified 250 times.

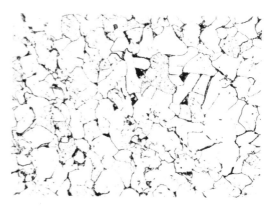

Fig. 15-11. Photomicrograph of electrolytic iron powder, pressed at 50 tsi (689,500 kN/m^2 or 689.5 MN) and sintered at 2100°F (1177°C) for 30 minutes. Shows similarity to structure of commercial pure iron. Magnified 250 times.

oxides shows up very clearly in these structures. Small machine parts that do not require much strength and hardness can be made from such an iron powder. If surface hardness is required, parts made from this reduced iron can be case-hardened by carburization similar to that of low-carbon steels. Fig. 15-11 illustrates the microstructure of a pressed and sintered electrolytic iron powder. The structure of this iron looks similar to that of the regular commercial pure iron and is relatively free from oxides. Iron-carbon alloys similar to steel can be made by this process and can be used for the manufacture of parts that require a certain degree of hardness. Fig. 15-12 illustrates the microstructure of an iron-carbon alloy made from mixing iron and graphite powders, followed by pressing or briquetting at 50 tons/in^2 and sintering at 2100°F (1150°C) for 30 minutes. The structure is of a pearlitic nature similar to that found in annealed tool steel. A part made with this

Fig. 15-12. Photomicrograph of an iron-carbon alloy made by mixing iron and graphite powders, pressed at 50 tsi (689,-500 kN/m^2 or 689.5 MN) and sintered at 2100°F (1177°C) for 30 minutes. Shows a pearlitic structure similar to that of annealed carbon tool steel. Magnified 1000 times.

Metallurgy

type of structure can be heat-treated by hardening and tempering in the same way used for regular carbon steel.

Tungsten wire for filaments in the lamp industry is made from pure tungsten powder which is pressed and sintered to form a bar of tungsten. This bar is hot-swaged to about 0.10" (2.54 mm) or less in diameter and is then drawn through tungsten carbide dies to about 0.010" (0.25 mm dia) while at a temperature range of 1500° to 1750°F (815° to 955°C). The wire is then drawn through diamond dies to the finished diameter. Starting from a powder, the unit strength of the swaged and drawn tungsten wire reaches some 850,000 psi (5,960,800 kN/cm^2)—more than twice that of the hardest steel—and is the strongest material known.

Another important branch of powder metallurgy is the field of hard tool materials. Hard carbides, such as tungsten carbide (WC), titanium carbide (TiC), and boron carbide (B_4C), can be bonded together by the powder metallurgy process to form a very hard material that has proved extremely important as a cutting tool and die material. To bond these hard carbides together, a binder or metal matrix is employed, usually cobalt. Nickel is also used in this respect. The carbide and metal binder are made into powder form by the usual mechanical methods. The prepared powder can be cold-pressed and sintered or hot-pressed to form a solid of the required shape. Manufacture of hard carbide materials by the powder metallurgy process produces much more uniformity of structure and properties than obtained by the melting process as illustrated by Figs. 15-13 and 15-14. Car-

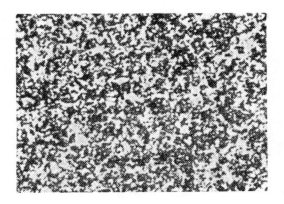

Fig. 15-13. Photomicrograph of cemented tungsten carbide; magnified 750 times. (American Society for Metals)

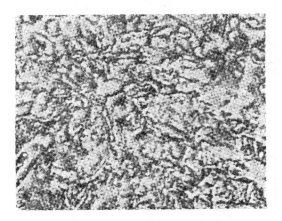

Fig. 15-14. Photomicrograph of cast tungsten carbide; magnified 750 times. (American Society for Metals)

bide tools and machine parts are exceedingly hard, although brittle, and find many applications in the field of cutting tools and dies.

Besides machine and tool applications, the powder metallurgy process has been successfully employed in the manufacture

Powder Metallurgy

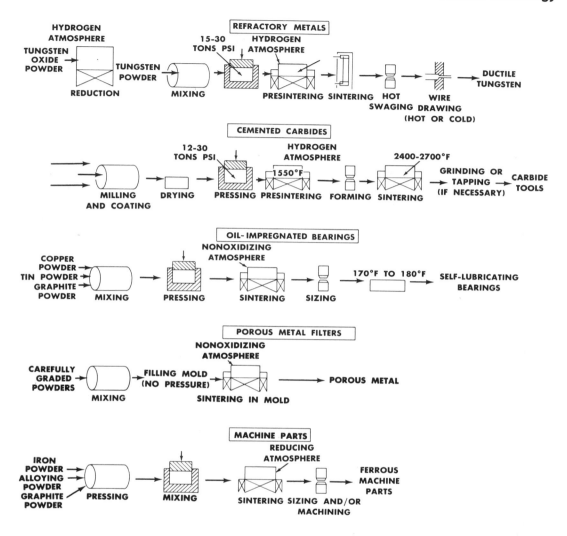

Fig. 15-15. Schematic of powder metallurgy processes. (Metals Handbook, American Society for Metals)

of contact points of tungsten, silver, or copper, clutch faces and brake linings where a mixture of a high coefficient of friction material is added to a metallic base, grinding wheels and drills made by combining diamond dust or carbide dust with softer metals, welding rods in which the necessary fluxes are incorporated with the special metallic material of the rod or bound to the surface of the metallic rod, Babbitt bearings made by bonding a bearing alloy to a metallic backing strip,

Metallurgy

and many other successful applications. There are many limitations to the application of the powder metallurgy process, but, with an understanding of these limitations, success for this process will be definite and continued. See Fig. 15-15 for the various powder metallurgy processes.

REVIEW

1. Define powder metallurgy.
2. What does the powder metallurgy process involve?
3. What was the first modern application of powder metallurgy?
4. What are recent applications of the process?
5. Is it possible to produce metallic alloys by the process?
6. How are metal powders produced?
7. What is the most troublesome impurity in powders?
8. How is this impurity caused and how can it affect the final product?
9. What five factors are important to the behavior of any powder?
10. What is the process of compacting the powders called?
11. Describe the process of molding metal powders.
12. What are the four requirements for satisfactorily molding in a mechanical press?
13. What lubricants are used in molding?
14. Describe the process of sintering.
15. What four factors influence shrinking?
16. Describe the atmosphere required in sintering furnaces.
17. Describe a self-oiling bearing.
18. How is tungsten wire made?
19. How are hard tool materials such as tungsten carbides made?
20. What industrial parts are made by powder metallurgy?

CHAPTER 16
PRODUCING NON-FERROUS METALS

Some metals are mined in the native state. Among these are silver, tin, mercury, platinum, gold and copper. Native metals need little refining to make them useful, either in pure form or alloyed with other metals. However, even in native metals there are impurities, such as rock, gravel, and sand, which require removal.

Most metals, however, are found as oxides (metal combined with oxygen). This metal oxide is called ore. Not all ores are found in oxide form; some of the most important occur as sulfide (metal combined with sulfur), or as carbonates (oxide of metal combined with CO_2). Carbonate ores of iron, copper, and zinc are important; some principal ores of copper and zinc occur as silicates. Ores of different metals are put through various refining processes to obtain the pure metal.

ORE DRESSING

One way in which the nonferrous metals differ from iron is in the manner of their occurrence. Iron oxide occurs in large and comparatively pure deposits; the other metals and compounds from which metals are derived are scattered through large volumes of rock, such as limestone or quartz. Since it would be difficult and costly to smelt large amounts of rock a method of concentration *(ore dressing)* by which the metals or metallic compounds are partially separated from the gangue (worthless material) is applied at the mine, before smelting. As the methods of ore

Metallurgy

dressing are rather general, we consider them here, rather than under the specific metals.

Gravity

The simplest method of ore dressing depends on the fact that in general the metallic compounds have a higher specific gravity than the gangue, and will settle faster in a stream of water. Gold panning is the simplest illustration of the procedure. On a larger scale, it is carried on in jigs where the ore is placed on a screen and a pulsating stream of water forced through the screen, causing the lighter gangue to be washed out. Another form of gravity concentrator is the table, consisting of a surface with longitudinal ridges, which is given a jerking end-to-end motion while a stream of water flows across it laterally. By this means the heavy ore is shaken over the end while the gangue washes off the front of the shaker.

Oil Flotation

Another method of ore separation is oil flotation. This process is based on the following: if a finely ground mixture of ore minerals and gangue is mixed with water, a little of certain oils added, and the mix stirred violently to produce a froth, the metallic mineral will be found in the froth. This method is capable of removing the last traces of mineral from the gangue, and hence is used to supplement the gravity process.

In discussing the metallurgical treatment of nonferrous ore, it should be understood that we refer to metallic mineral which has been concentrated by one of these methods. We shall also consider briefly the kinds of furnaces used for nonferrous smelting.

SMELTING AND REFINING

Reverberatory Furnace

The simplest kind of furnace is the reverberatory furnace. In this type the melting pot or hearth is long and narrow, and the charge is heated by a flame directed over the top of the material to be melted, so that much of the heating is indirect. The hearth has a slight tip toward the end of the furnace away from the firing end; molten metal and slag are drawn off here.

Slags are molten glassy materials which are purposely formed in certain metallurgical furnace operations for several reasons: first, impurities in the ore and the ash of the fuel must be removed and prevented from contaminating the metal. The slag layer over the metal also prevents excessive oxidation. Thus for lead ores, with silica as an impurity, iron oxide is added

Producing Non-Ferrous Metals

as a slag-former or flux. The silica and the iron oxide combine to form a glassy molten slag at the temperature of working. If the iron oxide were not added, the silica would combine with lead oxide, and lead would be left in the slag.

Reverberatory furnaces vary greatly in size and are used for a variety of purposes, such as copper matte smelting, lead fire refining, etc. They may be fired with gas, coal, oil or powdered coal, and may vary greatly in size.

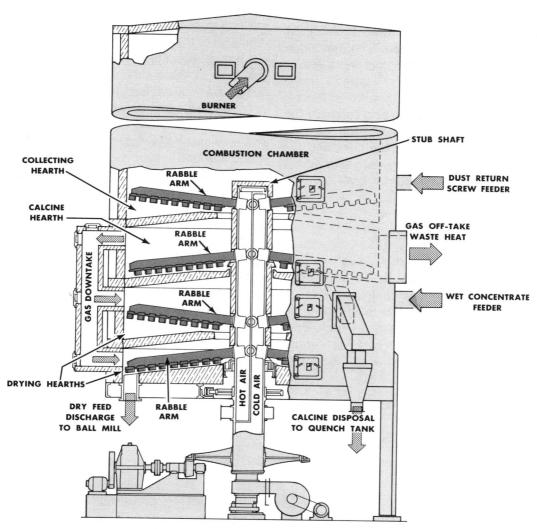

Fig. 16-1. Multihearth mechanical roasting furnace. (Wedge Mechanical Furnace Co.)

Metallurgy

Blast Furnace

The other type of furnace used principally in nonferrous smelting is the blast furnace, similar to the steel blast furnace except that it is much smaller and the blast is not heated. In the blast furnace the material to be melted is mixed with the fuel, usually coke, the heat for melting being obtained by combustion of the fuel when a blast of air is blown through the charge. Blast furnaces are taller than their cross-section, and the charge is fed in near the top. As this charge travels through the furnace, the coke is burned and the metal reduced and melted. The metal is drawn off at the bottom; the slag, formed from the ash of the coke and the impurities in the ore together with added fluxes, is drawn off from a layer immediately above the metal.

Roasting Furnace

In certain smelting processes it is necessary to oxidize the sulfide ore by roasting before smelting. Roasting consists in heating the ore in air until most of the sulfur is driven off and the sulfides are converted to oxides. For this purpose, various forms of furnaces are used. Fig. 16-1 shows a cutaway view of a multiple hearth furnace. In this, the ore is raked through the various hearths by means of rotating rabble arms, and the hot oxidizing gases travel upward through the furnace.

Pollution Control

In all smelting operations a considerable amount of fume is given off. The collection of this fume is important, both because of the damage it would do to surrounding vegetable and animal life, and because considerable metallic materials would be wasted. Bag houses or Cottrell treaters are used for collection of this fume.

Bag houses are rooms filled with long bags through which the smelter smoke passes. The solid material is retained by the bags, which are periodically shaken to remove the dust. The dust is treated to recover the metallic values.

The Cottrell treater takes advantage of the fact that if a gas with suspended solid matter is passed between two electrodes, between which is passing a high voltage discharge or corona, the dust will agglomerate and fall to one of the electrodes. This electrode is arranged so that it may be shaken and the dust recovered.

REFINING OF NON-FERROUS METALS

Copper

The most important use of metallic copper is wire and bars for electrical conductors. It is also used in tubes and boilers because of its properties in heat conduction and its corrosion resistance.

While 65 percent of U.S. copper pro-

Producing Non-Ferrous Metals

TABLE 16-1. COPER PRODUCTION.[1]

	SHORT TONS
World production	7,857,682
U.S. production (23 percent)	1,698,000
imported ores	170,000
scrap	420,000
consumption	2,196,950
USES: wire mills	1,530,000
brass	667,000

[1]*Metals Yearbook*, Vol. I., U.S. Department of the Interior, Bureau of Mines, 1973, GPO.

duction is used by the wire mills for the production of copper wire, some 667,000 tons (605,000 tonnes) of copper are used to produce brass, to account for the U.S. total of 1,530,000 tons (1,388,000 tonnes) in recent years. See Table 16-1.

The copper ore is first ground in massive ball mills to provide feed for later stages of production, Fig. 16-2. The ore is then concentrated by the flotation or gravity process, to yield 25 to 30 percent copper before it is refined.

Fig. 16-2. Copper ball mill processes 90,000 STPD. (Allis Chalmers)

Metallurgy

The most important copper ores are found in the form of a sulfide mixed with iron sulfide, although large amounts are found as oxides, carbonates, silicates, arsenides, and in other forms. Gold and silver are present in nearly all copper ores, and their extraction is one important step in the reduction of copper ores. Sulfide ores must be roasted to remove sulfur before smelting.

Smelting is accomplished either in a reverberatory or a blast furnace. This smelting does not result in pure copper, but in an alloy called *matte*, containing about 32 to 42 percent copper. The next operation, that of oxidizing the sulfur and iron, is carried out in a converter, Fig. 16-3, where air is blown through the molten matte. The product of the converter, known as *blister copper*, contains many impurities, including gold and silver, but is about 98 or 99 percent pure.

Fig. 16-3. View of two copper converters. (Asarco, Inc.)

Producing Non-Ferrous Metals

This blister or impure copper is further processed by partial refining in a furnace, after which the molten metal is cast in the form of plates, about 3' x 4' x 1" (91 x 122 m x 2.54 cm), called *anodes*. The metal anodes are then electrically refined by immersion in a solution of copper sulfate, after which they are placed in close to thin sheets of pure copper. An electric current is then passed through the solution. The copper dissolves from the anodes and is deposited on the thin plates or starting sheets. The impurities remain in the form of sludge, which falls to the bottom of the tank. Gold and silver are removed from this sludge by a separate process.

The deposited copper is stripped from the starting sheets from time to time, and new anodes are placed in the cell as the old ones are used up. Since this is a very slow process, requiring about one month for one anode to dissolve completely, tanks containing a very large number of anodes are used. Fig. 16-4 shows an electrolytic refinery based on the single cell concept also in Fig. 16-4.

The copper taken from the cathode sheets is remelted and cast into wire bar or cake for drawing into wire or tube, or rolling into sheet. See Fig. 16-5 for wire bar and sheet products. In this remelting process, the copper is oxidized to remove impurities. Such impurities as arsenic, antimony, iron and tin must not be allowed to remain as they lower the conductivity of the copper.

The copper is then reduced until it contains 0.04 to 0.06 percent oxygen. If further reduced, it becomes porous on casting. The remaining oxygen is in some cases removed by the addition of phosphor copper without seriously affecting the physical properties; however, this would lower conductivity. Fig. 16-6 diagrams a method of extracting and refining copper. Copper containing 0.05 to 0.06

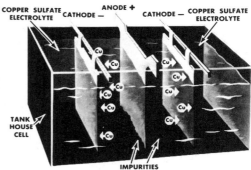

Fig. 16-4. Lifting cathodes of 99.99 percent pure copper from electrolytic refining tank at Great Falls, Montana refinery. (Anaconda Co.)

Metallurgy

Fig. 16-5. Casting copper wire bars. (Anaconda Co. and Chase Brass and Copper Co.)

percent oxygen is cast into wire bars and hot-rolled to ¼" (0.635 mm) rod.

Fig. 16-7 shows the process of drawing small diameter copper bar. The ¼" rod is pickled in sulfuric acid to remove the oxide, then reduced to wire of any size by drawing. Drawn copper wire is hard, and has a tensile strength of 46,500 to 70,000 psi (320,618 to 483,000 kN/m^2). The electrical conductivity is lowered some 2 to 3 percent; nevertheless, this kind of wire is preferred for line wire where strength is important. For other uses, the wire is annealed. This brings back its conductivity but decreases its strength to about 35,000 psi (241,000 kN/m^2). The annealing treatment is given in steam in order to leave the wire bright.

For castings and other uses where high electrical conductivity is not important, copper is usually deoxidized with phosphor copper to give a sound casting. For casting purposes as well as for fabrication processes other than wire drawing, copper alloys are to be preferred to pure copper.

Lead

Both lead and the alloys of which it is the major constituent are characterized by softness, pliability, and a low melting temperature. Pure lead melts at 621°F (327°C). Lead exhibits a high degree of resistance to atmospheric corrosion and at-

Producing Non-Ferrous Metals

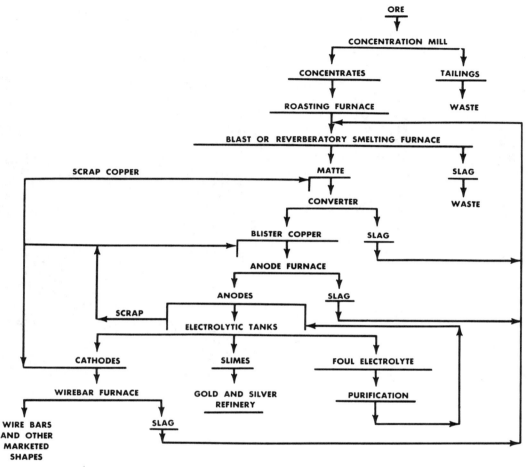

Fig. 16-6. Flowsheet of common copper extraction and refining. (Engineering Metallurgy by Stoughton and Butts, McGraw-Hill Book Co.)

tack by chemicals, particularly sulfuric acid. These properties determine some uses of metallic lead, namely: chemical piping and cable sheathing, where pliability combined with corrosion resistance is important; storage battery grids and chemical tanks, where resistance to sulfuric acid is important; and solder metals where low melting point is the major requirement.

Lead occurs in nature principally in the form of lead sulfide, called galena, a substance found in abundance in the United States, where over 600,000 short tons (545,000 tonnes) are produced annually. However, total U.S. consumption is nearly 1.5 million short tons (1,360,000 tonnes)

Metallurgy

Fig. 16-7. Small diameter soft copper bar being drawn on a special machine. (Chase Brass and Copper Co.)

per year, half of it for lead batteries. Metal production accounts for nearly 25 percent more: See Table 16-2.

Since scrap metal is a major source of lead, the reclaiming of lead batteries, which consumed nearly half of all lead produced in the past, has become important. See Fig. 16-8.

Three methods are used to refine the lead ores—the blast furnace, the reverberatory furnace, and the ore-hearth method. The blast furnace is the most important method at the present time.

TABLE 16-2. LEAD PRODUCTION.[1]

	SHORT TONS
World production	3,848,582
U.S. production	618,915
imports	347,000
consumption	1,485,254
USES:	PERCENT
metal products:	
cables, tube, sheet, ammunition	23
batteries	49
antiknock	19
paint	6

[1] 1973 *Metals Yearbook*

Producing Non-Ferrous Metals

Fig. 16-8. Recycling of lead batteries is a major source of reclaimed lead. (Lead Industries, Inc.)

Lead ores are subjected to a roasting-sintering process as a preliminary step in their reduction, an important procedure in the metallurgy of lead. The object of the roasting-sintering process is to reduce the percentage of sulfur by oxidation of the sulfur to sulfur dioxide (SO_2). Fig. 16-9 is a drawing of a Dwight-Lloyd sintering machine used in this process.

The lead blast furnace is similar to the blast furnace used in the production of copper matte. Its operation results in a crude lead containing both gold and silver. Both a metal and a matte from which lead is obtained are produced ordinarily in lead smelting. The crude lead is then cast into small bars to be further refined, Fig. 16-10.

The reverberatory process of refining lead by oxidation is no longer used to any extent in the United States. However, the ore hearth is used as the simplest method of smelting ores rich in lead. The ore hearth consists of a basin to hold the melted lead and to support the charge being smelted. The lead ore mixed with coke is fed on the hearth, and air is blown through the ore and passes above the surface of the molten lead. In this way the lead sulfide is reduced to metallic lead.

Lead obtained by any of these smelting methods contains the silver, copper, bismuth, arsenic, and antimony which were in the ore, and usually it must be purified. Lead from southeastern Missouri is an exception, being sufficiently pure to use without further treatment. Other leads are purified by fire or electrolytic refining.

Fire refining consists in oxidizing out the arsenic, antimony, and most of the copper by blowing air through the molten lead in a reverberatory furnace at a red heat. The lead so purified is then subjected to the Parkes process, to remove the gold and silver and the remaining copper. This process consists in adding a small amount of zinc to a kettle of molten lead. The zinc does not dissolve in the lead, but melts and stays on top of the molten lead. Most of the gold and silver enter this layer of zinc, and are skimmed off with it. The zinc-silver-gold alloy then is treated to recover the gold and silver, by the process described in the section on precious metals.

In electrolytic refining, the impure lead

Metallurgy

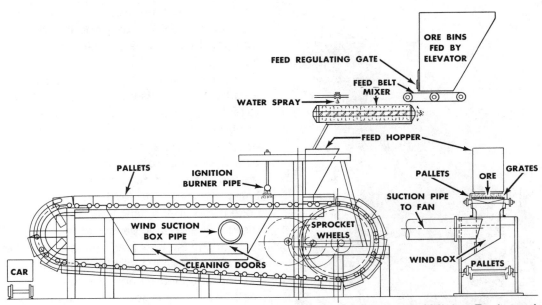

Fig. 16-9. Continuous roast-sintering machine. (American Institute of Mining Engineers)

Fig. 16-10. Casting of lead in a small foundry. (Lead Industries, Inc.)

or bullion is cast into anodes as described for copper, from which it is dissolved in a lead fluorosilicate solution and redeposited as pure lead through the action of an electric current. All the impurities, including bismuth and precious metals, are collected as a slime and treated to recover the metals. The deposited lead is then remelted and cast into pigs.

Producing Non-Ferrous Metals

Several grades of pig lead are available to the consumer: *common lead; corroding lead* which is used when a lead of exceptional purity is required; *chemical lead,* which is used in the chemical industries; and *antimonial lead,* containing between 6 and 7 percent antimony and used in the manufacture of storage batteries.

Zinc

Zinc rarely occurs in the native or pure metallic state. It is usually found in the form of zinc sulfide (ZnS), an ore called *zinc blende.* Other ores of zinc are: zinc carbonate ($ZnCO_3$), called *zinc spar,* and zinc silicate (Zn_2SiO_4) known as *Willemite.* Zinc blende is the principal ore worked in the United States. See Table 16-3.

The processes in the production of zinc are: concentration of the ore, roasting of the sulfide ores, reduction of the carbonates and silicates, and refining by distillation or electrolytic action. Zinc has a low melting point of 787°F (420°C) and vaporizes at 1661°F (905°C). The low vaporization temperature makes distillation a very economical refining process. Fig. 16-11 shows one such refining process.

In the distillation process, the powdered zinc oxide from the roasters is mixed with finely ground hard coal and placed in air-free, vertical retorts. The retort is heated externally, raising the temperature of the coal to the kindling point. The coal obtains most of its oxygen for combustion from the zinc oxide and metallic zinc is liberated. At the high temperature that results, the zinc is a vapor and is drawn off and passed through a condenser where it becomes a liquid. Liquid zinc runs from the condenser into molds and forms the solid *pig* used in commerce.

In electrolytic refining, the zinc oxide from the roasters is leached in tanks of sulfuric acid where it becomes zinc sulfate in a water solution. This solution is pumped into electrolytic tanks equipped with lead anodes and aluminum cathodes. When direct current electricity is passed through the solution, zinc is deposited on the cathode, as in Fig. 16-12. After about fifty pounds of zinc has been deposited on the aluminum cathode, the zinc is stripped off, melted and pigged.

Zinc is brittle in cast form but may be rolled into sheets or drawn into tubes. The principal uses of metallic zinc, however, are in *galvanizing* and as an alloying metal. Zinc is used as a metallic coating to protect iron and steel from corrosion. This process is called *galvanizing.* The zinc may be applied to a steel surface by electroplating but it is more commonly applied by *hot dipping.* The steel articles: pails,

TABLE 16-3. ZINC PRODUCTION (MINE).[1]

	SHORT TONS
World production	6,377,392
smelter	5,795,352
U.S. production	478,850
consumption	1,503,938
imports (ore)	199,053
zinc slab	588,712
1974 World production	5,920,673
	PERCENT
USES: Galvanizing	36
Brass production	14
Die casting alloy	40
Zinc oxide (paint)	4

[1] 1973 *Metals Yearbook*

Metallurgy

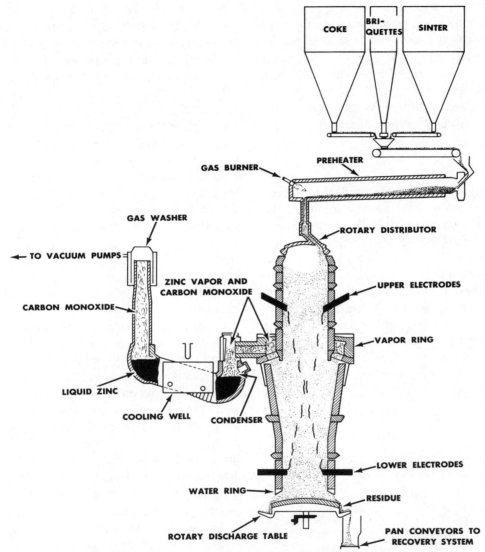

Fig. 16-11. Electrothermic zinc metal furnace, schematic.

buckets, tubs, etc., are fabricated and then dipped into a tank of molten zinc. As zinc has a good wetting property, (ability to cover or plate readily) it covers the metal surfaces and penetrates into the inner seams for 100 percent coverage and rust protection. Zinc coated mild steel is commonly referred to as *galvanized iron*.

Producing Non-Ferrous Metals

Fig. 16-12. Purifying zinc through electrolytic production.

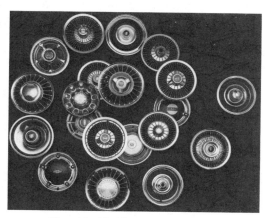

Fig. 16-13 Zinc die castings.

The major use of zinc as a structural material is in the form of zinc die castings, examples of which are shown in Fig. 16-13. Zinc die casting alloys are low in cost and have good strength. They can be cast to close dimensional limits, require minimum machining and possess good resistance to surface corrosion. Zinc die castings are limited to maximum operating temperatures of 200°F (93°C) because of their rapid loss in strength and hardness above this temperature. They also become brittle at below freezing temperatures (32°F; 0°C).

Other uses of zinc are as cathodes in dry cell batteries, zinc oxide pigment in white paint, zinc oxide in surgical dressings, printing plates, die casting alloys and bearing metals when alloyed with copper and aluminum. Zinc is alloyed with brass.

An important zinc base alloy called *Kirksite* is made of 4 percent aluminum, 3 percent copper and 93 percent zinc. This

Metallurgy

alloy has moderate hardness and tensile strength and a very high impact strength. It can be readily cast into die shapes for either blanking or drawing a wide variety of aluminum and mild steel parts in industry.

One of the large users of Kirksite is the aircraft industry which uses a great variety of light metal stampings in small quantities. Production runs as high as 25,000 small pieces have been recorded, but Kirksite dies most generally are used for short production runs because of a tendency of the die surface to wash out, lose the sharpness of detail in surface embossments. As the dies are easily melted and recast with nearly 100 percent recovery of the metal, Kirksite is an economical die material to use. The automobile industry uses Kirksite dies for experimental and pre-production runs of body stampings. Up to 200 fenders, hoods or door panels have been made from one set of Kirksite dies.

Cadmium

Cadmium, which is obtained as a by-product in zinc and lead smelting, is closely related to zinc in its properties. It is a silvery-white crystalline metal, symbol Cd, resembling tin. Its ore is a sulfide called *Greenochite*, CdS, and contains 77.7 percent cadmium. Cadmium has a specific gravity of 8.6 and is very ductile. It can be readily rolled into thin sheets. See Table 16-4.

Cadmium is sold as anodes for electroplating. Electrolytic cadmium is 99.95 percent pure and is used as a corrosion-resistant plated covering on steel fasteners. Cadmium melts at 608°F (320°C) and is a constituent of low melting point alloys used as temperature fuses in sprinkler heads. Wood's metal is a cadmium alloy which melts at 154°F (68°C). An alloy of copper with 1 percent cadmium is used in the trolley wires of bridge cranes as it has high strength combined with good wearing properties and good electrical conductivity.

TABLE 16-4. CADMIUM PRODUCTION.[1]

	SHORT TONS
World production	18,747
U.S. production	3,714
consumption	6,228
	PERCENT
USES: batteries, plating	50
copper alloys	50

[1]1973 *Metals Yearbook*

Nickel

Nickel is a silvery-white metal with a faint yellow cast. Its principal ore is Pyrrholite or Magnetic pyrite which is a mixed sulfide of iron and nickel. It has a specific gravity of 8.84, melts at 2446°F (1341°C) and is magnetic up to 680°F (360°C). It is resistant to corrosion and to most acids, except nitric.

Nickel owes most of its use to its resistance to atmospheric and chemical corrosion, and its ability to take a high polish. For these reasons it is used for vats and vessels in the chemical industry, and as an electroplating on a great variety of hardware. A greater amount of nickel; however, is used as an alloy in *nickel steel* and in the important nonferrous alloys such as *nickel silver, nichrome,* and *monel metal.*

Nickel is derived principally from mixed nickel sulfide ores by a process of roasting followed by reduction with carbon in a blast furnace—the Orford or *tops-and-bottoms process.* The product as cast from the blast furnace solidifies in two layers, the top containing copper sulfide, and the bottom layer nickel sulfide. The cast nickel is then further refined electrically, or by the Mond or *carbonyl process*, which consists of vaporizing the nickel in carbon monoxide and redepositing. See Table 16-5.

Nickel is produced in various forms: nickel pellets, produced by decomposition of nickel carbonyl gas without fusion; electrolytic cathode sheets; nickel shot or blocks, made by casting nickel pig molds or pouring into water without deoxidation; malleable nickel, a deoxidized cast nickel; nickel cubes, reduced from oxides; nickel salts; and nickel powders.

TABLE 16-5. NICKEL PRODUCTION.[1]

	SHORT TONS
World production	226,014
U.S. production	18,272
imports	191,073
consumption	197,723
exports	22,070
	PERCENT
USES: stainless steel	28
hi-nickel	26
alloy steel	12
plating	18
castings	3

[1] 1973 *Metals Yearbook*

Aluminum

Aluminum is almost a 20th century metal. The earth's crust contains more aluminum than any other metallic element. Aluminum occurs most commonly as a silicate or an oxide. It is found in almost all rocks, except sandstone and limestone, and in all clays. Aluminum, as an element, was discovered in 1727 but did not become commercially available due to the extreme difficulty in reducing its oxide form. This was accomplished in 1886 by a young American, Charles Hall, and a Frenchman, Paul Heroult, who simultaneously but independently discovered the electrolytic process. A singular fact about these two young scientists is that they were born in the same year, discovered the process in the same year and both died in the same year.

Aluminum is never found free in nature in its metallic state. Its ore is a clay, an hydrated oxide of aluminum, which is known as *bauxite*. The ore is usually dug in open-pit mines in Arkansas, Georgia, Alabama and the West Indies. (Fig. 16-14). Aside from producing aluminum, the ore is also used for manufacturing chemicals and high temperature insulating materials, and for grinding wheels and stones.

There are several different methods for refining aluminum; however, the most commonly used process was discovered by Karl Bayer. In this process, the bauxite is first crushed to a powder and then mixed in large pressure tanks with a hot solution of caustic soda, Fig. 16-15. The caustic soda dissolves the aluminum hydroxide, but not the impurities. This solution is then filtered to remove the impuri-

Metallurgy

Fig. 16-14. Large power shovel scoops up bauxite and loads heavy-duty truck. Bauxite is carried to rail head and goes by train to coast. Bauxite in Jamaica lies close to surface; in mining operations the topsoil is first scraped to one side and preserved, then replaced for land rehabilitation. (Kaiser Bauxite Company)

Fig. 16-15. Digestion area of alumina plant. The plant refines bauxite from Jamaica, which is then shipped to reduction plants for primary aluminum production.

Producing Non-Ferrous Metals

ties and then pumped into large tanks. On slow cooling, the aluminum hydroxide settles out in the form of fine crystals. These crystals are then washed to remove the caustic soda.

The aluminum hydroxide crystals are now fed into large revolving kilns, Fig. 16-16 and heated until they are white hot. This heat drives off the water in the form of steam, the residue being a white powdery chemical of aluminum oxide, called *alumina*.

The last step in this process is to send the alumina through reduction plants where the aluminum is released from the oxygen with which it is now firmly combined. This separation is effected by means of an electrolytic cell. A material called cryolite fills the cell. When the electric current has melted the cryolite, the powdery alumina is dissolved in the cryolite bath. Passage of the electric current breaks up the combination of aluminum and oxygen, the oxygen being freed at the carbon electrode, while the aluminum deposits in a molten layer at the bottom of the furnace, Fig. 16-17. This deposit is poured into molds and cools as a metallic

Fig. 16-16. Rotary kilns bake the hydrated alumina crystals, forming pure alumina (aluminum oxide). (Kaiser Aluminum & Chemical Corp.)

Metallurgy

Fig. 16-17. Special vacuum crucible draws molten aluminum from the bottom of an electrolytic pot. (Aluminum Company of America)

Fig. 16-18. Removing 10,000 pound ingot from direct chill-type casting station. (Kaiser Aluminum and Chemical Corp.)

Producing Non-Ferrous Metals

aluminum called pigs. The pigs are remelted to remove any remaining impurities, and then poured to form aluminum ingots, Fig. 16-18.

Aluminum is the lightest of the common metals, having a specific gravity of 2.70. Aluminum does not rust and strongly resists corrosion. Due to its high electrical and heat conductivity, as well as its chemical resistance, it finds many everyday uses.

Because of its lightness, the greatest single use of aluminum is in the field of transportation. Aluminum is also used to a very great extent for electric cables, for cooking utensils, and for protective foil for wrapping foods and other products. Its use in the building industry is of no small importance. The total volume of aluminum production is indicated in Table 16-6.

Aluminum lends itself readily to shaping. It is easily cast into molds of all shapes and kinds. It can be rolled hot or cold into thick plates, flexible sheets, or the thinnest of foil. In addition, it can be rolled into bars, rods, or drawn into the finest of wire.

Aluminum alloys are in common use. It alloys easily with copper, silicon, manganese, and chromium. Such alloying, with subsequent heat treatment, results in greatly increased strength and hardness.

Magnesium

Magnesium is the lightest metal now used commercially, having a specific gravity of 1.74. It is produced from minerals used as ores, the most common being magnesite and carnallite, and from concentrated brines and sea water. See Table 16-7.

In one of the many processes, the metal is obtained by first producing fused magnesium chloride, and then electrolyzing

TABLE 16-6. ALUMINUM PRODUCTION.[1]

	SHORT TONS
World production	13,359,000
U.S. production	4,529,000
consumption	5,685,000
primary production (1972)	4,177,190
imports	466,765
old scrap	188,594
new scrap	755,762
TOTAL	5,588,311
USES (1972):	PERCENT
plate, sheet, foil	57
rolled	16
extruded	23
forgings	1.3

[1] 1973 *Metals Yearbook*

TABLE 16-7. MAGNESIUM PRODUCTION.[1]

	SHORT TONS
World production	261,110
U.S. production	140,000
imports	3,283
USES:	PERCENT
batteries	49
metal products	23
ammunition, cables, tubes, sheet, antiknock agents	19
pigments	6

[1] 1973 *Metals Yearbook*

Metallurgy

the salt. It is also produced by heating the carbonate, dissolving the oxide thus formed in a fused salt bath, and then electrolyzing.

In producing magnesium from brine, brine containing magnesium chloride, sodium chloride, and calcium chloride is first treated to bring about a separation of these chlorides. The magnesium chloride from this treatment is then subjected to a drying treatment to remove most of the water. This material is then electrolyzed, and magnesium, which collects on the cathode side, is dipped from the cells and cast into ingots.

Magnesium is now produced in great abundance from ordinary sea water. There are three essential steps in this processing. First, lime water is mixed with sea water to convert its soluble magnesium salts into milk of magnesia. This milk of magnesia is then treated with acid, converting it into a magnesium chloride solution. The last step involves electrolyzing this molten product, yielding solid magnesium at one electrode, and chlorine gas at the other.

Like most metals, magnesium is seldom used in the pure form for metal parts. It is most commonly alloyed with aluminum, although several other alloys are produced.

Magnesium can be forged, rolled, and extruded. It displays its best mechanical properties in the extruded form, where it attains a tensile strength of more than 25,000 psi (172,000 kN/m_2). As a cast material, its tensile strength is about half this amount. Cold working results in a rapid increase in hardness; therefore, it is advisable to carry on such operations at temperatures ranging between 500° to 650°F (260° to 345°C). Magnesium can be heat treated thus, resulting in a substantial increase in mechanical properties.

Magnesium stands high in the electrochemical series, its corrosion resistance being only fair for ordinary purposes. Machining qualities of this metal are excellent; however, care should be taken in handling small chips or filings as they will oxidize and burn very readily. The danger of such burning decreases as the size of the piece increases.

The greatest single use of magnesium is the field of aircraft and general transportation where its extreme lightness is the dominating factor. Its metallurgical values as desulfurizer, deoxider, and alloying element are well known. Like aluminum, it finds many other important everyday uses.

Tin

Tin is a relatively soft, silvery-white metal with a brilliant luster. It has been used from earliest times. It is soft and very malleable and can be rolled into sheets as thin as 0.0002" (0.0051 mm). The principal ores are *Cassiterite*, Sno_2, called tin stone, and *Stannite* or tin pyrites, SnS_2. The principal tin producing countries are the East Indies, the Malay peninsula and Bolivia.

The preliminary concentrating and smelting of tin ore is similar to that of copper. After smelting, usually in a reverberatory furnace, the tin must be separated out from its impurities such as iron, lead, antimony, bismuth, arsenic and sulfur.

Producing Non-Ferrous Metals

Fig. 16-19. Transfer of 10 tons of tin from reverberatory furnace to refining kettles. (Tin Research Institute)

This is done in a thermal separation process called *liquation*. In the liquation process, the impure tin pigs are melted carefully on a sloping hearth, taking care not to raise the temperature above the melting point of tin, 450°F (238°C). The almost pure tin will flow down the slope to a collecting point. The impurities will remain, unmelted at the top of the slope. Fig. 16-19 shows molten tin being transferred from the smelting furnace to the refining process.

Small traces of impurities have an influence on tin. Lead softens the metal, while arsenic and zinc harden it. The purposes for which tin is used determines which metals must be removed and which are added.

Tin is used in industry to coat thin sheets of cold rolled steel, producing *tin plate* from which tin cans are made. Fig. 16-20 shows hot dipped tin plate in profile. When alloyed with 75 to 82 percent lead, the alloy is coated on thicker gage steel to produce *terne plate*. The resulting coated metal is highly resistant to atmospheric corrosion and terne plate is used to make automobile gasoline tanks, gas and brake lines and other automotive components. Bundy tubing is a special lap rolled tubing made from terne plate which is then sintered to produce a tubing which is both corrosion and leak proof. Tin is also used in the alloys of bronze, which are treated in Chapter 18. It has been used since the bronze age began.

369

Metallurgy

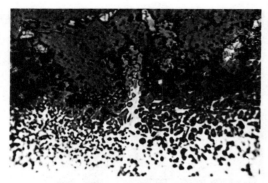

Fig. 16-20. Hot-dipped tin plate (tin coated steel), etched in picric acid and magnified 1500 times. (Tin Research Institute)

Manganese

Since manganese is a very hard metal; it is not used as a pure metal, but it is frequently used as an alloy in steel. Most manganese ore is an oxide and is found in Russia, Brazil, Africa and India. American ore is generally low grade, containing only 5 to 10 percent manganese. One to two million tons of manganese are imported annually into the United States.

Manganese is used in the steel industry as a deoxidizer and desulfurizer and nearly all steel contains small amounts of manganese. Manganese, when added to steel in large amounts produces a hard, wear-resistant steel used as armor plate and in rock crushers.

Manganese ores are used to produce ferromanganese for alloying steel and bronze. It contains 35 percent or more manganese. A special ferromanganese compound called *spiegeleisen* (mirror iron) was used extensively in the past in making manganese steel by the Bessemer process.

Chromium

Chromium is a high melting-point metal similar to nickel. It is used extensively in nickel alloys, alloy steels and in chromium plating. Chromium occurs in nature only in combination with other elements. The principal chromium ore is *Chromite* containing an oxide of chromium combined with oxides of other metals such as iron and aluminum. See Table 16-8.

Chrome ore is mined principally in Rhodesia; other sources are in India, Russia and Turkey. The metal is obtained by a reduction (smelting) process and refined by electrolysis. The metal is very hard and inert to nitric acid. Chrome plating is used for bearing surfaces, wear plates and tool room gages. For decorative purposes, chromium plating as thin as 0.0002" (0.0005 mm) may be used. Chrome plating is usually a final coating over undercoats of copper and nickel, as in the case of auto bumpers.

TABLE 16-8. CHROMIUM PRODUCTION.[1]

	1000 SHORT TONS
World production	7,507
U.S. imports	931
consumption	1,387
USES: ferrochromin (stainless steel) metallurgy refractories chromium chemicals	

[1] 1973 *Metals Yearbook*

Tungsten

Tungsten has a number of uses, principal among them its use in lighting as an incandescent lamp filament. Its value here is due to its very high melting point, about 6100°F (3370°C), and its ability to be drawn into fine wire. It is used also for phonograph needles and as heating elements in very high-temperature furnaces. However its chief use, from a tonnage standpoint, is as a constituent of high-speed steel. It is available on the market as tungsten powder, wire, and as ferro-tungsten, Table 16-9.

Tungsten is a heavy metal, its density being approximately 2½ times that of iron. It is never found free in nature, but in an ore from which an oxide is obtained. A metallic tungsten powder is made by the reduction of this tungsten oxide, either with carbon or hydrogen. It is not practical to melt down the powder to convert it to solid metal form, because of the extremely high melting point. Instead, a procedure known as *powder metallurgy* is used to change tungsten powder to the solid metal, a procedure which is discussed in Chapter 15.

Cobalt

Cobalt is similar to nickel in nearly all its properties. It occurs with both iron and nickel, and is used as a constituent of certain special steels.

Antimony

Antimony is a brittle metal with a silvery luster. The pure metal has almost no industrial uses. However, it is an important constituent of several alloys, principally those containing lead. It is available on the market as metallic antimony, much of which is imported from China, and also as antimonial lead—an alloy of lead with 15 percent antimony, obtained as a by-product in lead smelting. Table 16-10 indicates current uses.

TABLE 16-9. TUNGSTEN PRODUCTION.[1]

	1000 POUNDS
World production	85,320
consumption	84,504
U.S. production	17,096
consumption	17,984
	PERCENT
USES: cutting tools	62
mill production	18
special tool steels	11
hardfacing	9

[1] 1973 *Metals Yearbook*

TABLE 16-10. ANTIMONY PRODUCTION.[1]

	SHORT TONS
World production	76,419
U.S. production	41,000
consumption	20,613
USES: antimony lead castings ammunition primers bearing metals	

[1] 1973 *Metals Yearbook*

Metallurgy

Bismuth

Bismuth is a metal similar in appearance to antimony. It is a by-product in electrolytic lead refining. Its only use is as a constituent of the low melting alloys mentioned in the discussion of cadmium. See Table 16-11.

TABLE 16-11. BISMUTH PRODUCTION.[1]

	MILLION POUNDS
World production	8.3
U.S. consumption	2.3
imports	1.3
USES: pharmaceuticals	1.0
fusible alloys	1.3

[1] 1972 *Metals Yearbook*

Ferroalloys

Several of the metals which are difficult to produce in pure form, or can be more conveniently used when alloyed, are available as ferroalloys. These are used as additions to steel. The metals thus available are molybdenum, vanadium, tungsten, titanium, and silicon.

Precious Metals

The uses of gold and silver in coinage and jewelry are well known. Gold is a standard of value and the basis of world monetary systems in the past. These metals are thus used because of their permanence and their rarity. Most of the gold and silver produced today are recovered as by-products of lead, zinc, and copper refining.

In lead refining, the gold, silver, and platinum are obtained in the zinc crust of the Parkes process. This crust is placed in a retort and the zinc distilled off. The residue is an alloy containing gold, silver, and platinum, in addition to the lead. This alloy is then cupelled (heated in an oxidizing atmosphere until the lead is oxidized), leaving an alloy of the precious metals. This alloy is known as *dore bullion*. By treating with nitric acid the silver is dissolved, leaving the gold and platinum, or all three metals may be separated electrolytically.

Silver for purposes of coinage and jewelry is alloyed with 8 to 10 percent copper. In England the silver standard was set at 925 parts per 1000 of silver; this is known as *sterling silver*. American coinage formerly contained 900 parts/1000 of silver.

Gold for coinage purposes and for jewelry is usually hardened by adding copper. For coinage, an alloy of 90 percent gold and 10 percent copper is used. The purity of such alloys in jewelry is expressed in carats. A carat is 1/24 part; hence 24-carat gold is pure, while 14-carat gold is 14 parts gold and 10 parts copper.

Platinum, palladium, and iridium also belong among the precious metals and are even more rare and expensive than gold. Platinum, due to its resistance to chemical attack and its high melting point, finds use in chemical equipment. Palladium is principally of use in dental alloys; iridium alloyed with platinum is used as pen points because of its hardness.

REVIEW

1. Name some common metals found free in nature.

2. What are the common impurities in native metals that must first be removed?

3. Describe gangue.

4. Name two methods of ore dressing.

5. Describe the method of ore treatment called oil flotation.

6. Name three types of furnaces used in refining metals.

7. What is the most important use of metallic copper?

8. What properties determine the principal uses for metallic lead and its alloys?

9. What is the Parkes process which is used in purifying lead?

10. What are the two principal uses of metallic zinc?

11. List some of the important uses of nickel.

12. Describe briefly the Bayer process for refining aluminum.

13. Name some of the important uses of magnesium.

14. What are the principal tin-producing countries?

15. List the most important uses of tungsten.

16. What are ferroalloys? And how are they used?

CHAPTER 17 — LIGHT METALS AND ALLOYS

As the world moved into the *Air Age*, it soon became apparent that light and stronger metals were needed. As progress was made in the development of processes to produce and shape the light metals, their use followed in the aircraft industry and the transportation equipment field. The use of light aluminum alloys in aircraft gave increased capacity with less weight. Highway trucks with aluminum vans and high-strength magnesium alloys in the running gear increased the revenue pay load. Titanium is used as the sheet metal covering in super-sonic aircraft where it can withstand the high heat of wind friction without a reduction in tensile strength.

High-strength, light-weight alloys are becoming increasingly important in reducing weight in vehicle structures, engines, transmissions and running gear. Light weight is proving to be the means to obtain economies in the operation of transportation on land, on water, in the air and in space travel.

The metals commonly classified as light metals are: aluminum, magnesium, beryllium and titanium. All have densities less than steel. Each of these metals will be discussed in the following pages.[1]

ALUMINUM AND ITS ALLOYS

Pure aluminum resists corrosion because a hard, tough surface coat of aluminum oxide forms immediately on the exposure of the hot metal to air. This prevents further oxidation and corrosion. However, when pure aluminum is alloyed

---

1. Since expressions of psi (pounds per square inch) in units of tens and hundreds of thousands are not absolute quantities, the conversion units of kN/m^2 (kilo-Newtons per meter squared) have been rounded off appropriately.

with other elements, it no longer will form the protective coat of pure aluminum oxide, but instead will form complex oxides of aluminum and the alloying elements which are not hard and tough and which do *not* protect the surface against further corrosion.

Five elements are commonly alloyed with aluminum to change its properties. These are: chromium, silicon, copper, manganese and magnesium. Although these alloys look alike, the physical properties and characteristics of each differ. Their workability, strength, heat-treat temperatures, ability to age and to harden and many other properties are vastly different. They cannot be worked or handled in the same way.

Aluminum is one of our foremost industrial metals and can be manufactured nearly pure, ranging from 99.5 to 99.9 percent aluminum. Nearly pure aluminum may be used in the following applications: to deoxidize steel, cooking and chemical apparatus, in the form of wire as an electrical conductor, as a pigment in paint, aluminum foil, welded to high-strength aluminum alloys as a covering, as fuel tanks for aircraft, etc. However, over 50 percent of the aluminum produced is used in the alloy form, making a series of alloys with much higher physical properties than those found in pure aluminum.

Properties of Aluminum

Aluminum is a very soft and ductile metal with a tensile strength from 9,000 psi (62,100 kN/m^2) in a single crystal form, to 24,000 psi (165,000 kN/m^2) in the fully work-hardened condition. Its excellent ductility and malleability allow the rolling of aluminum into thin sheets and foil, and the drawing of aluminum into wire. Hardness of aluminum varies from 25 Brinell in the annealed condition, to 40 Brinell in the fully work-hardened state. Good electrical conductivity is one of its most valuable properties, being 60 percent that of copper and 3.5 times that of iron.

Aluminum has nearly five times the thermal conductivity of cast iron. Due to its high reflectivity and low radiating power, aluminum is useful in the field of insulation. Aluminum has very good resistance to corrosion by weathering. The aluminum ornamental spire on the Washington Monument, Washington, D.C., has successfully withstood many years of weathering.

Aluminum can be successfully welded by the blowtorch, electric resistance, and electric arc methods. The soldering of aluminum is difficult and not recommended.

Clad Aluminum Products

In order that the corrosion of aluminum alloys, particularly in the aircraft field, be reduced to a minimum, a patented product called *Alclad* has been developed. Alclad is made by coating an aluminum alloy ingot with high purity aluminum and reducing the duplex ingot by rolling and shaping to the desired dimensions. The bond of high purity aluminum has proved very efficient, and the protection offered by the coating is very good even at the edges where the coating, being electronegative, protects the exposed edges by galvanic action. Aircraft sheets, for example, made from strong wrought alloys such as 2017 and 2024 are coated on both sides with a thin layer of pure aluminum.

Anodic treatment

Anodic oxidation methods are now available by which an oxide coating is artificially built up on the surfaces of aluminum alloys for increasing their resistance to corrosion. In a process called *anodizing*, the part to be treated is made the anode of a cell whose electrolyte may contain chromic acid, sulfuric acid, a nitride, or some other oxidizing agent. The anodic coating is only somewhat resistant to abrasion but does improve the corrosion resistance. In most cases, the coating is somewhat porous, thus producing a suitable surface for paints or dyes and further protection against corrosion.

Electroplating methods are also available, and in most cases are used where an improvement in wear resistance is desired. Electro-deposits of zinc, chromium, and nickel have been successfully produced.

Machining of Light Metals

In general, the light alloys of aluminum and magnesium are easy to machine and offer no real difficulties. Factors that reduce the machinability of these alloys are greater plasticity and high thermal expansion. The softer grades of these alloys, with their greater plasticity, are more difficult to machine than the harder alloys. Magnesium alloys with their lower plasticity, as compared with the aluminum alloys, are more easily machined. In general, satisfactory machinability of the light alloys requires special cutting tools, such tools having greater rake and clearance angles than steel-cutting tools and made sharper by grinding and polishing to a smoother finish. Lubricants of kerosene and mixtures of kerosene and lard oil are used with the machining of aluminum alloys, whereas magnesium alloys may be satisfactorily machined without lubricants, using sharp tools and high machining speeds.

A free-cutting aluminum alloy has been developed containing 5.5 percent copper, 0.5 percent lead, and 0.5 percent bismuth, desirable because it develops a heterogeneous structure containing lead and bismuth inclusions. The free lead and bismuth aid in machining, similar to the action of lead in free-cutting brass, by acting as chip-breakers and thus producing fine chips.

Aluminum alloys may be classified into two groups: (1) casting alloys and (2) forging alloys.

Casting Alloys

Aluminum is alloyed for casting for two main reasons: (1) to improve strength and (2) to improve casting properties. Because most of the elements alloyed with aluminum increase the strength and hardness of the resultant alloy, the added element is called a *hardener*. These hardeners may be added to the aluminum in the melting furnace, or added to the ladle of molten aluminum. The final casting may be made

Light Metals and Alloys

by pouring or casting the molten metal into a sand mold and allowing the metal to solidify.

However, many aluminum castings are made by casting the molten metal into a permanent or metal mold, or by a practice of die-casting in a die-casting machine. The properties and characteristics may be greatly influenced by the practice used in the production of the casting. Many of the defects found in commercial castings may be attributed to the practice used.

TABLE 17-1. NOMINAL CHEMICAL COMPOSITION — WROUGHT ALLOYS.

The following values are shown as a basis for general comparison of alloys and are not guaranteed.

Refer to Standards Section, Table 6.2 same issue for composition limits.

Alloy	PERCENT OF ALLOYING ELEMENTS—Aluminum and Normal Impurities Constitute Remainder									
	Silicon	Copper	Manganese	Magnesium	Chromium	Nickel	Zinc	Lead	Bismuth
EC			99.45 percent minimum aluminum							
1050			99.50 percent minimum aluminum							
1060			99.60 percent minimum aluminum							
1100		0.12	99.00 percent minimum aluminum							
1145			99.45 percent minimum aluminum							
1175			99.75 percent minimum aluminum							
1200			99.00 percent minimum aluminum							
1230			99.30 percent minimum aluminum							
1235			99.35 percent minimum aluminum							
1345			99.45 percent minimum aluminum							
2011	....	5.5	....	....	....	....	....	0.40	0.40
2014	0.8	4.4	0.8	0.50	....	....	....	....	....
2017	0.50	4.0	0.7	0.6	....	....	....	....	....
2018	....	4.0	....	0.7	....	2.0	....	....	....
2024	....	4.4	0.6	1.5	....	....	....	....	....
2025	0.8	4.5	0.8	....	....	....	....	....	....
2117	....	2.6	....	0.35	....	....	....	....	....
2124	....	4.4	0.6	1.5	....	....	....	....	....
2218	....	4.0	....	1.5	....	2.0	....	....	....
2219[2]	....	6.3	0.30	....	....	....	....	....	....
2618[3]	0.18	2.3	....	1.6	....	1.0	....	....	....
3003	....	0.12	1.2	....	....	....	....	....	....
3004	....	....	1.2	1.0	....	....	....	....	....
3005	....	....	1.2	0.40	....	....	....	....	....
3105	....	....	0.50	0.50	....	....	....	....	....
4032	12.2	0.9	....	1.1	....	0.9	....	....	....
4043	5.2	....	....	....	....	....	....	....	....
4045	10.0	....	....	....	....	....	....	....	....
4343	7.5	....	....	....	....	....	....	....	....
5005	....	....	....	0.8	....	....	....	....	....
5050	....	....	....	1.4	....	....	....	....	....
5052	....	....	....	2.5	0.25	....	....	....	....
5056	....	....	0.12	5.1	0.12	....	....	....	....
5083	....	....	0.7	4.45	0.15	....	....	....	....
5086	....	....	0.45	4.0	0.15	....	....	....	....
5154	....	....	....	3.5	0.25	....	....	....	....
5252	....	....	....	2.5	....	....	....	....	....
5254	....	....	....	3.5	0.25	....	....	....	....
5356[1]	....	....	0.12	5.0	0.12	....	....	....	....
5454	....	....	0.8	2.7	0.12	....	....	....	....

Metallurgy

TABLE 17-1. CONTINUED.

The following values are shown as a basis for general comparison of alloys and are not guaranteed. Refer to Standards Section, Table 6.2 same issue for composition limits.

Alloy	Silicon	Copper	Manganese	Magnesium	Chromium	Nickel	Zinc	Lead	Bismuth
5456	....	....	0.8	5.1	0.12	....	....	....	....
5457	....	....	0.30	1.0	....	....	....	....	....
5652	....	....	....	2.5	0.25	....	....	....	....
5657	....	....	....	0.8	....	....	....	....	....
6003	0.7	....	....	1.2	....	....	....	....	....
6005	0.8	....	....	0.50	....	....	....	....	....
6053	0.7	....	....	1.3	0.25	....	....	....	....
6061	0.6	0.27	....	1.0	0.20	....	....	....	....
6063	0.40	....	....	0.7	....	....	....	....	....
6066	1.3	0.9	0.8	1.1	....	....	....	....	....
6070	1.4	0.30	0.7	0.8	....	....	....	....	....
6101	0.50	....	....	0.6	....	....	....	....	....
6151	0.9	....	....	0.6	0.25	....	....	....	....
6162	0.6	....	....	0.9	....	....	....	....	....
6201	0.7	....	....	0.8	....	....	....	....	....
6253	....	....	....	1.2	0.25	....	2.0	....	....
6262	0.6	0.27	....	1.0	0.09	....	....	0.55	0.55
6463	0.40	....	....	0.7	....	....	....	....	....
6951	0.30	0.25	....	0.6	....	....	....	....	....
7001	....	21.	....	3.0	0.30	....	7.4	....	....
7005	....	....	0.40	1.4	0.13	....	4.5	....	....
7072	....	....	....	....	....	....	1.0	....	....
7075	....	1.6	....	2.5	0.30	....	5.6	....	....
7079	....	0.6	0.20	3.3	0.20	....	4.3	....	....
7178	....	2.0	....	2.7	0.30	....	6.8	....	....

[1]Titanium 0.13 percent   [2]Titanium 0.06; Vanadium 0.10; Zirconium 0.18   [3]Iron 1.1; Titanium 0.07

(Aluminum Standards and Data, 1974–75 Aluminum Association of America)

Many different alloys with widely different compositions are made, and Table 17-1 may be referred to for study of some of the more common aluminum alloys that are made by sand-casting techniques. The two principal casting light metal alloys are the aluminum-copper and the aluminum-silicon alloys. Perhaps the aluminum-copper may be considered the most common of all the casting alloys. These usually contain less than 15 percent copper alloy. A study of the aluminum-copper alloy diagram, Fig 17-1, indicates that the microstructures of these alloys consist of mixtures of solid solution (α), and the eutectic, which is made up of the solid solution (α) plus the compound $CuAl_2$. The eutectic is a fine mechanical mixture of the alpha and the hard $CuAl_2$ compound. It is questionable whether or not the hard intermetallic compound $CuAl_2$ is a solid solution instead of a compound. However, although the presence of the pure compound $CuAl_2$ is questioned, the term, $CuAl_2$ is widely used in referring to these alloys and will be used in the following discussion.

It will be seen in Fig. 17-1 that the eu-

Light Metals and Alloys

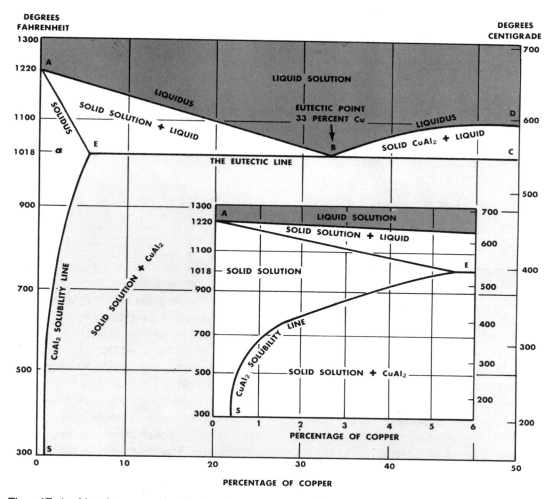

Fig. 17-1. Aluminum end of aluminum-copper diagram. (After Stockdale, Dix and Richardson)

tectic in these alloys increases in amount with the increase in copper content until a 100 percent eutectic is formed which is composed of 33 percent copper and 67 percent aluminum. The microstructure of a cast-aluminum alloy which contains 8 percent copper is shown in Fig. 17-2. The eutectic is plainly visible in the light background of the alpha solid solution. The eutectic constituent, being the last to freeze, is found between the dendritic branches of the primary crystals.

Metallurgy

Fig. 17-2. Cast aluminum-copper alloy. Etched ½ percent HF, magnified 100 times.

These alloys develop exceptionally high physical properties and are readily machinable. Tensile strengths ranging from 19,000 psi (131,000 kN/m^2) up to as high as 23,000 psi (159,000 kN/m^2) in the as-cast condition are common, with elongations from 1½ to 2 percent in 2 inches. These alloys may be welded and heat-treated, although heat treatments are more commonly applied to the forging types of aluminum alloys. Aluminum casting alloys are used extensively in the field of transportation and wherever a light structural alloy casting exhibiting the properties of these alloys is called for.

The aluminum-silicon casting alloys constitute an important series of casting alloys; the silicon in them may vary from 5 to 12.5 percent. These alloys are easier to cast than the aluminum-copper alloys due to greater fluidity at the casting temperature. The properties of certain compositions of these alloys may be improved by heat treatments, or by a modified practice in the foundry just prior to casting the molten metal in the mold. If the casting is made by the usual practice and contains about 12 percent silicon, the casting develops a coarse grain and is relatively brittle. By the use of a modified foundry practice, consisting of the addition of a small amount of metallic sodium to the ladle of molten metal, the casting freezes with a much finer and tougher grain structure. Through the modified practice, the foundryman can produce superior castings having distinctly higher strengths and higher elongations. The difference in grain structure of the cast metal made by the usual practice as compared with the modified practice is seen in Figs. 17-3 and 17-4.

The aluminum-silicon alloys exhibit excellent casting properties. This characteristic together with greater ductility and resistance to shock than exhibited by the aluminum-copper alloys, added to an even better resistance to atmospheric corrosion, make this type of casting alloy one of the most widely used in this country. Aluminum-silicon alloys are used in architectural and ornamental castings, castings for outboard motors, marine fittings,

Fig. 17-3. Microstructure of an unmodified 13 percent silicon casting, ½ percent HF etched, magnified 250 times. (Aluminum Company of America)

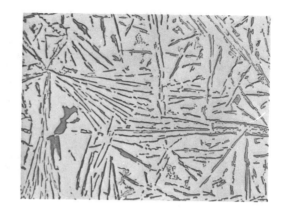

Fig. 17-4. Microstructure of a modified 13 percent silicon casting, ½ percent HF etched, magnified 250 times. (Aluminum Company of America)

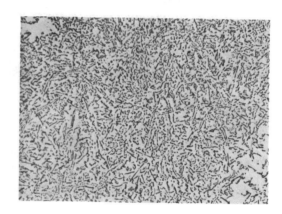

and wherever water-tightness and resistance to corrosion are considered important factors in the selection of a casting.

In addition to the aluminum-copper and aluminum-silicon casting alloys, some alloys are cast containing suitable proportions of magnesium and zinc. Alloys containing magnesium are said to be even more resistant to corrosion than the aluminum-silicon alloys and develop higher mechanical properties than the low-silicon casting alloys. Zinc may be used as an alloy in aluminum to increase its strength and hardness. It makes a cheaper alloy than copper, with approximately the same properties, but the aluminum-zinc alloys are heavier and have been reported to be less resistant to corrosion.

Wrought-Aluminum Alloys

Aluminum alloys of the forged or wrought type make up by far the greatest tonnages. The higher physical properties developed in these alloys are due to the beneficial effects obtained from the me-

Metallurgy

Fig. 17-5. 168-inch rolling mill. (Kaiser Aluminum & Chemical Corp.)

chanical treatment of rolling, Fig. 17-5, extruding, and forging, Fig. 17-6. This treatment refines the grain structure and makes the alloys more homogeneous. Due to the action of breakdown of the cast structure in any forging operation, a forging may always develop properties superior to those of a casting. In general, lower percentages of alloying elements are used in the wrought alloys. The operations in the production of wrought aluminum alloys include: (1) casting of a blank or ingot; (2) hot and often cold-working or shaping; (3) heat treatments.

Hot and cold-working. Hot-working of aluminum alloys is carried out between 600° and 900° F (around 315° to 480° C), and many desirable shapes of nearly pure aluminum and aluminum alloys may be produced by hot-working methods, including sheets, plates, tubes, bars, and structural shapes. However, it is more dif-

Light Metals and Alloys

Fig. 17-6. Large 8000-ton hydraulic press produces high-strength aluminum alloy forgings for aircraft, defense and commercial applications. (Kaiser Aluminum & Chemical Corp.)

ficult to hot-work aluminum and its alloys than it is steel. This is perhaps due to the lower temperatures used in the hot-working of aluminum compared to the temperatures used when hot-working steel.

The cold-working of aluminum and its alloys results in the same work-hardening effects resulting from the cold working of any metal or alloy, Fig 17-7. Shaping by cold-rolling, drawing, or by any of the methods used in our modern metalworking plants, may be used for the double purpose of getting the desired shape in the aluminum and at the same time causing improvement of physical properties through the changes brought about by the cold-working or work-hardening.

In the normal production of cold-worked shapes of aluminum, the aluminum is first cast into the ingot form, the ingot form is then reduced by hot-working methods to the desired size and shape for the cold working operations. The cold-working starts where the hot-working stops, although it should be recalled that annealing or other heat treatments may be given the metal before any cold reduction is started.

Aluminum finished by cold-rolling may be purchased with varying amounts of cold-reduction and therefore with varying degrees of hardness.

Metallurgy

Fig. 17-7. New sheet facility produces coil weights up to 40,000 pounds (18,181 kilograms) per roll. (The Aluminum Association)

Heat treatment of aluminum and its alloys. The commercially pure aluminum 1100 may be subjected to annealing after cold-working. The annealing practice carried out with cold-worked 1100 consists of heating the metal slightly in excess of 650°F (345°C), where complete softening is almost instantaneous, followed by cooling to room temperature. Although it is well to establish a standard practice of annealing, the exact temperature used and the time at this temperature, along with the cooling rate, is important only provided the annealing temperature selected exceeds the recrystallization temperature of the metal. The annealing heat treatment allows the strain-hardened crystal state of the aluminum to recrystallize into new grains. These new-formed grains grow to a desired size, removing all of the strain-hardened condition of the aluminum and restoring the original soft and plastic state of the metal.

The wrought-aluminum alloys may be

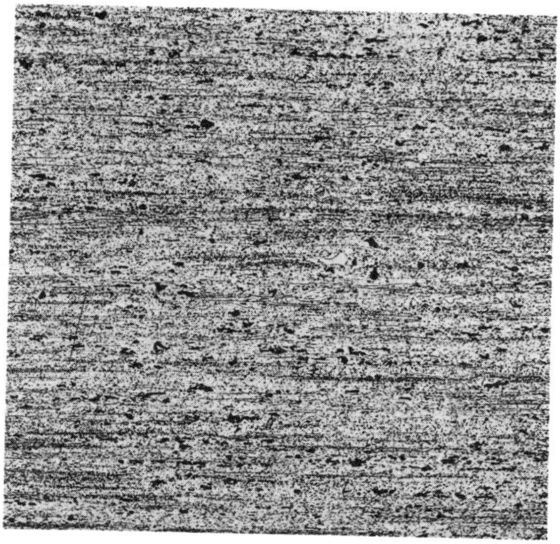

Light Metals and Alloys

Fig. 17-8. Hard-rolled duralumin type alloy sheet, with distortion of structure due to cold working. Magnified 100 times. (After Dix and Keller)

subjected to annealing heat treatments similar to the pure aluminum, but with the alloyed aluminum a very slow rate of cooling from the annealing temperature is required if the maximum temperature used in the annealing operation exceeds 650° F (about 345° C) by much more than ten degrees. The reason for this may be obtained from a study of the aluminum-copper diagram, Fig. 17-1.

Examination of the aluminum-copper diagram reveals that an alloy of 95 percent aluminum and 5 percent copper freezes as a solid solution alloy known as duralumin. Upon slow cooling, the solubility of copper (probably as $CuAl_2$), decreases along the *E-S* line of the diagram, so that at 300° F (149° C) the solubility of copper in aluminum is only 0.5 percent copper. Therefore, upon slow cooling of this alloy of aluminum-copper from around 970° F (520° C), the copper is gradually precipitated within the grains of aluminum as rather coarse particles of $CuAl_2$. This condition may be considered as the annealed or normal state of the aluminum-copper alloy; the structure is in a state that may be subjected to cold-forging and consists of the relatively hard $CuAl_2$ constituent embedded in a soft aluminum matrix. This structure may be cold-worked as indicated in Fig. 17-8 and if we wish to carry out an annealing treatment after cold-working, a temperature not exceeding 650°F (345° C) is sufficient to recrystallize the structure and remove the strain-hardening resulting from the cold working. A temperature of 650° F (345° C) will recrystallize the aluminum phase in the aluminum-copper alloy without markedly dissolving the $CuAl_2$ phase as indicated by the *S-E* line in Fig. 17-1. The structure developed by this treatment is illustrated in Fig. 17-9. The rate of cooling from the annealing temperature is not important if the copper as $CuAl_2$ is not dis-

Metallurgy

Fig. 17-9. Showing same section as Fig. 17-8, after annealing; many small particles of $CuAl_2$ present in structure. Magnified 100 times. (After Dix and Keller)

solved; however, if the copper becomes dissolved, rapid cooling will keep it in solution and the alloy is not considered to be in its best annealed condition. If the alloy is to be annealed completely for severe cold-forming operations, heating to a temperature of 750° F to 800° F (about 400 to 425° C) for about two hours should be used, followed by cooling at a rate not to exceed 50° F per hour (28° C/hr) down to 500° F (260° C).

Solution heat treatment. Several of the aluminum alloys may have their properties greatly enhanced by suitable heat treatments. The 2017 type of aluminum-copper alloy is one of the oldest and is still one of the most widely used alloys that will respond effectively to simple heat treatments. The possibility of carrying out heat treatments depends upon the changing solubility of aluminum for the constituent $CuAl_2$ during heating and cooling, within the solid state, as indicated by the curve S-E, Fig. 17-1.

If we reheat a normal, annealed aluminum-copper alloy containing 4.5 percent Cu to a temperature approximately 950° F (510° C) for a period of 14 hours, most of the $CuAl_2$ phase present in the original annealed condition of this alloy is dissolved. If we cool this solid solution structure very rapidly, such as by quenching in water, most of the $CuAl_2$ phase will be retained in solution in the aluminum. The best results of this treatment are obtained if a drastic quench is used, although sometimes a slower quenching medium such as oil is employed.

After this treatment, examination of the structure of the alloy will reveal very little of the $CuAl_2$ phase present. This condition is illustrated in Figs. 17-10 and 17-11. Apparently, the balance has been retained in solution by the severe rate of quenching from the solution temperature. This treatment has been referred to as a *solution heat treatment*. Improvement in corrosion resistance and increase in tensile strength are obtained by this treatment. Cold working may be carried out after the solution heat-treatment, thereby increasing the strength without any effect upon the corrosion resistance of these alloys. However, properties of the solution-heat-treated alloy may spontaneously change with the passage of time at room temperature. After 30 minutes to one hour at room temperature, the solution-treated alloy may noticeably increase in strength, a phenomenon known as *precipitation hardening*.

Light Metals and Alloys

Fig. 17-10. Structure of duralumin type alloy sheet after heat treatment; practically all the soluble $CuAl_2$ constituent is dissolved. Magnified 100 times. (After Dix and Keller)

Fig. 17-11. Showing same structure as Fig. 17-10, magnified 500 times. Light particles with black boundaries are undissolved $CuAl_2$. (After Dix and Keller)

Precipitation hardening. The increase of strength that takes place with the passage of time at room temperature in a solution-heat-treated alloy is thought to be caused by the gradual precipitation of the $CuAl_2$ phase in very fine submicroscopic particles. This precipitation takes place at the aluminum grain boundaries and along crystallographic planes within the aluminum crystals. The very fine particles of $CuAl_2$ precipitate act as keys and build up resistance to slip, thereby greatly increasing the strength of the alloy. After the solution heat treatment, the strength of the 2017 alloy is about 44,000 psi (303,000 kN/m^2), with an elongation of 20 percent in 2 inches. Upon aging at room temperature, the strength increases rapidly within the first hour, and after 4 to 5 days the strength has increased to approximately 62,000 psi (422,000 kN/m^2). A very singular fact is that the ductility, as measured by elongation, does not change during the aging period, but remains about 20 percent in 2 inches. This treatment produces an alloy that is almost as strong and as ductile as structural steel, with about one-third the weight.

Instead of aging the quenched alloy at room temperature in order to obtain the precipitation hardening, the alloy may be caused to precipitate the $CuAl_2$ phase more rapidly by heating somewhat above room temperature; however, rapid precipitation does not develop so high a tensile strength. Some compositions of aluminum alloys that respond to heat treatments, such as solution or precipitation treatments, do not develop any increase in tensile strength at room temperature after the solution treatment. However, in these

Metallurgy

alloys, heating to slightly above room temperature allows the precipitation to take place, producing an increase in strength. The precipitation hardening that takes place at room temperature in these alloys may be prevented by cooling solution-heat-treated alloys to below 0° F (-18° C), and holding this low temperature. This may be accomplished through the use of dry ice (CO_2). It is assumed that the low temperature increases the rigidity of the structure enough to prevent the precipitation of the $CuAl_2$ phase; or that the atomic structure is so lacking in atomic mobility that any change, such as the precipitation of the $CuAl_2$ constituent, cannot take place.

The capacity for precipitation hardening, also called age hardening, to take place at room temperature depends on the presence of a small quantity of magnesium such as is found in the composition of the duraluminum type (2017) alloys.

The aluminum alloys, without magnesium, age-harden much less at room temperature. The addition of silicon to aluminum-magnesium alloys produces little age hardening at room temperature but exerts pronounced precipitation hardening effects at high temperatures. The recommended heat treatment for the Al-Mg-Si type alloys consists in quenching from 960° to 980° F (about 515° to 525° C) in water and aging at 310° to 340° F (156° to 171° C) for as many as 18 hours in order to get the maximum precipitation hardening effects.

There are a considerable number of different types of precipitation-hardening aluminum alloys that are commercially used, some of which are indicated by the alloy combinations Al-Cu-Mn, Al-Cu-Mg-Si, Al-Cu-Mg-Ni, Al-Mg_2Si, and Al-Mg_2Zn.

Heat treatment of casting alloys. Cast alloys of the aluminum series, as well as the forging alloys, may be subjected to heat treatments which markedly improve their physical properties. As a general rule, the casting alloys require much longer soaking periods at the temperatures used for the solution heat treatment, and the temperatures used for precipitation heat treatments, in order to bring about the desired changes in structure and in properties.

DIE-CASTING

Die-casting is considered the simplest method of converting raw material into a finished product. The advantage of die-casting is that it allows low-cost part production, close tolerance, and extremely smooth surfaces, which require little or no machining, Fig 17-12.

In die-casting, the metal must be heated to a molten state. It must not be formed while it is in a plastic state, which was required for hot-forging operations. The molten metal is forced into a die at a high pressure. A die consists of two matching halves whose facing sides contain a

Light Metals and Alloys

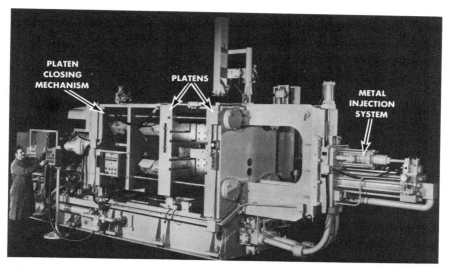

Fig. 17-12. Modern automatic die-casting machine. (Greenlee Tool Co.)

smooth cavity formed exactly like the final part desired. After a casting is made, the two halves are parted to enable removing the finished product. The die is permanent; it is not destroyed with each casting cycle as is the case with sand molds.

Die-Casting Machines

Such machines are made in two types, which are known as (a) hot-chamber machines and (b) cold-chamber machines. Both types are generally made completely automatic. They have all their major components automatically operated.

Hot-chamber machine. For die-casting zinc, use is made of a machine that is constructed as shown in Fig. 17-13. It contains a hemispherical cast-iron pot fitted into a furnace. The zinc is maintained in a molten state by a gas burner. The life of the pot depends somewhat on the thickness of its walls and also upon the extent that it is overheated. A pot has a life of from 2 to 6 months. Some improvement in pot life can be obtained by applying a moderately heavy oil to its warm surface followed by application of heat to carbonize the oil.

A cylinder with a ram-plunger and an attached gooseneck is submerged in the molten metal as indicated in Fig. 17-13. When the plunger is retracted, metal enters through the indicated port that is below the metal level. As the plunger is forced downward, the molten metal is forced through the gooseneck into the die. The movable platen, which is actuated by a plunger, presses the die-seat against the nozzle to prevent leakage of the metal. After the casting is formed and has solidi-

Metallurgy

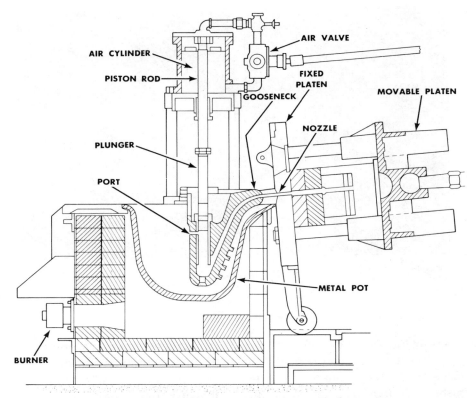

Fig. 17-13. Hot chamber type of die-casting machine.

fied, the die is opened. The sprue breaks away between the die and the nozzle to free the casting, which can now be removed.

In such die-casting machines, the control for metal injection is interlocked with the die operating mechanism so that metal cannot be injected unless the die is securely locked.

Cold-chamber machine. Some metals have a considerably higher melting point than zinc. Common metals falling in this class are aluminum, copper, and magnesium. It is undesirable to die-cast these materials by using a hot-chamber machine. In such a machine, these metals kept in a molten form in an iron pot would pick up iron. This iron pick-up would eventually cause the plunger to seize in the cylinder and would also cause a short pot life. The properties of the core produced would also be affected.

Because of these detrimental effects, higher melting alloys (aluminum, copper, and magnesium) are cast in cold-chamber machines. The typical construction of such a machine is shown in Fig. 17-14. The cylinder, which forms the cold cham-

Light Metals and Alloys

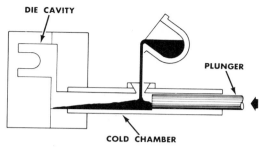

Fig. 17-14. Cold chamber type of die-casting machine.

ber, is horizontal and is filled through the pouring slot. The two halves of the die are locked together and the plunger is advanced, forcing the metal into the die. A slug is left in the end of the cylinder, which is attached to the casting. Next, the die-holder is unlocked and the die is opened. Because the plunger is still under pressure, it continues its stroke and retains the slug and the attached die-casting in contact with the movable die.

Die-Cast Machine Materials

The kind of steel from which dies are made depends upon the kind of material that is to be cast. For casting lead, tin, and Babbitt materials, mild steel is adequate. For casting zinc, an intermediate grade of steel is recommended that is heat-treated for long usage. For casting metals having a higher melting point, such as aluminum, special steel is required for making the dies, which must be heat-treated. The same steel serves the purpose for magnesium castings. For copper-base die castings, very special alloy steels are required, which also must be heat-treated. Die-life is short for copper-base materials, considerably better for aluminum and magnesium, while an extremely long life may be expected from a die used for making zinc castings.

Common Die-Casting Materials

The metals most commonly used for die-casting are zinc, aluminum, copper, and magnesium. The extent each is used depends upon (a) cost of the alloy, (b) ease of casting, (c) adaptability for a particular use, (d) ease of machining, and (e) cost of dies.

For die-casting, zinc has many favorable properties and it is therefore extensively used for the following reasons: (1) cost per pound is low; (2) it is cast at a low temperature (having a melting point of 718°F or 381°C) which minimizes the cost of fuels, low initial die cost and die upkeep; (3) it lends itself to casting within close dimensional limits; (4) it is easily machined, where machining is necessary; (5) it is readily finished by electroplating or organic coating; (6) it has greater strength than any other non-copper alloy; and (7) it has resistance to corrosion.

Since zinc has a low melting point, it is casted with a hot-chamber machine shown in Fig. 17-13. It is essential that only a very high grade of zinc, 99.99 percent pure, be used. In the cast state, zinc has the properties indicated in Table 17-2. If impurities such as lead, cadmium, or tin are used in excess, the desirable physical and mechanical properties are affected.

Metallurgy

TABLE 17-2. PROPERTIES OF ZINC FOR DIE CASTING.

ELEMENTS		PERCENT
Copper, max	SAE903	0.10
	SAE925	0.75–1.25
Aluminum		3.5–4.3
Magnesium		0.03–0.08
Iron, max		0.1
Lead, max		0.007
Cadmium, max		0.005
Tin, max		0.005
Zinc (99.99+)		remainder
MECHANICAL PROPERTIES	SAE 903	SAE 925
Tensile strength (ultimate)	41,000 psi	47,600 psi
Brinell hardness	82	91
Compression strength	60,000 psi	87,000 psi
Melting point	717.6°F	717.1°F
Specific gravity	6.6	6.7

MAGNESIUM AND ITS ALLOYS

As stated previously, magnesium is the lightest of the commercial metals, having a density of 1.74 at 68°F (20°C). Its weight per cubic foot is 108.5 pounds, (49 kilos) as compared with 175 (80 kilos) for 99.0 percent aluminum, and 557 pounds (253 kilos) for copper. It has a melting point of 1202°F (650°C) and may be cast. Magnesium may also be extruded and rolled hot or cold. Magnesium is plastic enough at 650°F (345°C) to be hot-worked by either rolling or extrusion methods. The metal work hardens rapidly upon cold-rolling and requires frequent annealing to maintain its plastic and workable nature. Although the mechanical properties of magnesium are relatively low, additions of alloying elements such as aluminum, zinc, and manganese greatly improve these properties.

The magnesium alloys are second in importance to the aluminum alloys as light structural alloys. The most prominent of the light magnesium alloys are the series of magnesium alloys containing 4 percent to 12 percent aluminum and from 0.2 percent to 1.5 percent manganese, the balance being magnesium. These alloys are sold in this country under the name *Dow Metal*.

The most outstanding characteristic of Dow Metal is extreme lightness, being about two-thirds as heavy as aluminum alloys. As cast, the magnesium alloys have a tensile strength of from 18,000 to 35,000 psi (124,000 to 241,000 kN/m^2). After forging, the strength may be increased to approximately 48,000 psi (331,000 kN/m^2). Although these alloys respond to heat treatments similar to those used on the aluminum-copper alloys, the forged and extruded shapes of Dow Metal do not require further heat treatments for the development of satisfactory properties. This is an important advantage for these alloys. However, casting alloys are heat-treated where severe service conditions demand maximum physical properties.

CAUTION:

Due to its high affinity for oxygen, magnesium is highly flammable. The stability of magnesium and its alloys has been proven by its resistance to normal atmospheric corrosion either in the natural state or after proper protective coatings have been applied to exposed surfaces. Lightness and easy machinability are two of the outstanding properties of the magnesium alloys.

Protective Coatings

When magnesium alloys are subjected to normal atmospheric exposure, very little surface change occurs other than the formation of a gray discoloration, and only slight loss of strength occurs after several years of exposure. However, if these alloys are exposed to salt water and are contaminated with iron and other impurities, very rapid corrosion occurs. In order to increase the corrosion resistance of magnesium alloys, care should be exercised during their manufacture to avoid metallic impurities and flux inclusions. To further increase their corrosion resistance, magnesium alloys can be given a chemical treatment that forms a protective surface film on exposed surfaces. Much of the protection offered by chemical treatments lies in their ability to clean and remove impurities from the metal surfaces. Two commonly used chemical treatments are the chrome-pickle treatment and the dichromate treatment.

The chrome-pickle treatment consists in immersing the parts for one minute at room temperature in a bath of the following composition:

Sodium dichromate
 ($Na_2Cr_2O_7 \cdot H_2O$).......................... 1.5 lb
Concentrated nitric acid
 (specific gravity 1.42).................... 1.5 pt
Water to make up to....................... 1.0 gal

After immersion, the part is rinsed in cold water, then dipped in hot water, and allowed to dry. The resultant coating on clean surfaces is of a matte to brassy, iridescent color. This treatment is largely used for protection of magnesium alloys during processing, storage, and shipment.

In the dichromate treatment, the parts are first cleaned and carefully degreased and immersed for 5 minutes in an aqueous solution containing 15 to 20 percent hydrofluoric acid (HF) by weight, the solution being at room temperature. The parts are then rinsed in cold running water. This treatment is followed by boiling the parts for 45 minutes in an aqueous solution containing 10 to 15 percent sodium dichromate. After boiling, the parts are rinsed in cold running water, followed by a dip in hot water to facilitate drying. The addition of calcium fluoride or magnesium fluoride to the dichromate bath will improve corrosion resistance and promote film formation and uniform coating.

Metallurgy

Magnesium Casting Alloys

Aluminum is the principal alloying element used to improve the mechanical properties of magnesium. Zinc and manganese are also added because they improve the mechanical properties and resistance to corrosion. Both aluminum and zinc are soluble in magnesium in the solid state to a limited extent. The magnesium end of the magnesium-aluminum alloy system is illustrated in Fig. 17-15 and it will be seen from a study of this diagram that about 2 percent aluminum is soluble in magnesium when at room temperature. Upon heating magnesium to 817°F (436°C), the solubility increases to about 12 percent aluminum. This change in solid solubility with the temperature change makes it possible to subject these alloys to heat treatments comparable to those given to aluminum alloys; i.e., solution and precipitation heat treatments. The solution heat treatment usually requires heating to 760° to 780°F (about 405° to 415°C) for 16 to 18 hours, followed by water quenching. The aging or precipitation treatment consists of heating the solution-heat-treated alloys to 350° to 360°F (177° to 182°C) for 4 to 18 hours. The precipitation hardening resulting from this treatment raises the strength and hardness but lowers the ductility and toughness characteristics of these alloys.

Magnesium alloy castings require special care in foundry practice in order to produce satisfactory castings, magnesium being very sensitive to oxidation conditions and having a low specific gravity with a large shrinkage ratio. The factors of special foundry technique include: protective flux during melting and pouring, a specially treated molding sand to prevent reaction between the sand and metal, large gates and risers to compensate for the extreme lightness of the metal, and pattern design allowance for shrinkage.

The high mechanical properties of the 6 percent Al : 3 percent Zn type, combined with its superior corrosion resistance, have made it the leading casting alloy. This type of magnesium alloy is heat treatable and develops its maximum properties only upon proper heat treatment.

Magnesium alloy castings are used in the manufacture of household appliances, foundry equipment, portable tools, aircraft gear and engine castings, high-speed machinery, and in the automobile industry.

Wrought Magnesium Alloys

Alloys of magnesium containing aluminum, manganese, and zinc are now available in many of the standard wrought structural shapes, such as bars, rods, tubes, angles, etc. Although these alloys do not lend themselves readily to cold forming or forging methods, they may be satisfactorily shaped by hot methods such as extrusion, press forging, bending, etc. They work-harden so rapidly upon cold shaping that only small amounts of cold work can be carried out. However, at a temperature range between 550° to 600°F (about 290° to 315°C) most of the alloys may be successfully shaped by the usual mechanical methods. If only slight or light working is required to shape a part, temperatures as low as 400°F (205°C)

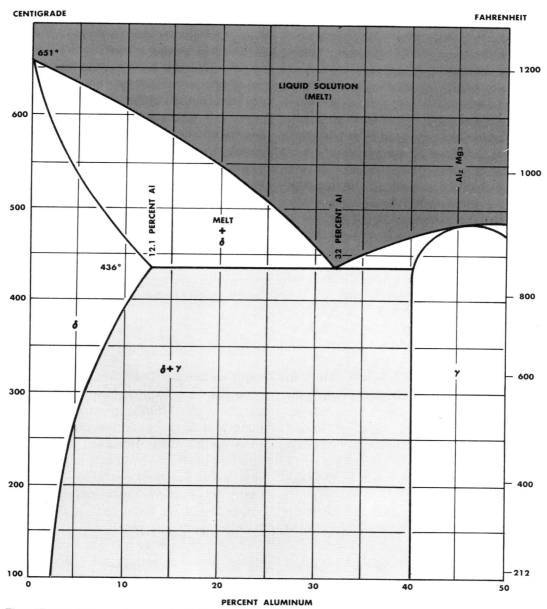

Fig. 17-15. Magnesium end of the magnesium-aluminum diagram. (After Hanson, Gayler, Schmidt, and Spitaler)

Metallurgy

may be used. An alloy with about the best cold-forming characteristics is the ASTM No. 11, containing 1.5 percent Mn. This is recommended for use where maximum cold-forming properties are needed, along with resistance to salt water corrosion.

BERYLLIUM

This high-strength, light-weight metal has a density near to that of magnesium. Until the 1950's its use was principally as an alloying agent which hardened copper. Beryllium has a grayish-steel color and does not corrode easily as an oxidized film which forms immediately on the surface protects it from further oxidation.

Beryllium is available in blocks produced by hot pressing of beryllium powder, in sheets, plates and extrusions. Minimum tensile strengths, from 70,000 psi (482,700 kN/m^2) for hot pressed parts to over 200,000 psi (1,4000,000 kN/m^2) for drawn wire, gives beryllium an outstanding strength-to-weight ratio. A forged beryllium alloy with a strength of 128,000 psi (882,000 kN/m^2) is equivalent to a steel forging with a strength of 555,000 psi (3827 kN/m^2).

Beryllium has a high specific heat which makes the metal useful for electronic heat sinks, aircraft brakes and aero-space re-entry heat shields. A pound of beryllium will absorb as much heat as two pounds of aluminum or five pounds of copper in a one degree temperature change.

Beryllium has three industrial disadvantages in its use. These are its high cost, potential toxic hazards from inhalation of beryllium dust during machining operations and a tendency for brittleness at room temperature. However, its production has been slowly rising, reaching current levels indicated by Table 17-3. High purity beryllium from electron-beam zone refining has been found to be more ductile than arc melted beryllium.

Solid beryllium is non-hazardous if no dusts, fumes or vapors are produced by processing (these are toxic when inhaled). Beryllium can be forged, extruded, drawn and rolled. It can also be welded, using the TIG or MIG welding process.

TABLE 17-3. BERYLLIUM PRODUCTION.[1]

	SHORT TONS
World production	4,740
U.S. imports	3,345
consumption	7,781
USES: BeCu alloys, BeO ceramics	

[1] 1973 *Metals Yearbook*

TITANIUM

Titanium-bearing ores are plentiful, but the technical difficulties in recovering, melting, and processing titanium make the metal costly. However, since titanium has a very high strength-to-weight ratio at temperatures between 300°F and 700°F (150° to 370°C), it is widely used in airframes and aircraft engines at prices 30 to 40 times that of stainless steel.

Aeronautic engineers claim that a one-pound weight reduction in a jet engine results in an eight to ten-pound reduction in air-frame weight. Titanium thus saves 500 pounds (227 kg) in each of the eight jet engines of a heavy bomber and makes possible a reduction of 40,000 pounds (18,143 kg).

No other structural metal presents the problems encountered in producing titanium. The liquid metal seems to be a universal solvent, which either dissolves or is contaminated by every known refractory. The contaminants acquired during the manufacturing process are generally fatal to titanium's physical properties. The metal, when molten, unites with the oxygen and nitrogen in the air with such speed that all reduction and ingot-melting processes must be carried out in a vacuum or in an inert atmosphere. No method has been found to remove contaminants acquired during processing.

The simple and efficient continuous Kroll-type reactor has lowered the cost of titanium and is still in use. In the reactor, titanium tetrachloride is reacted with magnesium to form spongy titanium and magnesium chloride. The process produces many tons of titanium.

Some progress has been made in obtaining commercial titanium by reacting purified sodium with tetrachloride, by continuous thermal decomposition of the iodide and by arc dissociation of halides (such as titanium tetrachloride). However, major efforts are being expended on electrolytic methods. The electrolysis of molten-salt electrolyte seems to promise recovery of nonmassive deposits in pure enough form to melt into ductile metal.

The first step in winning titanium metal from its ore is to chlorinate an oxide-carbon mixture to secure titanium tetrachloride. Titanium tetrachloride is reduced to titanium sponge by reduction of the chloride by magnesium. The reaction produces titanium sponge and magnesium chloride. The magnesium chloride is passed through an electrolytic cell and magnesium and chlorine are recovered and may be used again.

Production

In the molten state, titanium reacts vigorously with the oxygen and nitrogen of the air. If the amount of these elements is kept below 0.5 percent, they act as hardening agents, but in amounts over 0.5 percent oxygen or 0.25 percent nitrogen they

Metallurgy

drastically reduce the ductility of titanium. Hydrogen embrittles titanium, reducing its ductility. Adequate protection from moisture and the atmosphere are therefore important. Vacuum technology or inert atmospheres of pure argon or helium can accomplish this.

The ordinary techniques of melting and casting are ruled out. The method usually used is the von Bolten process of arc-melting in an inert atmosphere at near atmospheric or greatly reduced pressures. Electrodes may be of tungsten, carbon, or consumable electrodes of titanium. A difficulty of tungsten and carbon electrodes is that they are subject to spalling and melting, particularly if they are spattered with titanium. Melting is usually done in a water-cooled copper crucible, building up each ingot individually, or by extracting the ingot through the bottom of the crucible. By using a reduced pressure and double melting with consumable electrodes, contamination is avoided and hydrogen content of the metal is kept at a low level.

Forging, rolling, and drawing present no special problems. Mills capable of handling the stainless steels and the alloy steels can handle titanium.

Properties

Silver-gray titanium lies between aluminum and steel in strength, density, elasticity, and serviceability at elevated temperatures. It has a density of 0.16 lb/in^3 (0.26 kg/cm^3). It is 60 percent heavier than aluminum and only 56 percent as heavy as an alloy steel, which has a density of 0.286 lb/in^3 (0.458 kg/cm^3). The alloys of titanium are stronger than all aluminum alloys and most steel alloys. Titanium also has excellent ductility. The titanium alloys are superior in strength-to-weight ratios over the usual engineering metals and alloys. Titanium has an endurance ratio well above heat-treated alloy steels and nonferrous materials. Its endurance limit is always more that 50 percent of its tensile strength.

The coefficient of thermal expansion of titanium alloys is about 40 percent less than for austenitic stainless steel. Titaniums's low expansion sometimes introduces thermal stresses when alloyed with other metals. Thermal conductivity of titanium alloys is about the same as austenitic stainless steel. It may vary up to 25 percent depending upon alloying constituents. The electrical resistivity of titanium is approximately the same as austenitic stainless steel and increases with more alloying elements, making it easy to heat by electrical means for working.

Fatigue resistance. Titanium and its alloys have good fatigue resistance in the unnotched condition. With very sharp notches, titanium appears to be greatly inferior to steel. The deficiency seems to disappear as the notch is made rounder and, under some notch conditions, titanium may be superior to other metals. Where notch fatigue is important, the user should conduct specific tests to determine titanium's desirability. The fatigue limit of titanium improves as the temperature is lowered to sub-zero. Not much research has been done on titanium and its alloys at low temperatures. However, what little research has been done indicates that hardness and tensile strength are in-

Light Metals and Alloys

creased as temperature is lowered. Elongation decreases at low temperatures, and notch sensitivity appears to increase.

Elevated-temperatures. Titanium has excellent weight strength ratios between 300° and 700°F (150° to 370°C), but it is inadequate at high temperatures. The ultimate strength and yield strength of titanium drop fast above 800°F (425°C) Titanium readily absorbs oxygen and nitrogen from the air above 1200°F (650°C), embrittling the metal. There is hope for development of alloys which will withstand temperatures up to 1400°F (760°C).

Chemical corrosion properties. Titanium is almost immune to most corrosives, but when it is subject to attack, the rate of corrosion is usually severe. One chemical concern tested titanium with sulfuric acid concentrations up to 22 percent at high heat and pressure, and titanium remained unaffected.

Titanium is exceptionally immune to sea water and marine atmosphere. It is the only structural metal that has a corrosion behavior in salt water identical to that in air. This places a great demand on titanium for ship building. Unalloyed titanium retains its original luster and appears to be absolutely unaffected when exposed to rural and marine atmospheres for a period of five years. Titanium should be insulated when used with other metals by organic coating or electrical insulation, because it causes severe corrosion to metals with which it is in contact.

Commercially Pure Titanium

Titanium of the highest purity possible at a reasonable production cost is called commercially pure titanium, which is 99.5 percent pure. The importance and need for a standard was recognized by consumers, so commercially pure titanium was established as a composition of 0.2 percent maximum iron, 0.10 percent maximum nitrogen, a trace of oxygen, less than 0.70 percent carbon, and 0.02 percent tungsten, the remainder being titanium.

Type Ti-75A (Timet Corp.) has the following characteristics: hardness, 85-95 Rockwell B; ultimate strength, 80,000 psi (552,000 kN/m^2) min; yield strength, 70,000 psi (483,000 kN/m^2) min, and elongation in two inches, 20 percent min.

Commercially pure titanium does not respond to heat-treatment but can be cold-worked to above 120,000 psi (827,000 kN/m^2) tensile and 100,000 psi (690,000 kN/m^2) yield strength with a consequent drop in ductility. Commercially pure titanium and titanium base alloys are available in the form of sheet, strip, plate, wire bar, rod, forging billets, and forgings.

Titanium Alloys

Laboratories have explored thousands of titanium alloys of various combinations. Some having very impressive physical properties are commercially available. One difficulty experienced was reproducing the alloys from heat to heat. This problem was solved by improving melting practices. The variations in raw sponge and the tremendous effect of contaminants which are readily picked up by titanium made it very difficult, if not impossible, to duplicate previous heats.

Metallurgy

The size of the titanium atom is such that it fits well with other metallic atoms, indicating that it may make many good solid solution type alloys. Titanium also makes a physical transformation at 1625°F (885°C), thus indicating a possibility of producing various age-hardenable alloys with alloying elements whose solid solubilities decrease with temperature.

Titanium shows strong tendencies toward forming interstitial solid solutions with *metalloids* (elements having both metallic and nonmetallic properties). Oxygen, nitrogen, and carbon dissolve in molten titanium to form solid solutions or non-metallic inclusions (foreign matter). *Binary* (two-constituent) alloys of these elements do not respond to heat-treatment but are readily work-hardened. Iron, chromium, and molybdenum stabilize the high temperature beta phase of titanium, producing alloys which harden when quenched. The *quaternary* (four-constituent) alloys formed from titanium, iron, chromium, and molybdenum have better combinations of high strength and ductility than have been obtained from *binary* or *ternary* (three-constituent) alloys at the same level.

In addition to the commercially pure grade (Ti-75A) and the extra soft variation special grade, the Timet Corp. markets several different titanium-base alloys, having nominal compositions and annealed physical properties of sheet as follows: Type Ti-100A, an oxygen-nitrogen alloy, is used primarily for sheet, plate, strip, and wire nonheat-treatable but work-hardenable. Mill products of this alloy have physical properties as follows: hardness 30-C Rockwell max; tensile strength, 100,000 psi (690,000 kN/m^2) min; yield strength, 90,000 psi (620,000 kN/m^2) min; and elongation, 15 percent min.

Type Ti-140A is a two-phase, high-strength alloy sold primarily as bar and forgings, but also in experimental production as sheet and strip. This alloy has high-temperature stability and excellent impact values. It may be spot-welded, but fusion welds show only indifferent ductility values. In bar and forgings the annealed properties are as follows: hardness, 300 to 304 Brinell hardness number; tensile strength, 130,000 psi (896,000 kN/m^2) min; yield strength, 120,000 psi (827,000 kN/m^2) min; and elongation, 12 percent min.

Type Ti-150A is an alloy sold primarily as bars and forgings-heat-treatable and work-hardenable. It is moderately responsive to heat treatment is supplied as annealed at about 1250°F (675°C). Physical properties are: hardness, 311-364 Brinell hardness number; tensile strength, 135,000 psi (930,000 kN/m^2) min; yield strength, 120,000 psi (827,000 kN/m^2) min; and elongation, 12 percent min.

Type Ti-155A is a very high-strength forging alloy, beta stabilized by iron, chromium, and molybdenum additions and with sufficient aluminum to maintain high strength at temperatures up to 1000°F (538°C). This alloy is only in experimental production, but has already been welcomed in the jet-engine industry as an ideal composition for compressor wheels, particularly for later stages where high temperatures build up as the air is compressed. This alloy also exhibits excellent creep and impact performance. Annealed Ti-155A bars and forgings have the following physical values: hardness, 300 to 370 Brinell hardness number; tensile strength, 155,000 psi (1,070,000 kN/m^2); yield strength, 140,000 psi (965,000 kN/m^2); and elongation, 12 percent.

Uses

Virtually all titanium metal is being used today so as to take advantage of its favorable ratio of strength-to-weight between 300° and 700°F (150 to 370°C). Titanium would be used more if it cost less.

Aircraft gas turbines. Titanium bar stock and forgings are used in making compressor disks, spacer rings, rotating and stationary compressor blades, and vanes, through bolts, turbine housings, and liners, and miscellaneous hardware for turbo-prop engines. Titanium sheet is used for fire shields, brackets, and shroud stock.

Airframes. Titanium and titanium alloy sheet are used in airframes for both structural and nonstructural applications, primarily surrounding engines, where service temperatures are in the range from 300° to 700°F (150° to 370°C). Titanium is used in both military and commercial aircraft where even its high cost is justified by the savings in weight and the accompanying extra pay load.

Fasteners. Titanium and titanium alloy rivets, nuts, bolts, and screws have been manufactured in a variety of sizes, principally for evaluation. The high-shear type of rivet, on which stainless steel or Monel clips or collars are *swaged* (pressed, pulled, or hammered), has great possibilities. Shear-type fasteners may be important in design because of the relatively high-shear-tensile ratio of titanium. Titanium alloy screws, and particularly bolts, also offer possibilities for saving weight.

Other Applications. Because of superior corrosion resistance, titanium and titanium alloys are being used experimentally as seats and disks in globe valves, and metering disks in displacement-type fuel systems on ships; also, for military equipment which must be light for mobility or air transportability, and in the chemical industry for pipe and fittings to carry highly corrosive chemicals. Titanium oxide is an important pigment in long-lasting titanium white paint.

REVIEW

1. Name the uses to which nearly pure aluminum can be put.

2. Name the properties that make aluminum so useful.

3. How do you account for the resistance of aluminum to corrosion?

4. Describe briefly how the patented product *Alclad* is made.

5. What is the anodic treatment used to increase the corrosion resistance of aluminum alloys?

6. What factors reduce the machinabil-

Metallurgy

ity of light alloys of aluminum and magnesium, and how are these overcome?

7. Into what two groups are aluminum alloys classified?

8. Cast aluminum is alloyed to improve what two properties?

9. Why are aluminum-silicon alloys more easily cast than aluminum-copper alloys?

10. What effect does the addition of a small amount of metallic sodium have on an aluminum-silicon casting?

11. What four outstanding characteristics are exhibited by aluminum-silicon castings?

12. Name several uses of aluminum-silicon castings.

13. What is the outstanding characteristic of aluminum alloys containing magnesium?

14. Why is zinc alloyed with aluminum?

15. Why do wrought aluminum alloys have physical properties superior to those of castings?

16. What three operations are needed to produce wrought aluminum alloys?

17. Are the hot-working temperatures used on wrought aluminum lower or higher than those used on steel? Does this make such work on aluminum easier or harder than work on steel?

18. Describe the process of annealing commercially pure aluminum.

19. What precaution is observed when cooling wrought-aluminum alloys during annealing?

20. At what temperature does recrystallization occur during annealing?

21. What improvement is obtained for certain aluminum alloys by a solution heat treatment?

22. Describe the phenomenon known as *precipitation hardening*.

23. What effect does precipitation hardening have upon elongation?

24. How may precipitation hardening be prevented?

25. Upon what does the capacity for precipitation hardening at room temperature depend?

26. What is *Dow Metal*? What is its outstanding characteristic?

27. Name and describe briefly the two chemical treatments given magnesium alloys to increase their corrosion resistance.

28. Name the principal alloying element used to improve the mechanical properties of magnesium.

29. Why are zinc and manganese added to magnesium?

30. What results are obtained from the *precipitation heat treatment* of magnesium alloys?

31. Name the special foundry techniques for magnesium alloy casting.

32. List the uses for magnesium alloy castings.

33. Why is it that wrought magnesium alloys do not lend themselves readily to cold-forming or forging methods?

34. In what ways is winning titanium from its ore different from winning other metals from their ores?

35. What effects do contaminants have upon the physical properties of titanium? Contrast these effects with the effects of contaminants upon other metals.

36. What are the effects of oxygen, nitrogen, and water and what means must be taken to eliminate them while producing titanium?

37. Discuss the properties of titanium.

CHAPTER 18

COPPER AND ITS ALLOYS

Where copper was first produced is not known. It was known to the Romans as the metal from Cyprus, from which the name copper was derived. Many of the ancient peoples used it. The Chaldeans had developed the art of working copper as early as 4500 B.C. Later, the mines of Cyprus yielded copper to the Egyptians, and from Egypt the use of copper spread into Europe. In America copper was used by the pre-Columbian peoples and the Indians long before any contact with Europeans. Native copper is found on the earth's surface in many places.

Copper seems to have been used first for ornaments; later it was used for tools and arms. Discovery of the hardening effects of tin, combined with copper to produce bronze, gave various peoples important advantages in war. Where, when, and how bronze and brass were discovered is not known, but the early, long-used method for brass was by direct reduction from a mixture of copper, or some of its ores, and calamine which is zinc carbonate.

It was only in 1781 that Emerson invented the process of direct fusion of metals to produce a copper-zinc alloy we call brass. In America, the first rolling of brass sheet was done in Waterbury, Conn., and Connecticut has continued to be a center of American brass manufacture.

Copper ranks next to iron and steel in commercial importance as a metal. Due to its electrical properties it is of particular use to electrical engineers. Copper is easily rolled and drawn into wire. It exhibits great resistance to weathering, has good mechanical properties, and is of moderate cost. In ductility copper is definitely surpassed only by gold and silver. Its high ductility allows the shaping of copper without trouble from cracking, and with relatively small expenditure for power and for wear and tear of machinery. Cast copper has a tensile strength of 24,000 psi

(165,000 kN/m²),[1] while in the cold-worked condition the tensile strength may exceed 70,000 psi (483,000 kN/m²). Copper is a soft metal with a Brinell hardness of 35 in the *as-cast condition*. The hardness of copper may be increased to approximately 100 Brinell in cold-working operations. It has been believed by many that in ancient times copper was made approximately as hard as hardened tool steel by some method of heat treatment, a treatment that has often been referred to as *the lost art of hardening copper*.

Present knowledge leads us to believe that some of the ancient tools made from copper may have contained some impurities which contributed to its hardness and, coupled with cold-forging, resulted in a harder variety of copper. The only way to harden commercially pure copper is by cold work-hardening methods.

Copper is an extremely tough metal in that it has a remarkable resistance to fracture from sudden shock loads. The fact that its elastic limit upon loading is only 50 percent of its ultimate strength may be taken as an indication of its abililty to deform without rupture when loaded beyond its elastic limit.

Commercial Grades of Copper

Some of the commercial grades of copper include:

1. Electrolytic and low-resistivity lake copper, both of which run around 99.9 percent copper with silver, are used for electrical conductors. Structure of these types of copper is illustrated in Fig. 18-1.

Fig. 18-1. High-conductivity copper, annealed. Magnified 75 times. (Revere Copper and Brass, Inc.)

2. Arsenical copper containing from 0.25 percent to 0.50 percent arsenic. This type of copper is used where its high recrystallization temperature is an advantage, such as in stay bolts for locomotives.

3. Fire-refined copper, other than lake copper, which has not been electrolyzed. It contains about 99.10 percent copper and silver with a maximun of 0.10 percent arsenic; it is used for mechanical purposes.

4. Deoxidized and oxygen-free copper with no analyzable oxygen. It possesses exceptional plasticity and good welding properties and is not subject to embrittlement resulting from exposure to a reducing atmosphere.

5. Tough-pitch copper which may contain as much as 0.070 percent oxygen. If this type of copper is heated in a reducing atmosphere above 750° F (about 400° C),

---

[1] Since expressions of psi (pounds per square inch) in units of tens and hundreds of thousands are not absolute quantities, the conversion units of kN/m² (kilo-Newtons per meter squared) have been rounded off appropriately.

Copper and Its Alloys

Fig. 18-2. Electrolytic-pitch copper; copper 99.90 percent; annealed. Magnified 75 times. (Revere Copper and Brass, Inc.)

the reducing gases react with the oxide particles at the grain boundaries and form cracks which cause the section to become brittle. The presence of small amounts of oxygen is essential for sound castings from the ordinary refining furnaces and results in good mechanical properties. The structure of this type of copper is illustrated in Fig. 18-2.

Impurities in copper. Some of the impurities found in copper are oxygen as Cu_2O, sulfur, bismuth, antimony, arsenic, iron, lead, silver, cadmium, phosphorus, and others. Some of these impurities show a pronounced effect by lowering the electrical conductivity of the copper, whereas others have little effect. Arsenic and phosphorus have a marked detrimental effect on the conductivity of copper. Cu_2O in the concentration found in commercial copper has little effect on the mechanical properties. Silver is one of the most common impurities found in copper. It has little effect upon the properties of copper with the exception of raising the recrystallization and annealing temperatures. Lead in amounts over 0.005 percent reduces the hot-working properties but has little effect on the plasticity at room temperature.

Alloys of Copper

Copper is alloyed with several different metals because the resulting alloys are superior in many ways.

1. Copper alloys are stronger and harder than pure copper, and they may be further improved in their mechanical properties by cold working and, in some instances, by heat treatments.

2. Commercial grades of copper do not make satisfactory castings. Alloying of copper improves the casting characteristics, and these alloys, such as brass and bronze, are used to make castings.

3. The alloys of copper, in general, are much easier to machine than commercial copper, which is too soft and tough for easy machinability.

4. The corrosion resistance of many of the copper alloys is superior to that of commercial copper.

5. Alloys of copper-zinc (brass) are cheaper than copper due to the low cost of zinc and are superior for many uses.

6. The higher elastic nature of the copper alloys makes them superior to commercial copper, the latter being almost totally lacking in elastic properties unless it has been subjected to severe cold work hardening.

Table 18-1 may be used by the student as a reference table. It covers characteristics of common forms of copper and copper alloys.

TABLE 18-1. COPPER AND ITS ALLOYS IN WROUGHT FORM

METAL	TYPE COMPOSITION					PROPERTIES (HARD AND SOFT)			FORMS	PROPERTIES AND USES	METHODS OF WORKING
	Cu	Zn	Pb	Sn	Ni	TENSILE STRENGTH (psi)	PERCENT ELONGA-TION	ELASTIC LIMIT (psi)			
Copper (electrolytic)	99.9					51,000 / 32,500	4 / 47	48,000 / 12,000	sheet; bar; tube; rod plate	corrosion resistance; ductility; high conductivity. roofing; bus bar; high conductivity tubing	stamp; draw; weld; solder; forge; form
Copper (lake)	99.9				7 oz silver per ton	51,000 / 32,000	4 / 47	48,000 / 12,000	sheet; strip; rod	high annealing point; auto radiator fins; lock seam tubing	stamp; draw; solder; form
Copper (phosphorized)	99.9				0.04 phosphorus max	55,000 / 35,000	5 / 45	44,000 / 16,000	sheet; strip; tube	draws and coils better than electrolytic. Water, refrigerator and oil-burner tubing	stamp; draw; forge
Copper (arsenical)	99.9				0.04 phosphorus 0.30 arsenic	60,000 / 36,000	4 / 40	55,000 / 7,000	sheet; tube; plate	high strength; resists heat and flaking; condenser tubes	stamp; draw; forge
Copper (cadmium)	99				1.00 cadmium	80,000 / 35,000	4 / 45	68,000	rod	high strength; high-strength parts; trolley wire	draw; forge
Copper (beryllium)	98				2.00 beryllium	175,000 / 75,000	6 / 45	134,000 / 31,000	sheet tube; rod	very high strength; hardness; high conductivity; springs; cutting tools	stamp; draw; forge; form

BRASSES

METAL	Cu	Zn	Pb	Sn	Ni	TENSILE STRENGTH (psi)	PERCENT ELONGATION	ELASTIC LIMIT (psi)	FORMS	PROPERTIES AND USES	METHODS OF WORKING
Gilding metal	95	5				55,000 / 35,000	5 / 38	39,000 / 11,000	sheet; strip; tube	ductility; reddish gold color; primers; detonator fuse caps; jewelry; forgings	stamp; draw; forge; form
Commercial bronze	90	10				67,000 / 37,000	3 / 40	53,000 / 11,000	sheet; strip; tube	ductility; used for color match; stamped hardware; bullet jackets; jewelry; caskets; screen cloth	stamp; draw; forge; form; perforate
Rich low brass	85	15				75,000 / 42,000	4 / 43	52,000 / 15,000	sheet; strip; tube	corrosion resistance; brass pipe; jewelry; badges; name plates; etchings; tags; dials	stamp; form; draw; blank; etch: weld
Low brass	80	20				85,000 / 43,000	4 / 50	65,000 / 15,000	sheet; strip;tube	corrosion resistance; yellow color; jewelry (for gold plating); fulton bellows.	stamp; form; draw; spin
Seventy-thirty or cartridge brass	70	30				86,000 / 45,000	4 / 50	65,000 / 15,000	sheet; strip; tube	high ductility; deep drawing; pins; rivets; eyelets, radiators; cartridge shells; spun articles	stamp; spin; deep drawing
High brass	66	34				90,000 / 48,000	4 / 50	70,000 / 15,000	sheet strip	high ductility; deep drawing; brass pipe; auto reflectors; stampings; radiator fins	stamp; spin deep drawing
Leaded high brass	65	33.5	1.5			80,000 / 45,000	5 / 60	60,000 / 15,000	sheet strip	forming by bending; free machining; engravers' brass; lighting fixtures; clock and watch backs; gears; keys	stamp; form; bend; punch
Free cutting rod	62	35	3			62,000 / 47,000	20 / 60	25,000 / 15,000	rod	typical brass rod; free machining; extruded shapes; screw machine parts	machine; thread; extrude
Forging rod	60	38	2			70,000 / 50,000	10 / 45	31,000 / 15,000	rod	hot forgings; faucet handles; shower heads	forge; extrude; machine;
Muntz metal	60	40				80,000 / 57,000	9.5 / 48	60,000 / 15,000	sheet; plate; tubes	condenser tubes and heads; ship sheathing; perforated metal; brazing rod	draw; punch; forge

Name	Cu	Zn	Pb	Sn	Other	Tensile/Yield	Elong/Rockwell	55k/15k	Form	Uses	Operations
Architectural bronze	56	41.25	2.75			70,000 / 50,000	10 / 20	55,000 / 15,000	sheet; strip	strength; hardness; free cutting; extruded shapes; forgings; interior ornamental bronze	extrude; forge; machine

SPECIAL BRASSES

Name	Cu	Zn	Pb	Sn	Other	Tensile/Yield	Elong/Rockwell	Strength	Form	Uses	Operations
Silicon brass	78	20			2.0 silicon	110,000 / 55,000	4 / 61	83,000 / 12,500	sheet; strip	high strength; weldability; refrigerator evaporators; fire extinguisher shells	resistance weld; stamp; draw
Aluminum brass	78	22			2.0 aluminum	83,000 / 62,000	17 / 52	75,000 / 16,000	tube	resistance to corrosion and erosion; self-healing skin. condenser tubes	draw; extrude
Admiralty	71	28		1		95,000 / 45,000	5 / 60	92,000 / 18,000	sheet; strip; tube	resistance to corrosion, especially of sea water; condenser tubes	stamp; draw; extrude
Naval brass	60	39.25		0.75		75,000 / 54,000	15 / 45	39,000 / 15,000	sheet; rod; tube	resistance to corrosion in sea water; tube heads; marine shafting; bolts; forged parts; window anchors	draw; forge

BRONZES

Name	Cu	Zn	Pb	Sn	Other	Tensile/Yield	Elong/Rockwell	Strength	Form	Uses	Operations
Phosphor bronze	98.75			1.2	0.05 phosphorus	65,000 / 40,000	4 / 48	50,000 / 15,000	sheet; strip	resilience; strength; hardness; corrosion resistance springs; bearings; small parts	stamp; form; weld
Phosphor bronze	92			8	0.05 phosphorus	110,000 / 55,000	3 / 55	85,000 / 25,000	sheet; strip; rod	similar to above; welding rod	stamp; form; weld
Silicon bronze	96.25			0.50	3.25 silicon	110,000 / 60,000	5 / 55	100,000 / 25,000	sheet; tube; rod	strength; weldability; corrosion resistance; tanks; bolts; screws; lags; chain; locomotive hub liners; welding rod.	stamp; draw; forge; weld; extrude; cast
Aluminum bronze	95				5.0 aluminum	105,000 / 57,000	5 / 55	80,000 / 24,000	sheet; tube; rod	corrosion resistance; strength; golden color; condenser tubes; gift articles	stamp; extrude; draw
Manganese bronze	59	39	0.75		1.25 Fe, 0.05 Mn	75,000 / 60,000	5 / 35	50,000 / 15,000	sheet; strip; rod	resistance to wear and corrosion; welding rod; perforated coal screens; extruded wearing parts	extrude; perforate; weld

NICKEL SILVERS

Name	Cu	Zn	Pb	Sn	Other	Tensile/Yield	Elong/Rockwell	Strength	Form	Uses	Operations
Nickel silver (typical)	65	20			15	93,000 / 58,000	5.5 / 45	75,000 / 15,000	sheet; strip; rod	resistance to corrosion; strength; extruded shapes; table silver; instruments; key stock; springs	forge; extrude; stamp

CUPRO-NICKELS

Name	Cu	Zn	Pb	Sn	Other	Tensile/Yield	Elong/Rockwell	Strength	Form	Uses	Operations
Cupro-nickel (eighty-twenty)	80				20	80,000 / 49,000	3 / 42	78,000 / 17,000	tube	resistance to corrosion, erosion, heat and chemical attack; condenser tubes	
Cupro-nickel (seventy-thirty)	70				30	84,000 / 49,000	4 / 50	83,000 / 18,300	tube	same as above but more resistant to corrosion; condenser tubes	
Cupro-nickel (zinc alloy)	75	5			20	85,000 / 50,000	5 / 35	77,000 / 23,000	sheet; tube; rod	same as 80-20 above but less resistant to corrosion; condenser tubes	

Reprinted by permission of American Society for Metals

Metallurgy

Annealing of Copper

Annealing is the only method of heat treatment used on pure copper although other heat treatments are used in connection with some of the copper alloys. The purpose of annealing copper is to restore the original ductility and softness to work-hardened copper resulting from any cold-working operation.

The process of annealing cold work-hardened copper involves heating to a temperature of 1100° F (about 595° C), holding at this temperature for a certain period of time, and then allowing the metal to cool to room temperature. The rate of cooling from the annealing temperature of 1100° F (about 595° C) is without effect; therefore, fast cooling rates such as a water quench may be used. The annealing temperature of 1100° F (about 595° C) is above the recrystallization temperature of the cold-worked state and results in a complete change in the original grain structure without promoting undue grain growth. Annealing at temperatures higher than 1100° F (about 595° C) has no material effect on the strength of copper, but the ductility is reduced somewhat due to increase in size of grains.

BRASS

The alloys of copper and zinc are commonly classified as *brasses;* however, the term *commercial bronze* may be used in reference to some compositions of copper-zinc. These terms as used in the brass industry may be very misleading to the student. The copper-zinc alloys are the most important of the copper alloys due to their desirable properties and relatively low cost. Zinc readily dissolves in copper in both the liquid and solid states, forming a series of solid solutions; with less than 36 percent zinc, a solid solution is formed that is referred to as alpha (α) solid solution. The alpha phase is a strong and very ductile structure. If the combination contains above 36 percent zinc, a solid solution known as beta (β) is formed, which is relatively hard and much less ductile than the alpha brass. When zinc exceeds 50 percent in the alloy, a gamma (γ) solid solution is formed which is hard and brittle and of no value industrially except for decoration.

A study of the copper-zinc alloy diagram, Fig. 18-3, illustrates the areas where the alpha (α), beta (β), gamma (γ), etc., solid solutions exist. The copper-zinc

Copper and Its Alloys

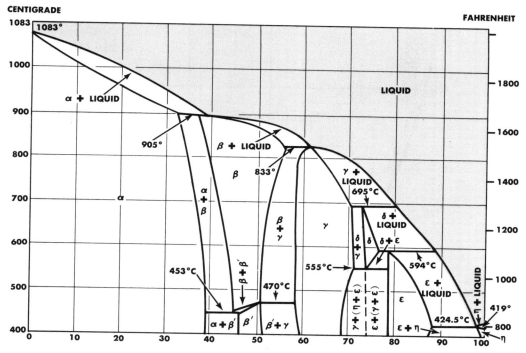

Fig. 18-3. Diagram of copper-zinc alloys. (Metals Handbook, American Society for Metals)

diagram may be readily understood and interpreted if the student will refer back to the chapter dealing with types of alloy systems, and divide the copper-zinc alloy diagram into sections resembling the more simple diagrams described there.

Brasses containing over 62 percent copper consist of only one phase, the alpha solid solution, which is very ductile and has a face-centered cubic type of crystal structure. The alpha solid solution brasses are used mostly where the parts are wrought to shape. Many of the commercially important brasses are of this type. The mechanical properties of these alpha brasses depend largely upon the zinc content and the degree of cold working they receive. Fig. 18-4 illustrates the effect of composition upon the tensile strength and ductility of alloys containing up to 50 percent zinc. It will be seen from Fig. 18-4 that the tensile strength and ductility are both improved with additions of zinc up to approximately 30 percent zinc, above which composition the tensile strength continues to increase up to about 45 percent zinc, with a marked drop in the ductility when the zinc content exceeds 30 percent of the total weight.

When the percentage of zinc exceeds

409

Metallurgy

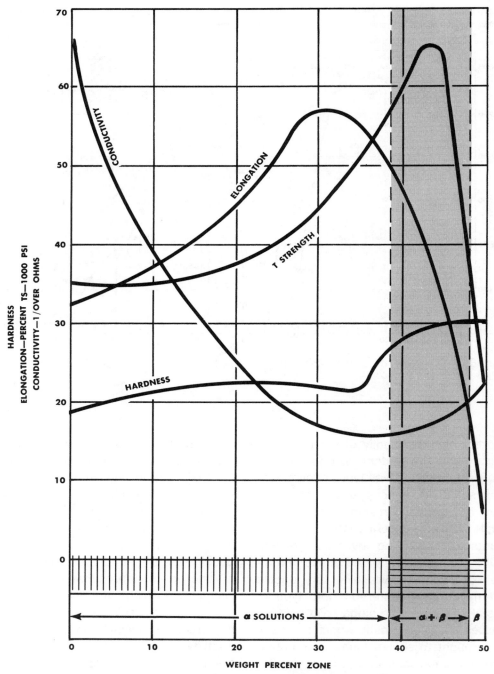

Fig. 18-4. *Properties of the high-copper brasses.*

approximately 38 percent, the structure of the alpha solid solution changes to the beta phase. This beta phase is much harder and stronger but less ductile than the alpha phase. Also, the beta phase differs from the alpha phase in that it is a body-centered cubic crystal structure, and upon cooling to approximately 880° F (about 670° C), the beta phase undergoes a structural transformation changing the beta (β) to a phase designated as β'. The two beta phases differ in that the β has its solute atoms haphazardly arranged in the solvent lattice, whereas the β' phase has the solute atoms arranged orderly in its atom structure of solvent atoms.

The color of brass varies as its composition from a copper red for the high copper alloys, to a yellow color at about 38 percent zinc. The color changes and is slightly more reddish in the β' phase.

Red Brasses

There are four types of wrought brasses in this group: *gilding metal*—95 to 5, *commercial bronze*—90 to 10, *rich low brass*—85 to 15, and *low brass*—80 to 20. These brasses are very workable both hot and cold. The lower the zinc content, the greater the plasticity and workability. These brasses are superior to the yellow brasses or *high brasses* for corrosion resistance and show practically no bad effects from dezincing or *season cracking*. Due to their low zinc content, they are more expensive and are used primarily when their color or greater corrosion resistance or workability are distinct advantages. The structure of annealed red brass with its alpha phase is illustrated in Fig. 18-5.

Fig. 18-5. Annealed red brass. Copper 85 percent; zinc balance. Magnified 75 times. (Revere Copper and Brass, Inc.)

The applications for red brass include valves, fittings, rivets, radiator cores, detonator fuse caps, primer caps, plumbing pipe, bellows and flexible hose, stamped hardware, gaskets, screen cloth, etc. These alloys may be shaped by stamping, drawing, forging, spinning, etc. They have good casting and machining characteristics and are weldable.

Yellow or Cartridge Brass

The yellow alpha brass is the most ductile of all the brasses, and its ductility allows the use of this alloy for jobs requiring the most severe cold-forging operations such as deep drawing, stamping, and spinning. Fig. 18-6 illustrates the structure of annealed yellow or cartridge brass,

Metallurgy

Fig. 18-6. Annealed cartridge brass. Copper 70 percent; zinc balance. Magnified 75 times. (Revere Copper and Brass, Inc.)

which is a copper containing from 27 to 35 percent zinc. This alloy is used for the manufacture of sheet metal, rods, wire, tubes, cartridge cases (from which it gets its name), and many other industrial shapes.

Unless the brass is exceptionally free from lead, it may not respond well to hot working, and all fabrication should be done cold. Lead may be used to improve the cold-working characteristics.

A rolled and annealed brass has a tensile strength of about 28,000 psi (190,000 kN/m^2). By cold rolling, however, its tensile strength may be increased to 100,000 psi (690,000 kN/m^2). Brass is obtainable in varying degrees of hardness by cold rolling, usually designated as *quarterhard*, *half-hard*, and *hard*. The hard brass is brass with the maximum degree of cold rolling that may be considered practical for a given thickness of section and from a workability viewpoint.

The effect of cold working upon the mechanical properties of this alloy is illustrated in Fig. 18-7. The only heat treatment given to this alloy is annealing to remove any cold work hardening and to remove residual stresses as the result of cold-working operations.

Annealing of Brass

The only heat treatment applicable to alpha brass is that of annealing after cold working. Cold working, as we have seen, produces a distortion of the crystal state, which is accompanied by an increase of hardness and strength and a loss of plasticity. The cold-worked condition, with its greater strength and lower ductility, renders the metal less workable and may result in failure due to rupture of the less plastic condition of the crystal state. Restoration of the original properties may be

Copper and Its Alloys

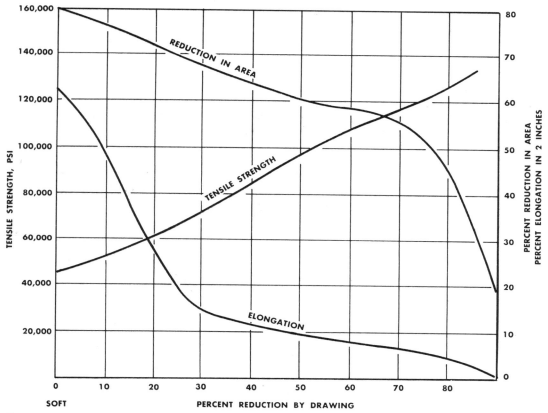

Fig. 18-7. Effect of cold drawing on the physical properties of the high brass. (Metals Handbook, American Society for Metals)

accomplished by an annealing operation; i.e., heating of the cold-worked brass to above its recrystallization temperature. During the heating of the cold-worked brass, any internal stresses which may be present in the brass from the cold-forming operations may be removed. Three changes take place during the heating of the cold-worked brass in an annealing operation: (1) the relief of internal stresses (2) the recrystallization of the cold-worked crystal structure into new and very fine grains and (3) the growth of the fine new-formed grains into larger and fewer grains or crystals. The annealing operation completely removes any trace of the original cold-worked state of the brass, and restores the original ductility, with a lowering of the tensile strength and hardness to the normal values.

The effect of annealing temperatures ranging from room temperature up to

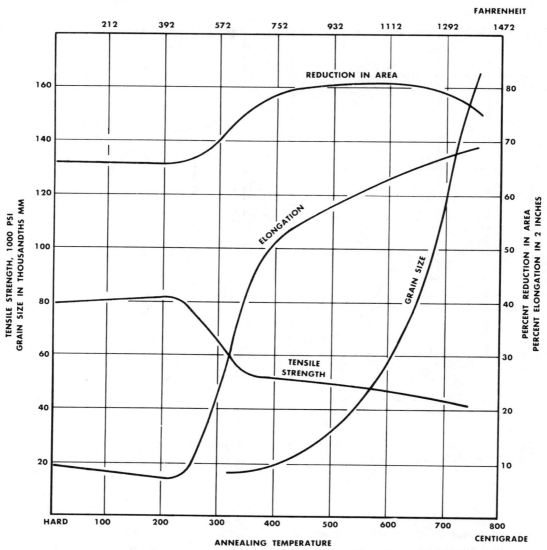

Fig. 18-8. Effect of annealing temperatures on the properties of high brass. (Metals Handbook, American Society for Metals)

1478° F (800° C) on the properties of cold work-hardened brass of the yellow (high brass) composition is illustrated by Fig. 18-8.

Annealing may be carried out by heating the cold-worked brass to within a temperature range of 1100° F to 1200° F (about 595 to 650° C), followed by cooling

Copper and Its Alloys

at any convenient rate. The rate of heating and cooling is almost without effect on the size of the new-formed crystals. The size of the annealed crystals is influenced largely by the amount of cold work the brass received before the annealing operation, and the maximum temperature used in the annealing operation. (See the discussion of recrystallization.) The atmosphere in the annealing furnace may often be controlled to prevent excessive oxidation or even discoloration of the surface of the brass to be annealed.

Season Cracking

Highly stressed brass and bronze may be sensitive to failure by cracking in service under conditions of a corrosive nature. This failure is spontaneous and may occur without any added strain from service. It constitutes one of the most serious defects or failures in cold-worked brass. The conditions leading to season cracking may be eliminated by careful control of the cold-working operations, or by removal of the internal stresses before subjecting the brass to service. Annealing should remove any danger of season cracking. Also, a low-temperature baking or annealing treatment, which may not lower the tensile strength below satisfactory values, may be used that will prove safe as a treatment to prevent season cracking. Low temperature annealing within temperatures of 500° to 800° F (about 260° to 425° C) may prove satisfactory as a means of removing the internal stresses and maintaining the tensile strength within acceptable limits.

Sensitivity to season cracking may be determined by exposure of a cold-worked brass, for approximately 10 minutes, to a mercurous nitrate solution containing nitric acid. Satisfactory brass will show no evidence of a crack after a period of one hour following the treatment in this bath.

Alpha-Beta Brasses, Muntz Metal

When the content of zinc in brass is increased from 40 percent to 45 percent zinc, the alloy is called Muntz metal. This alloy contains both the alpha and beta constituents in its structure. Muntz metal may be hot-worked even when it contains a high percentage of lead. Such brass is accordingly useful for screws and machine

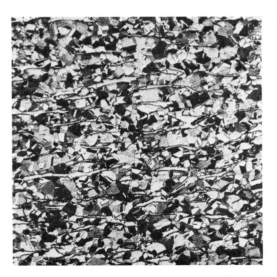

Fig. 18-9. Muntz metal, annealed. Copper 60 percent; zinc balance. Alpha plus beta structure. Magnified 75 times. (Revere Copper and Brass, Inc.)

Metallurgy

parts, where ease of working and particularly ease of machining are more important than strength.

A photomicrograph of this alloy, Fig. 18-9, illustrates the type of structure found in Muntz metal. As stated previously, both the alpha and beta phases are present in this alloy. The lighter constituent is the harder and less ductile beta phase, and the darker constituent is the softer and more plastic alpha phase. Some of the applications for Muntz metal include sheet form for ship sheathing, condenser heads, perforated metal, condenser tubes, valve stems, and brazing rods.

Fig. 18-10. Annealed arsenical admiralty metal, magnified 75 times. Copper 70 percent; tin 1 percent; arsenic 0.03 percent; zinc balance. (Revere Copper and Brass, Inc.)

Admiralty Metal

Admiralty metal is a copper-zinc alloy containing approximately 1.0 percent tin and 0.03 percent arsenic. Its resistance to corrosion is superior to that of ordinary brasses that run free from tin. The structure of this type of alloy is shown in Fig. 18-10. This composition is used for tubing condensers, preheaters, evaporators, and heat exchangers in contact with fresh and salt water, oil, steam, and other liquids at temperatures below 500° F (260° C).

BRONZE

Copper-Tin Bronze

Bronze is an alloy containing both copper and tin although commercial bronzes may contain other elements besides tin; in fact, they may contain no tin at all. Bronze for "copper" coins, medals, etc. contains 4 to 8 percent tin to increase its hardness and resistance.

Tin forms compounds with copper, the compounds being dissolved in the copper up to 5 percent tin, forming a solid solution (α) similar to the alpha solution of the brass. However, tin increases the strength, hardness, and durability of copper to a much greater extent than zinc. Over 5 percent tin produces a heterogeneous structure consisting of an alpha sol-

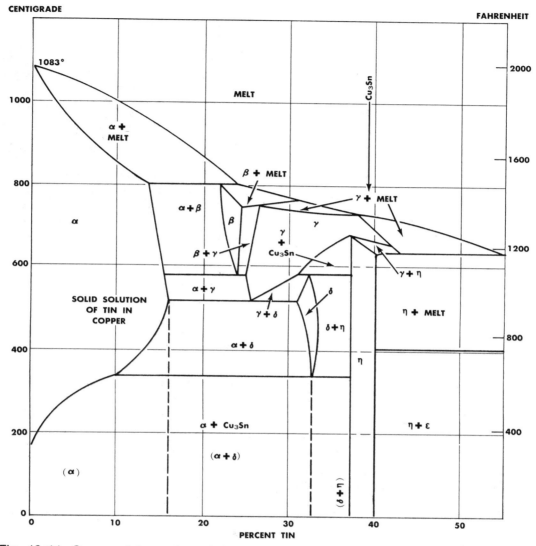

Fig. 18-11. Copper-rich portion of the copper-tin constitution diagram. (Metals Handbook, American Society for Metals)

id solution plus a compound, Cu_3Sn, that acts as a hardener in the alloy.

The alloy diagram of the copper-tin system is shown in Fig. 18-11. From a study of this diagram, it will be seen that the solubility of the compound Cu_3Sn decreases with a lowering of the temperature from the alpha solid solution area to room tem-

Metallurgy

perature. In commercial copper-tin alloys that are cast and cooled relatively fast, the solubility is about 5 percent tin and for slowly cooled alloys about 10 percent tin; i.e., the capacity to hold tin in solid solution is a function of the cooling rate.

Alloys with increasing amounts of tin contain, upon cooling to room temperature, an alpha solid solution phase plus the compound Cu_3Sn. This compound is extremely hard, brittle, and wear resistant and results in a harder alloy. The copper-tin alloys containing from 1.25 percent to 10.50 percent tin are more expensive than copper-zinc alloys but are harder, stronger, and more resistant to corrosion than the copper-zinc brasses. They are used for many parts, including high-strength springs, clips, snap switches, electric sockets and plug contacts, fuse clips, flexible tubing, hardware, bushings, bearings, welding rods, etc.

Fig. 18-12. Phosphor bronze as-cast, copper 92 percent; tin 8 percent, magnified 75 times. (Revere Copper and Brass, Inc.)

Phosphor Bronze

Phosphorus is added to bronze containing from 1.5 to 10 percent tin (see Table 18-1), for deoxidation purposes during melting and casting operations. The phosphorus also increases the fluidity of the molten metal, thereby increasing the ease of casting into fine castings and aiding in the production of sounder castings. Also, with the higher phosphorus content, a hard compound Cu_3P is formed which combines with the Cu_3Sn compound present in these alloys, increasing the hardness and resistance to wear. The structure of a cast phosphorus-treated bronze is shown in Fig. 18-12. The alpha constituent appears as dendritic crystals in a matrix consisting of a mixture of alpha phase and the hard, brittle compound Cu_3Sn. These alloys are used largely for gears and bearings.

Leaded Bronze

Lead does not alloy with copper but may be mixed with copper while the copper is in the molten state by agitation or mechanical mixing, and under suitable conditions this may be satisfactorily cast in a mold, resulting in the lead being well distributed throughout the casting in small particles. Lead may be added to both bronze and brass for the purpose of increasing the machinability and acting as a self-lubricant in parts that are subjected to sliding wear, such as a bearing. The lead particles, with their soft, greasy nature, reduce the frictional properties of the alloy. Lead is a source of weakness and is usually kept below 2 percent although some bearing bronzes may contain as high as 50 percent lead.

Aluminum Bronze

Alloys of copper containing aluminum in place of tin are known as *aluminum bronze*. They may also contain other elements such as silicon, iron, and nickel, frequently added to increase the strength of the alloy. Typical structures found in these alloys are illustrated in Figs. 18-13, 18-14, and 18-15. The influence of cooling rates is shown by the structures developed in these illustrations; i.e., the extruded material is practically 100 percent alpha phase, the forged specimen has retained some of the beta phase and shows a two-phase structure, alpha plus beta, whereas the specimen water-quenched from a temperature of 1700° F (925° C) shows a marked two-phase, alpha-plus-beta type of structure.

The structure of the slow-cooled alloys of this composition may completely change to the alpha phase upon reaching room temperature; however, at elevated temperatures, the alpha phase changes to a beta phase, which may be retained, in part, at room temperature by means of rapid cooling. It is possible to improve the hardness and other properties of many of these alloys by heating to the beta phase

Fig. 18-13. Aluminum-silicon bronze, extruded. Aluminum 7 percent; silicon 2 percent; copper balance. Alpha plus beta structures. (Revere Copper and Brass, Inc.)

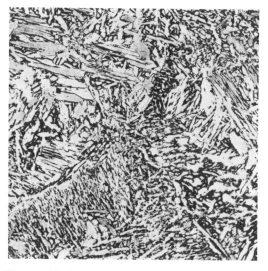

Fig. 18-14. Aluminum-silicon bronze, forged. Aluminum 7 percent; silicon 2 percent; copper balance. Alpha plus beta structures. (Revere Copper and Brass, Inc.)

Metallurgy

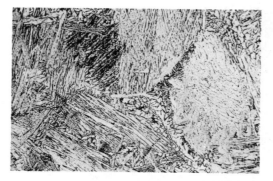

Fig. 18-15. Aluminum-silicon bronze, heated to 1700°F (about 925°C) and quenched in water. Aluminum 7 percent; silicon 2 percent; copper balance. Alpha plus beta structures. (Revere Copper and Brass, Inc.)

not unlike those found in brass and bronze. The mechanical properties of aluminum bronze are superior to those developed by ordinary bronze. The tensile strength in the as-cast condition of an alloy containing 10 percent aluminum is about 65,000 psi (448,000 kN/m^2). It is nearly as ductile as brass and has double the ductility of tin-copper bronze. Many special aluminum bronzes are on the market, which develop very high hardness and strength; some of these are subjected to heat treatments similar to aluminum-copper alloys in order to develop their maximum properties. It is more difficult to obtain uniformity in aluminum bronzes than in the more common copper alloys.

range of temperature, approximately 1650° F (900° C), followed by rapid cooling. Upon reheating to 700 to 1100° F (about 370° to 595° C), the retained beta phase becomes unstable and undergoes a transformation with the separation of the beta phase to a fine alpha and delta crystal form. Such a change causes a marked increase in hardness and strength at the expense of ductility. Heat-treated aluminum bronzes may be used for hand tools, such as chisels, where nonsparking characteristics are essential to avoid fires and explosions. They are also widely used in the oil refineries and in other industries handling inflammable or explosive gases and liquids. Other applications include aircraft engine parts, valve seats and guides, spark plug inserts, and bushings.

With percentages below 10 percent aluminum, the copper dissolves the aluminum and forms solid solutions, α, β and γ,

Miscellaneous Bronzes

Other bronze alloys include *manganese bronze, bell bronzes, nickel bronze,* and *silicon bronze*. These alloys find a field of usefulness in industrial, marine, and household applications. The manganese bronze is fundamentally a brass of approximately 60 percent copper, 40 percent zinc, and manganese up to 3.5 percent and is well known as an alloy for ship propellers. It has excellent resistance to corrosion and good mechanical properties. The bell bronze contains 20 to 25 percent tin and is a hard, brittle alloy which gives a tone quality to bells that no other composition yields.

Addition of nickel to bronze and brass improves mechanical properties and is used to improve the hardness and wear resistance of gears and bearing bronzes. Silicon bronze contains, with other elements, 1 to 4 percent silicon, which is added to improve cold work-hardening characteristics, developing greater strength upon cold rolling or drawing. In addition to high capacity for work hardening, the silicon bronzes exhibit excellent resistance to corrosion caused by many organic acids, sugar solutions, sulphite solutions, etc. Their uses include electrical fittings, marine hardware, water wheels, boilers, pumps, shafting, and many parts that operate under corrosive conditions.

OTHER COPPER ALLOYS

Beryllium-Copper Alloys

The answer to the *"lost art of hardening copper"* is found in some of the modern copper alloys such as the newer beryllium-copper alloys containing 1 percent to 2.25 percent beryllium. Although beryllium makes a very costly alloy, its initial cost is offset by the remarkable properties developed by this new alloy. It may be cast, worked hot and cold, welded, and given heat treatments similar to those given the aluminum-copper alloys; i.e., solution and precipitation heat treatments.

As may be seen from the alloy diagram of beryllium-copper, Fig. 18-16, the solubility of beryllium in copper increases with temperature from less than 1 percent beryllium at room temperature to more than 2 percent at 1500° F (815° C). Also, as the beryllium content in copper increases, the alpha solid solution, which forms first, changes to a eutectoid structure consisting of a mixture of alpha and gamma phases. Any gamma phase present in the alloy at room temperature changes to a beta phase upon heating to above 1050°F (565° C). Due to these changes in solubility and the change from one phase to another, the beryllium alloys are susceptible to heat treatments which greatly alter their mechanical properties with improvement in both strength and hardness. Heat treatments include annealing, heating, and quenching, known as solution heat treatment, and aging or precipitation treatments.

In the best annealed condition, obtained by quenching the alloy from a temperature of 1472° F (800° C), an alpha solid solution structure is developed. The tensile strength of this quenched alloy is about 60,000 psi (414,000 kN/m^2). In the quenched state, the solution alloy can be readily cold worked, and may develop a tensile strength in excess of 100,000 psi (690,000 kN/m^2). After cold working, the alloy may be subjected to a precipitation hardening treatment by heating to a temperature between 475° F and 575° F (247 to 302° C). This treatment precipitates a hard gamma crystal phase of the alloy, rendering it much harder and less ductile. By means of these treatments the hardness may be varied from 215 Brinell to 400 Brinell, with a tensile strength, in the

Metallurgy

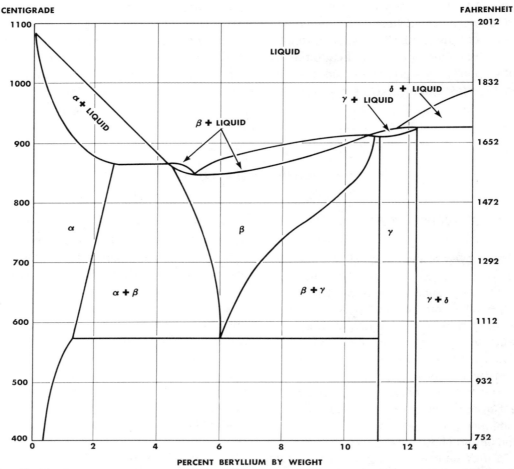

Fig. 18-16. Diagram of copper-beryllium alloys. (Metals Handbook, American Society for Metals)

higher Brinell hardness range, as high as 200,000 psi (1,380,000 kN/m^2).

These alloys exhibit good electrical properties and excellent corrosion resistance. Because they possess a combination of high Brinell hardness and tensile strength, these alloys find many applications as springs, tools, surgical and dental instruments, watch parts, firing pins, and welding electrodes. The main objection to these alloys is cost.

Copper-Manganese-Nickel Alloys

A substitute for the beryllium alloys has been developed that contains approximately 22 to 24 percent manganese, 22 to

24 percent nickel, and the balance copper. This alloy exhibits age-hardening effects resulting from heat treatments and may be cast and wrought to shape. Alloys of this type may be greatly improved in strength and hardness by quenching from 1200° F (650° C) followed by tempering or aging at 660° to 840° F (about 350° to 450° C) for approximately 24 hours. With proper heat treatments, hardness values of Rockwell C54 and higher are obtainable with tensile strength values in excess of 200,000 psi (1,379,000 kN/m^2). These particular alloys may be used in applications similar to the beryllium-copper and the aluminum bronzes and are used where hardness and nonsparking characteristics are needed, such as in some applications for hand tools.

Copper-Nickel Alloys

Copper and nickel form a very simple alloy in that both metals dissolve in each other in all percentages to form a solid solution. Nickel added to brasses and bronzes has been found to improve their electrical properties and toughness. A familiar copper-nickel alloy is found in the United States five-cent piece, which is coined from alloy sheets containing 75 percent copper and 25 percent nickel. All alloys of copper and nickel with more than 20 percent nickel are white in color. Alloys containing 30 percent nickel are much used in marine service where strength and resistance to corrosion are important factors in the selection of an alloy.

Monel Metal

Monel metal is a so-called *natural alloy* in that it is made by smelting a mixed nickel-copper ore mined in Sudbury, Ontario. It contains approximately 65 percent nickel, 28 percent copper, the balance consisting chiefly of iron, manganese, and cobalt as impurities. Monel metal is harder and stronger than either copper or nickel in the pure form; it is also cheaper than pure nickel and will serve for certain high-grade purposes to better advantage than the pure metal. Monel is a substitute for steel where resistance to corrosion is a prime requisite.

Monel metal can be cast, hot and cold worked, and welded successfully. In the forged and annealed condition it has a tensile strength of 80,000 psi (552,000 kN/m^2) with an elongation of 45 percent in 2 inches, and is classified as a tough alloy. In the cold-worked condition it develops a tensile strength of 100,000 psi (690,000 kN/m^2), with an elongation of 25 percent in 2 inches. With excellent resistance to corrosion, pleasing appearance, good strength at elevated temperatures, Monel metal finds many uses in sheet, rod, wire, and cast forms. Included in the applications for monel metal are sinks and other household equipment, containers, valves, pumps, and many parts of equipment used in the food, textile, and chemical industries.

A Monel metal containing 3 to 5 percent aluminum, known as K-Monel, is available. This alloy is of interest because it can be made to develop very high properties by means of heat treatment as a substitute for cold-work hardening in the other

Metallurgy

nickel-copper alloys. A Brinell hardness of 275 to 350 Brinell may be obtained by means of a precipitation heat treatment. This hardness may be further increased by cold-work hardening.

Nickel-Silver

Nickel-silver, also called German silver, is a brass alloy containing 10 to 30 percent nickel and 5 to 50 percent zinc, the balance copper. These alloys are very white, similar to silver, hence the name. They are resistant to atmospheric corrosion and to the acids of food stuffs. Their use depends largely upon their resistance to corrosion and pleasing appearance.

Nickel-Iron Alloys

The nickel-iron alloys, aside from the nickel steels, are of considerable importance. An alloy of 36 percent nickel is known as *Invar,* and has a very low coefficient of expansion so that it is used for standards of length. The alloy of iron and nickel which contains 46 percent nickel is known as *Platinite,* since it has the same coefficient of expansion as glass and may be used in place of platinum to seal into glass.

The alloy of 78½ percent nickel and 21½ percent iron is known as *Permalloy* and has very high magnetic permeabilities at low fields so that it has been used in the construction of submarine telegraph cables.

The alloys of nickel and chromium, as well as the ternary alloys of iron, nickel and chromium, have the property of withstanding high temperature without scaling so that they are very useful for furnace parts, annealing boxes, etc. These alloys also have a high electrical resistance and are hence useful as windings for resistance furnaces, electric irons, and toasters. They are known by various commercial names, depending on their composition. Some of the best known are nichrome, chromel, etc.

REVIEW

1. Name the properties of copper that make it so important commercially.
2. Name the approximate strengths of cast copper before and after cold working.
3. Can copper be hardened by heat treatment?
4. Describe the difference in qualities of the alpha, beta, and gamma brasses. How are they formed?
5. With what metal is copper alloyed to form brass?
6. Name the uses of red brasses.
7. What percentage of zinc is used in cartridge brass?
8. Name the approximate strength of

cartridge brass before and after cold rolling.

9. Brass is made in various degrees of hardness. How are these expressed?

10. What is Muntz metal and what particular uses are made of it?

11. What changes take place during heating of cold-worked brass in annealing?

12. What is meant by season cracking of brass and bronze, and how is danger of such failure removed?

13. Why is tin used in bronze?

14. What percentage of tin is used in bronze for electric sockets?

15. Why is lead added to some brass and bronze?

16. What can be said of the mechanical properties of aluminum bronze?

17. What tensile strength is developed in beryllium-copper alloy? Name some of its uses.

18. What does the addition of nickel do for copper alloys? What special uses are made of these alloys?

19. What is Monel metal and how is it obtained?

20. Name the percentages of nickel and iron in Permalloy. What are some of its uses?

21. List some of the important impurities found in copper. How do they affect the properties of copper?

22. Why are copper alloys superior to commercial copper?

23. What is the purpose of annealing copper?

24. Describe the process of annealing cold, work-hardened copper.

25. Name the important red brasses.

Metallurgy

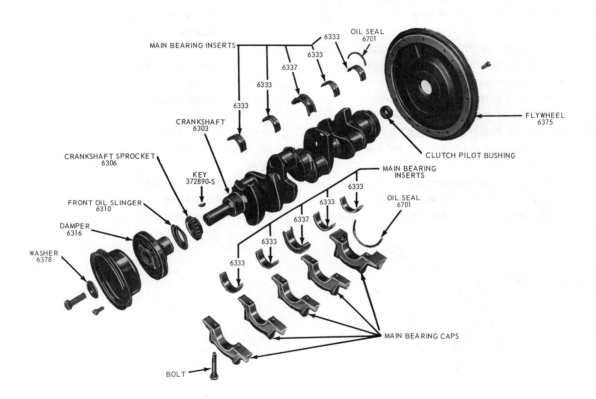

Bearing metals are a major part of automotive engine life. (Ford Motor Co.)

CHAPTER 19   BEARING METALS

We are now living in the *Alloy Age* and we may wonder at the many kinds and varieties of metals and alloys employed by industry. We soon discover that the reasons for the great variety of metals and alloys used are not difficult to find. Each has gained its commercial status because it meets certain requirements better than some other metal, or because its use provides some special satisfaction to human needs. In many cases, availability and economy dictate the selection of the metal or alloy for some specific use. Steel is used in rails, tungsten in lamp filaments; with economy the main factor in their selection, each has no competition.

Just because a metal is higher priced than some other is no economic reason to bar its use. Because of the large number of alloys that have been made and placed on the market, the engineer may make a selection from a great variety of alloys exhibiting combinations of properties that a few years ago were only wishful thinking.

It may seem to the student of metallurgy that too many combinations are already available, but engineers are always demanding new alloys, as yet undiscovered, that will exhibit properties now unavailable. A large number of the engineers' present inventions cannot economically be used until such time as alloys with combinations of properties needed for these new practices are made available.

BEARING METAL CHARACTERISTICS

Selection of a satisfactory bearing material is one of the major problems encountered in the design and construction of machinery, engines, or any part of a

Metallurgy

piece of equipment that requires motion of rotation or reciprocation. Many different materials may be chosen for use as the bearing surfaces of any moving part. The selection of any one type of metal or alloy will depend upon several factors, and the final selection may be determined by the trial and error method.

In many designs, roller and ball bearings are chosen where a minimum of frictional resistance is needed and where high bearing pressures are encountered. In the solution of bearing problems, the engineer has frequently depended upon the selection of a suitable metal or alloy to be used as the bearing, with the hope that the metal selected would exhibit low frictional characteristics without any tendency to weld or seize when the bearing surfaces came into contact with each other.

In most applications of metal bearing surfaces, the engineer depends upon a film of oil acting as a lubricating film to prevent welding, or seizure, and excessive wear. Theoretically, the type of metal or alloy that is chosen for the bearing surfaces should be of no consequence if a proper film of lubrication is maintained at all times between the bearing surfaces, thus preventing any metal-to-metal contact. However, there are times when there is an insufficient lubricating film present between the metal surfaces in any bearing, and a metal-to-metal contact occurs. It is then that the bearing metal should behave satisfactorily without seizure or burning out or excessive wear.

A number of metals and alloys, mostly of the white metal group, such as cadmium, lead, zinc, etc., are used as bearings of the flood-lubricated type. Bearing metals have received much attention from engineers, chemists, and metallurgists during the past few years, and as a result of diversity of viewpoints and interests, much literature covering bearing metals is available. The properties of a good bearing metal should include:

High compressive strength. This is essential if the bearing is not to squeeze out under high-compressive loads. The compressive strength should be high enough to support the load.

Plasticity. It must be sufficiently plastic to allow the bearing to deform in order to take care of misalignment; it also enables bearings to withstand shocks without cracking. If a bearing is finished to a high degree of dimensional accuracy and properly installed, a harder, stronger, and less plastic metal bearing alloy may be used.

Low coefficient of friction. The bearing metal of lowest coefficient of friction is not necessarily the best, as other factors affect behavior and may have more obvious effects. The frictional properties of a perfectly lubricated bearing would be of no account; however this condition is not attainable in practice. Some contact or rubbing of metal to metal takes place, and due to this a low frictional characteristic is desirable. In general, the harder the bearing metal, the lower the frictional characteristics.

Long life. Life of a bearing metal is largely a matter of resistance to wear, and this will depend upon the composition and hardness of the metal. In general, the bearing should be selected so that it will wear, rather than the shaft material. If a hard bearing metal is selected to increase the resistance to wear, it has the disadvantage that if there is any tendency to heat, it will heat more rapidly and consequently wear more rapidly.

High heat conductivity. Good thermal conductivity allows the rapid dissipation of heat generated and prevents the bearing metal from becoming overheated. Good contact between the bearing metal and the backing used to support the bearing is essential to insure good heat transfer. A good contact or adherence is often obtained by casting the bearing metal into contact with the backing. Testing the bond strength between the bearing metal and the backing is a good way to determine the efficiency of the bond.

Heat resistance. A bearing metal should maintain its compressive strength and hardness in the temperature range within which it may be expected to be used.

Besides these properties, a good bearing metal should have good casting properties and be easily workable. From this it should be clearly seen that bearing metals must meet varied requirements. They must be hard to resist wear, yet soft so that the bearing will wear rather than the shaft or spindle; tough and strong, so as to avoid failure by deformation or fracture, yet plastic and compressible so as to allow the bearing metal to conform to any slight misalignment or inaccuracy in dimensions or fit.

It is apparent that no commercially pure metal, or even homogeneous alloy, can successfully meet all of these requirements. As a result, practically all bearing metals are made up of several structural crystal constituents. The more common bearing metals have a strong and plastic background or matrix in which is dispersed one or more hard constituents. It may be assumed that the strong matrix allows plastic flow for the adjustments of fitting and prevents failure under shock loading. The hard crystals dispersed throughout this background add some strength, but largely increase the resistance to wear. It has been stated that there is an advantage in having the hardest component in the bearing metal the same or at least not any harder than the shaft or journal, so that mutual polishing rather than abrasion will take place.

Lead-Base Bearing Metals

Lead has some of the properties desired in a bearing metal, but due to its low strength, alloying of the lead with some hardener metal, such as copper, tin, antimony, calcium, etc., is often practiced. Members of one group of lead-base bearing alloys contain approximately 60 percent lead, the other 40 percent being made up of tin and antimony, with a little copper in certain of the mixtures. The structure of these alloys contains a matrix of a eutectic nature with hard crystals of an intermetallic compound embedded in the eutectic matrix. Lead-base bearing metal is the cheapest of all the white bearing alloys.

Other uses for these alloys of lead, tin, and antimony include solder, type metal, sheathing for telephone cables, and storage battery grids. In fact, due to the large tonnage of lead used in batteries and

Tin-Base Bearing Metals

The tin-base bearing metals are called *Babbitt metal.* A common alloy of this class contains 85 percent tin, 10 percent antimony, and 5 percent copper. Copper and antimony are essential alloys in these tin-base bearing metals. A typical structure found in these alloys is seen in Fig. 19-1, which consists mostly of a dark eutectic matrix containing embedded needle-like crystals of CuSn and cube-like crystals of SbSn. The hard crystals of CuSn and SbSn increase the hardness and wear resistance of the alloy. The physical properties of the tin-base alloys are superior to the lead-base alloys, having higher values for strength and hardness, and they are lighter. The lead-base alloys have a compressive strength of 13,-000 to 15,000 psi (90,000 to 103,000 kN/m^2), with a Brinell hardness of 15 to 20 Brinell; while the tin-base alloys, as a class, have equal or greater strength, with an elastic limit two to three times as high and a greater Brinell hardness. The term Babbitt metal can also refer to lead base bearings or other *cast-in-place metal bearings.*

Lead-Calcium Bearing Alloys

This alloy was a discovery by Francis C. Frary. When hard lead was wanted and it was found that there was a shortage of antimony as a hardener for shrapnel, Frary secured patents covering lead-base alloys, with barium and calcium as hardeners, that proved very successful. Out of this discovery a new alloy was created for bearings, containing a total of 2.5 percent calcium, tin, and other hardeners, and having an unusually high strength at temperatures near its melting point; consequently it is favored for high-speed gas engines. The structure of these alloys contains crystals of Pb_3Ca dispersed throughout a matrix or ground mass of tin and calcium in solid solution in the lead. This structure conforms to the usual structure of bearing metals.

Cadmium Bearing Alloys

A very popular bearing alloy consisting of 1.25 to 1.50 percent nickel, the balance cadmium, has been used very successfully as a flood-lubricated bearing metal. The structure of this alloy is shown in Fig. 19-2 and in the alloy diagram, Fig. 19-3.

Fig. 19-1. Babbitt bearing alloy. Magnified 100 times (dark matrix is mostly eutectic structure; white needle-like crystals are CuSn crystals; white cube-like crystals are SbSn crystals).

Bearing Metals

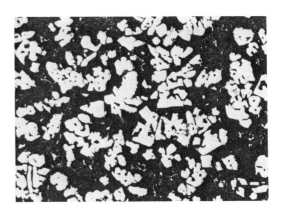

Fig. 19-2. Cadmium-nickel bearing alloy. Magnified 100 times (dark matrix is a eutectic structure; white crystals are $NiCd_7$ crystals).

The cadmium-nickel alloy has been found to have a higher melting point than the Babbitt metals, with its properties less affected by temperature; in fact, these alloys are operative at temperatures that would melt the Babbits.

Although the alloy diagram of the cadmium-nickel system, Fig. 19-3, may not show accurately the complete behavior of the two metals, it will serve to illustrate the influence of nickel in forming a eutectic mixture when 0.25 percent nickel has been added to cadmium. It also shows the formation of an intermetallic compound with the chemical formula $NiCd_7$. The solid solubility of nickel in cadmium is practically zero. The eutectic mixture solidifies at 605°F (318°C) and consists of nearly pure cadmium and crystals of the compound $NiCd_7$. The common bearing alloy of 1.25 to 1.50 percent nickel and the balance cadmium, consists structurally of a matrix of the eutectic structure with an excess of free cube-like crystals of the compound $NiCd_7$, shown in Fig. 19-2. The $NiCd_7$ compound is harder than the compound found in the tin-antimony-babbitt bearing alloys, but the eutectic matrix is slightly softer than that of the Babbitt alloys.

Other cadmium bearing alloys contain silver, copper, and indium. Silver is said to improve casting characteristics and to increase strength, while indium is used to combat the corrosive action of some lubricating oils.

Bearings of the following composition have been successfully used in several makes of American automobiles: 97 percent cadmium, together with 2.25 percent silver, and 0.75 percent copper. The structure of cadmium-silver-copper alloy is composed of a solid solution matrix with small quantities of embedded crystals of a harder intermetallic compound formed by the copper.

Copper-Lead Bearing Alloys

Bearing alloys of copper-tin-lead or copper-lead have been successfully used when heavy compressive loads and high

Metallurgy

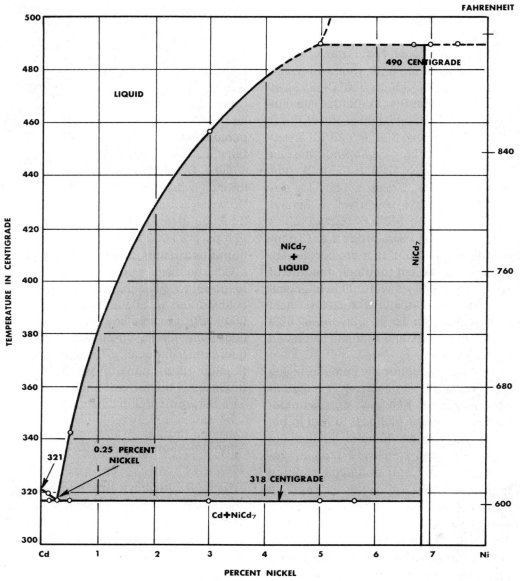

Fig. 19-3. Constitutional diagram of the cadmium-rich end of the cadmium-nickel system.

temperatures are encountered, such as in connecting-rod bearings in aircraft and automotive truck engines, rolling-mill bearings, and railroad service. The lead and copper form a mechanical mixture during the solidification process, inas-

Bearing Metals

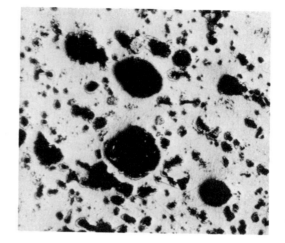

Fig. 19-4. Copper-lead alloy, polished section, magnified 100 times. Dark areas represent lead.

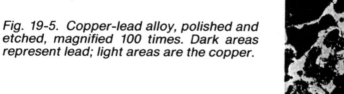

Fig. 19-5. Copper-lead alloy, polished and etched, magnified 100 times. Dark areas represent lead; light areas are the copper.

much as copper has very little or no solid solubility for the lead. The cast copper-lead alloy consists of lead freezing in particles throughout the matrix of copper. The addition of the lead to the copper increases the plasticity of the bearing metal and lowers the coefficient of friction, thus improving copper as a bearing metal. The lead weakens the metal but acts as a self-lubricant.

A representative structure is illustrated in Fig. 19-4, which shows a copper-lead alloy polished, the lead appearing dark and the copper, light. Fig. 19-5 shows the same alloy etched. The lead in these photomicrographs has largely eroded during

Metallurgy

the polishing operation, which accounts for the dark areas in these specimens. In the casting of bearings of copper-lead alloys, the lead is apt to segregate into large particles in certain areas of the bearing, and the alloy may form two layers, one copper, one lead. Such segregation may seriously impair the physical properties of the bearing alloy. Stirring of the molten alloy and rapid cooling during the casting operation serve to minimize this tendency to segregate. Also, the addition of nickel to the alloy seems to act as an aid in further reducing this tendency to segregate. A common binary (two element) alloy containing 25 percent to 36 percent lead and the balance copper gives satisfactory service results. Aircraft engine bearings have been made from a ternary (three element) alloy containing 72 percent copper, 25 percent lead, and 3 percent tin, with or without the addition of silver.

Bearings made from copper-lead alloys have less plasticity and greater frictional properties than the tin or cadmium base alloys and therefore require more accurate sizing and assemblies. A clearance between bearing surfaces of about twice that of the Babbitt metals is required, as well as special high-grade lubricants.

Composite Bearings

Bearings to be used at high speeds and for heavy loads are often manufactured by the casting or plating of a suitable bearing metal to a hard steel or bronze supporting backing or shell. The steel or bronze backing shell is lined with the bearing metal. The thickness of this lining of bearing metal may vary from 0.060'' to less than 0.0025'' (from 1.52 to 0.06mm).

The hard steel or bronze backing serves to improve greatly the mechanical properties of the bearing. Several methods may be used to cast or attach the bearing metal to the backing, such as die-casting, permanent mold casting, centrifugal casting, or electro-plating. A steel or bronze cylindrical or semicylindrical shell or strip may be used for the backing. If a cylindrical or semicylindrical shell is used, the bearing metal may be cast to the shell by the centrifugal casting method, whereas steel strips may have Babbitt or copper-lead alloys welded to them by moving a strip of the steel horizontally through the continuous casting machine. If the bearing metal is to be added to both sides of the steel strip, a vertical casting machine method may be employed. The final bearing may be made from the lined strip by cutting off short sections and then bending or forming them to the desired shape. One of the problems encountered in the manufacturing of a composite bearing by these methods is the problem of obtaining a good bond between the steel or bronze backing and the bearing metal. Controlling the structure of the cast bearing metal is an important factor influencing the characteristics of the bearing metal.

Zinc-Base Bearing Alloys

Zinc-base bearings of the following composition have been successfully used where the surface speeds are high and there is no shock loading or excessive vibration: 85 percent zinc, 10 percent cop-

Bearing Metals

per, and 5 percent aluminum. Alloy of this composition has a hardness of approximately 130 Brinell and compares favorably with some of the copper-base bearing metals. Some zinc-alloy die-castings can be made from a similar composition and require no additional bearing alloy in their finished assembly. Some applications for these bearing alloys include lathe and machine tool spindle bearings and electric motor bearings.

Porous Self-lubricating Bearings

Another type of bearing material consists of a porous bronze alloy made by the pressed powdered metals method known as *powder metallurgy*. Powdered copper and tin, or powdered tin and bronze are mixed with graphite, and the cold mixture is partially welded together by means of pressure. The pressure welding is accomplished in a steel die or mold which forms the powdered metal into the shape required. The cold-welded shape is then removed from the steel die or mold and placed in a furnace and heated to a temperature which further welds the pressed metal powder to form a solid. Accurate control of the density or porosity of the finished bearing is effected by controlling the pressure used in the cold welding operation and the temperature to which the pressed bearing is heated.

The heating of the pressed metal powder is a sort of annealing operation which is called *sintering*. If coarse metal powders are used and little pressure and low temperatures are employed, a porous bearing metal results. Porous bearing alloys are capable of absorbing substantial amounts of oil, which is held in the pores of the bearing metal until pressure causes it to flow out to the surfaces of the bearing, thus making it a self-lubricating bearing. Oil may be added to the porous bearing alloy from any outside surface, the oil working through the spongy metal as needed. These porous bearings have many applications and fill a real need in those applications where lubrication is difficult to accomplish or where very little lubrication can be used.

Silver-Lead-Indium Bearings

A bearing alloy widely used in aircraft engines and in other applications where a heavy-duty bearing metal is needed is the silver-lead-indium bearing alloy. This bearing alloy consists of a steel shell to which a coating of copper is electroplated. This copper layer is not a part of the bearing but merely assists in binding the steel to the silver layer which follows. This steel shell, with the thin copper surface, is then plated with about 0.02" to 0.03" (0.51 to 0.76mm) of silver. The bearing may then be machined to size, allowing for the additional thickness of the succeeding deposits of layers of lead and indium.

The machined silver surfaces are then slightly roughened by light sandblasting, and a thin layer of lead is deposited over the silver to a thickness of about 0.001" (0.025mm) by plating. This plated layer can be controlled as to thickness so as to avoid any further machining operations. The indium is then plated to the lead surface with a flash plating of only a few

Metallurgy

hundred-thousandths of an inch. The indium, which alloys readily with the lead and has a melting point of 311°F (155°C), is diffused into the lead by heating the bearing to 350°F (177°C) in an oil bath. The primary role of the indium is to reduce the corrosive action that lubricants would have on the unprotected lead surface, for indium is more corrosion resistant than either lead or silver.

Other bearing alloys. Included under the heading of bearing metals are aluminum alloys and cast iron. Some discussion of these alloys will be found in sections of this text dealing with these metals.

REVIEW

1. Name some of the factors that may dictate the selection of a certain metal for a job.
2. Why is high *compressive strength* necessary to a bearing metal?
3. Why is *plasticity* important in a bearing metal?
4. Why is a low *coefficient of friction* desirable in a bearing metal?
5. Discuss the length of life of a bearing.
6. Why is *high heat-conductivity* important in bearings?
7. Sum up the characteristics of a good bearing metal.
8. Why is it important to have good contact between bearing metal and its backing?
9. What is the advantage in having the hardest component in the bearing metal of the same hardness as the metal in the shaft or journal?
10. Name several uses of the lead-base alloys.
11. What peculiar property of lead-calcium alloys makes them suitable for use on high-speed gas engines?
12. In what respect are cadmium alloys superior to Babbitt metal for bearings?
13. How does silver improve the cadmium bearing alloys?
14. What type of bearing alloys is used when heavy compressive loads and high temperatures are encountered?
15. Describe what steps are taken to prevent segregation of lead in the copper-lead bearing alloys.
16. Describe composite bearings.
17. Name some of the applications for zinc-base bearings.
18. How are porous self-lubricating bearings made?
19. In what applications do porous bearings fill a real need?
20. Describe the widely used silver-lead-indium bearing.
21. What is the primary role of indium in the silver-lead-indium bearings?

CHAPTER 20

ZIRCONIUM, INDIUM AND VANADIUM

ZIRCONIUM

Pure zirconium metal was described as early as 1920, but not until the development of the Van Arkel process of iodide decomposition in 1925, in Leyden, Holland, was pure, coherent metal readily available.

The later Kroll process proved better than Van Arkel's for a large-scale operation. In that process, zirconium tetrachloride is broken down with magnesium to form magnesium chloride and zirconium metal. Hafnium, usually in quantities of 1.5 to 2.0 percent, is the chief impurity. The hafnium contamination of zirconium is in the same proportion as in the ore from which the metal was made.

Ore Supply

Zirconium was once considered one of the rare metals. Its ores, however, are widely distributed and no less available than those of better-known metals. The earth's crust is estimated to contain 0.04 percent zirconium, more than the combined percentages for copper, lead, zinc, tin, nickel, and mercury.

Almost all the zirconium used in American industry comes from the extensive ore deposits in Florida, Australia, and Brazil. The ores from Florida and Australia consist of seacoast sands from which zircon (zirconium silicate), rutile, ilmenite, and other minerals readily separated. Though the ore is lean, it is still cheaper to produce than if drilling, blasting, and crushing were also required. In Brazil, the chief zirconium ore is baddeleyite, an impure oxide with less silica than is contained in zircon.

Deposits of zircon have been operated profitably in Madagascar, India, and Australia. There are deposits in Oregon and

Metallurgy

North Carolina that could be worked if necessary. Table 20-1 shows current zirconium production.

TABLE 20-1. ZIRCONIUM PRODUCTION.[1]

		SHORT TONS
U.S.	imports	67,537
	exports	17,360
	consumption	168,000
		TONS
USES:	foundries	92,000
	refractories	25,000
	Zr oxides	18,000
	alloys	3,000
	other (welding rods)	30,000

[1] 1972 *Metals Yearbook*

Refining Methods

Zirconium first was refined from a crude metallic powder produced by reduction of either zirconium oxide by calcium, of the tetrachloride by sodium, or of potassium fluorozirconate by potassium. Such powders were often of very low purity, analyzing only 80 percent to 85 percent zirconium. They are converted to ductile metal bars by a refining operation known as the Van Arkel—de Boer process, which involves the formation and decomposition of zirconium tetraiodide in an air-free vessel. Iodine vapor in this vessel attacks the crude powder at a moderate temperature, and the tetraiodide formed is decomposed at a higher temperature.

By causing the deposition to occur on a wire heated by an electric current, a rod of zirconium crystals is built up, and the iodine released by decomposition of the tetraiodide then reacts with more of the crude zirconium in a cooler part of the vessel. The process is repeated until all of the crude zirconium is used up or the rod of refined zirconium becomes so thick that it can no longer be kept at the required temperature by the electric current available. Scrap metal or impure sponge, as well as a powder, may be used as the raw material in this process.

A serious limitation of the process is the very heavy electric current necessary to maintain a thick zirconium bar at the temperature (about 2372°F or 1300°C) required for decomposition of the iodide. Thus, only fairly thin bars can ordinarily be made. The starting wire, generally tungsten, remains inside the bar as an impurity, but this objection can be overcome by using a zirconium wire to start the deposition process.

Pure zirconium bars produced by the iodide process are small, costly, and of low strength. Swaged and annealed rods of this type are almost as weak as copper. Reported properties are as follows: tensile strength, 36,000 psi (249,000 kN/m^2); yield strength, 16,000 psi (110,000 kN/m^2); elongation in 1 inch, 26 percent; reduction of area, 32 percent; and Rockwell hardness of B-30.3.

For these reasons the Zirconium Metals Corporation bases its process for producing zirconium metal on the Bureau of Mines method, involving the reduction (breaking down into constituents) of zirconium tetrachloride by magnesium.

By a method developed by the Titanium Alloy Manufacturing Company, the ore is first treated in an arc furnace to convert the silicate to a carbonitride (more commonly called cyanonitride), which in turn is chlorinated to obtain the tetrachloride. The tetrachloride is then reduced to metal by magnesium in a closed retort. Extreme care is required in the later stages of this process to avoid contamination of the metal by the oxygen and nitrogen in the air. The metal is produced in the form of a gray crystalline sponge (porous) which is melted in graphite by induction, or in water-cooled copper by an arc, to obtain ingots. An atmosphere free from oxygen and nitrogen is essential for melting operations. Carbon pick-up from the graphite in induction melting is slight and not detrimental if superheating (heating above boiling point without converting into gas) and long holding in the molten state are avoided.

The ingots are forged or hot-rolled at about 1350° to 1400°F (730° to 760°C), and if higher temperatures are avoided, scaling and oxygen absorption are not sufficiently serious to impair the quality of the finished product. Annealing, prior to cold-working, may be done at 1300°F (705°C) in air. Scale-removal for cold-rolling or wire-drawing is difficult, and usually requires shot-blasting, followed by pickling.

Uses

Zirconium metal was first used for priming explosives and flashlight powders. The extreme combustibility of this powder makes it very suitable for this purpose, and decades ago zirconium was produced chiefly as a powder. The metal produces a very bright light when ignited. The reactivity of finely divided zirconium renders it effective for combining with the last traces of oxygen and nitrogen in evacuated vessels or tubes, and thus the complete removal of these gases from the interior atmospheres. These uses are still current.

Among the other uses that have been suggested are for grid wires in vacuum tubes and parts of discharge tubes exposed to high temperatures, as an electrolytic condenser for rectifiers and high-intensity electric lamp filaments, and electrodes in fluorescent tubes. It is used to clad fuel rods in nuclear reactors.

Not only is zirconium not attacked by the atmosphere or sea water at normal temperatures, but it is almost as resistant to acids as tantalum (a hard, ductile metallic element). It resists alkalies better than tantalum. Zirconium is far superior to tantalum in wear-resistance and low density, as well as in abundance and economy.

Physical Properties

Like titanium, zirconium undergoes a physical transformation at approximately 1590°F (about 855°C). The room temperature phase, called alpha, has a close-packed, hexagonal, crystal structure. Above 1590°F (about 855°C) the crystal structure is body-centered cubic, called beta zirconium. The hexagonal close-packed structure in metals is generally considered to cause poor ductility because of the presence of only one plane of easy slip, but zirconium, like titanium, is

Metallurgy

an exception to this general rule and can be worked with ease at room temperature.

Mechanical Properties

Metallic zirconium, hot-rolled to bar form, has the following tensile properties, as determined by specimens having a gage-length of 1" and 0.16" (25.4 to 4.0mm) dia: stress for 1 percent extension under load, 68,000 to 80,000 psi (469,000 to 552,000 kN/m^2); tensile strength, 97,000 to 113,000 psi (669,000 to 779,000 kN/m^2); elongation in one inch, 9 to 14 percent, and reduction of area, 10 to 30 percent.

The tensile and bend properties of zirconium sheet, cold-rolled to 0.035" to 0.11" (0.89 to 2.80mm) thick, and annealed at 1400°F (760°C), as determined on standard specimens, are as follows: yield strength at 0.2 percent offset, 56,000 to 72,000 psi (386,000 to 496,00 kN/m^2); tensile strength, 76,000 to 83,000 psi (524,000 to 572,000 kN/m^2); elongation in two inches, 13 to 19 percent, and minimum bend radius, 1.8 to 3.9 times thickness.

Zirconium sheet has no directional properties. Its transverse ductility is at least as good as in the lengthwise direction; its transverse yield strength is slightly better. The tensile strength of zirconium decreases with rising temperature to 50 percent at temperatures from about 600° to 700°F (315° to 370°C).

Hardness. The hardness of zirconium varies from about 175 to 275 Brinell, more than that of aluminum alloys but less than steel. Hardness is greatly dependent on the oxygen and nitrogen in the metal. The hardness-strength relation of zirconium differs somewhat from that of steel.

Work hardening. Cold-working of zirconium increases its strength and hardness, and cold-rolling produces a pronounced loss in ductility of the metal, as indicated by the reduction in the percentage of elongation that occurs in cold-rolled specimens. Fig. 20-1 shows the effects of cold-rolling on mechanical properties of commercially pure zirconium.

Effect of annealing on cold-worked zirconium strip. Cold-rolled zirconium sheet and strip thicker than about 0.045" (1.14mm) may be satisfactorily annealed at 1400°F (760°C) in air, without damage by oxidation. In annealing gage material, however, a vacuum or inert atmosphere is necessary to prevent embrittlement of the zirconium by gaseous contamination. The reduction in hardness which takes place when cold-rolled 50 percent strip is heated under a protective atmosphere of argon for one hour at various temperatures is indicated on Fig. 20-2. The effect of air is discussed in the section on corrosion resistance and illustrated by Fig. 20-3.

The graph in Fig. 20-2 indicates that complete stress-relief and at least some recrystallization occur at 1172°F (633°C), with little further change in hardness up to 1652°F (900°C). However, at a temperature between 1652 and 1832°F (900° to 1000°C) zirconium hardness drops very suddenly, so that full softening may be obtained by annealing at 1832°F (1000°C) for one hour. Annealing at this temperature is not practical in air and must be conducted in a protective atmosphere to avoid excessive oxidation of the metal. Hydrogen should not be used because it has been found to cause embrittlement.

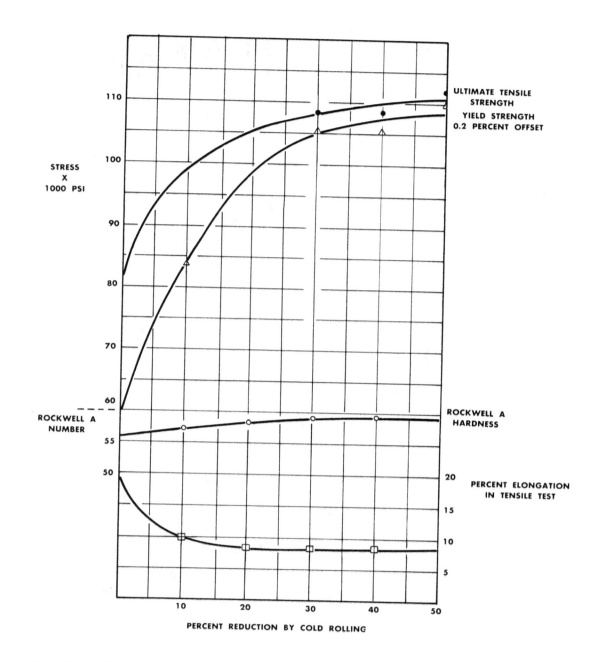

Fig. 20-1. Effects of cold rolling on the mechanical properties of commercially pure zirconium.

Metallurgy

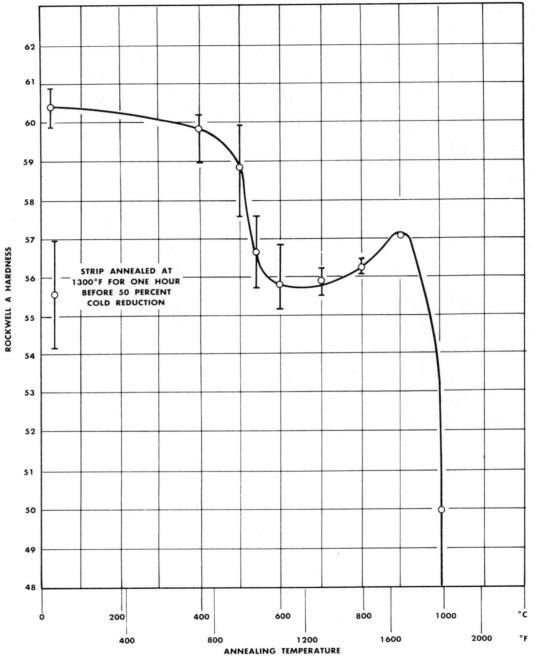

Fig. 20-2. The reduction in hardness when cold-rolled 50 percent zirconium strip is heated under a protective atmosphere of argon for one hour.

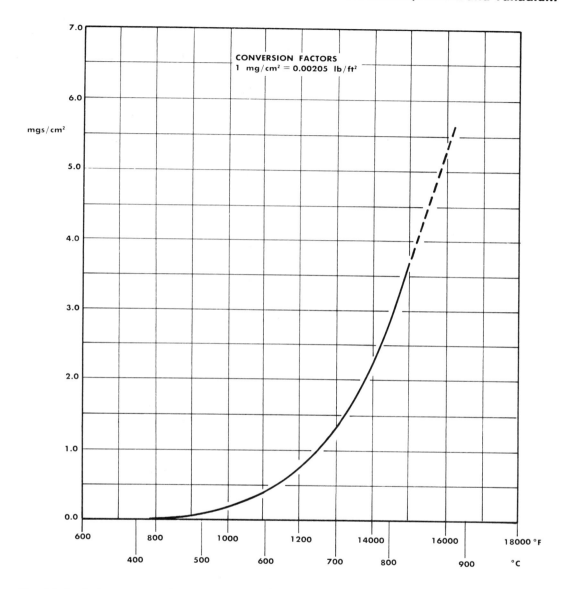

Fig. 20-3. Gain in weight of zirconium by oxidation in air in one hour.

The sudden drop in hardness at about 1652°F (900°C) is believed to be related to the physical transformation from alpha to beta zirconium that takes place in this temperature range and causes complete recrystallization. The protectivce atmosphere used for annealing thin zirconium strip or wire must be carefully purified to

Metallurgy

remove all traces of oxygen, nitrogen, water vapor, etc. Either argon or helium is sometimes passed over zirconium sponge at about 1600°F (871°C) for such purification.

Bending or forming. Zirconium sheet, strip, or wire may be deformed to a considerably greater angle in free bending if the initial deformation is performed at a temperature of about 400° to 600°F (205° to 315°C) instead of at room temperature.

Precision casting. Zirconium metal cannot be melted and cast by ordinary methods because of its reactivity with the atmoshphere and with refractories when molten. The purity of zirconium may be maintained in the molten state through melting in a closed, water-cooled copper vessel, either evacuated or filled with helium or argon. It does not wet the water-cooled copper. A method of melting in such a furnace, and pouring into an attached mold without contact with air, was developed by W.E. Kuhn.

This development work has shown that castings up to at least several pounds in weight can be made so that either molds or refractory investment molds (a process wherein molds are made by surrounding a wax pattern with a refactory ceramic material, producing molds which will sustain high temperatures) can be used. The fluidity of zirconium metal melted by an arc in this way is sufficient for rods 3" (7.62mm) long and only 5/32" (4.0mm) dia to be cast without defects. Intricate castings which produce the patterns precisely in size and detail can be obtained through the use of investment molds of zirconia cements.

Zirconium wire. Annealed, hot-rolled zirconium rods are readily cold-drawn to the form of wire, raising their strength to more than 100,000 psi (6,895,000 kN/m^2).

Notched-bar impact resistance of zirconium. The notch sensitivity or impact resistance of zirconium for both 45° *V* and standard keyhole notches has been determined by Charpy test at room temperature on hot-rolled and annealed [1300°F (704°C) in air] bar stock.

Impact tests on zirconium metal have been made at elevated temperatures, using *V*-notch Charpy specimens. The following results indicate that a transition from brittle to ductile fracture takes place on heating commercial zirconium metal in the vicinity of 900°F (about 480°C), where the impact value sharply increases from less than 15 ft-lb at 800°F (about 425°C), to more than 40 ft-lb at 1000°F (about 540°C).

It is possible that by special heat treatment, combined with vacuum annealing, the transition temperature may be lowered, or the transition on cooling eliminated, so that notched-bar impact values will be higher at room temperature.

Fire Precautions

Solid massive zirconium is not dangerous to handle. It will not burn when forged or rolled at red heat, and can be machined as safely as stainless steel. Turnings or chips of ordinary size can be handled in air with complete safety, but very fine turnings will sometimes ignite and burn slowly with an intense white light and considerable heat.

Zirconium, Indium and Vanadium

CAUTION

When either powder, sponge, or turnings start to burn, the fire should be extinguished only by smothering with dry sand, salt, or inert oxides such as titanium oxide or zirconium oxide. Carbon tetrachloride or carbon dioxide should not be used because they may cause explosions and more violent burning.

CAUTION

Zirconium *powder* in air is both pyrophoric (ignitable in air) and explosive, even when damp. Extreme caution should be exercised, when handling the dry powder, for friction, static electricity, or any form of heat concentration will cause the dry powder to ignite.

INDIUM

Indium is generally found in zinc blends. There are indications that it is found in a number of other ores, but the quantity of indium present in all ores will vary considerably. In most ores its extraction is economically impractical. Ore has been found in which indium occurs in sufficient quantities to make indium of commercial importance.

Indium may be extracted and purified in many ways; treatment of indium chloride with ether precipitation; by boiling with hydroscopic sulfite; treatment of indium and tin sulfates with hydrogen sulfide; separation of indium from tin with sodium hydroxide; solubilization of indium in an acid solution with sodium sulfate; treatment of indium-bearing lead or lead alloys with sodium hydroxide and sodium chloride; treatment of lead-residues from zinc ore smelting with sulfuric acid to form a calcine; separation of indium and gallium by the electrolysis method, and precipitation of indium as indium orthophosphate.

A complex ore containing lead, zinc, iron, copper, silver, gold, and indium is worked, separating the concentrates from the gangue (waste) by grinding and treatment by flotation process. Concentrates of zinc and lead-silver are obtained. Indium is found in the zinc concentrate. The zinc concentrate is roasted in the presence of sodium chloride in the first refining process. Since indium chloride will fume at roasting temperature, a Cottrell precipitator was used to catch these fumes. The fumes were dissolved and the indium plated out.

Today the zinc concentrate is roasted, and the soluble portion is dissolved in sulfuric acid. From this solution indium is plated out or thrown out by neutralization of the acid. Indium from this process is again dissolved, purified, and plated out. The new process produces indium of more than 99 percent purity.

Metallurgy

Physical Properties of Indium

Indium, which resembles tin, is a soft, white metal with a bluish tinge. It is ductile, malleable, softer than lead, crystalline, and diamagnetic. It appears between iron and tin in the electromotive series.

Indium does not react with water, even at boiling temperature. It will not combine with the gases in the air at normal temperatures but, when heated, it burns with a nonluminous, blue-red flame, producing indium oxide. Up to its melting point, the surface of indium remains bright but, at higher temperatures, a film of oxide appears. Indium readily combines with organic compounds which can be easily oxidized or combined chemically with water.

Indium has an atomic weight of 114.8. It is face-centered tetragonal crystal lattice in structure, very malleable and ductile, and has an approximate tensile strength of 15,980 psi (110,200 kN/m^2). It melts at 311°F (155°C) and boils at 2642°F (1450°C).

Uses

One major development in the use of indium has been the coating of metallic surfaces with indium, diffused into the base metal. This extensively prolongs the life of these moving parts by reducing friction and wear. The thicker the film of indium, the more resistant the surface is to wear. Indium has a low coefficient of friction and over a wide range of temperatures, is slippery and changes its viscosity very slightly.

A recent use is the combining of indium and graphite for lubrication and wear reduction, in moving parts for internal combustion engines, engine accessories, etc. The combination also reduces erosion and corrosion for electrical products.

Alloys of Indium

The addition of small quantities of indium hardens and strengthens the metal with which it is alloyed, increasing its tarnish or corrosion resistance.

Aluminum-indium. Small amounts of indium have an influence on the age hardenability of aluminum alloys. Additions of 0.1 percent of indium retard the age-hardening of duralumin, but increase both the rate and amount of hardening of 4 percent copper-aluminum alloys, from which magnesium is absent.

Beryllium-copper-indium. The addition of indium to beryllium copper alloys increases their hardness and tensile strength and lowers their melting and heat-treatment temperatures. The presence of indium increases the fluidity of the alloy.

Cadmium and indium dissolve completely in each other in the molten state in all proportions. A eutectic is formed at 75 percent indium and 25 percent cadmium with a melting point of 252.5°F (122°C). These alloys are used in the production of surfaces which are subjected to friction.

The addition of indium to gold provides sounder dental castings. Indium is softer than lead, more lustrous than silver, and as untarnishable as gold. Used as a deoxidizer and cleanser, it increases tensile strength 14 percent and ductility 24 percent.

Gold-indium alloys have some unique characteristics for brazing. The alloys of 77½ percent gold and 22½ percent indium have a working temperature of a little above 932°F (500°C). This is an ideal temperature for work on metal objects with glass inserts, since temperatures above 1112° F (600° C) are destructive to glass-metal seals, while temperatures of 932° F (500° C) must be endured, and indium prevents unwanted contamination of adjacent parts.

Lead-indium alloys are used widely in bearings and in solders. The addition of 1.0 percent of indium to lead doubles the hardness of the lead. The addition of indium to silver tends to limit tarnishing.

Low melting alloys with indium. In 1938, it was determined that the addition of indium to a Wood's or Lipowitz Metal caused its melting point to drop approximately 1.45° for every 1% indium present. The lowest melting point was 116°F (47°C), with an alloy containing 18 percent indium. The presence of indium decreases the tendency of such alloys to oxidize in the molten state. Development of low-melting solders available commercially are based on these facts. These solders possess superior bonding power, better wetting characteristics, and superior corrosion resistance.

Bearing alloys. The aviation industries were the first to recognize the value of indium-treated bearings. The Indium Corporation of America produced the first indium-treated aviation bearings.

Indium bearings have not been excelled in meeting the demands presented by high oil temperatures, generation of acid in oil, heavy loads, and the necessity of high wettability in aviation engines. Silver-lead indium bearings are the most widely used aviation bearings today.

Each of the three components of the silver-lead indium type bearing has a service to perform. Silver has the internal properties which resist failure due to fatigue, but externally silver lacks the quality of "oiliness" needed in a good bearing surface. To fill this requirement, a thin layer of lead is applied to the silver surface. Lead, unfortunately, is soluble in the organic acids present or formed in lubricating oils. To offset this difficulty, a thin layer of indium is deposited on and diffused into the layer of lead. The addition of indium to the bearing surface accomplishes the following things: it increases the strength of the bearing material; it prevents corrosion of the bearing surface without impairing the fatigue resistance or other bearing properties; it increases bearing wettability, and thus permits the bearing surface to retain its oil film more completely.

It was felt years ago that, if a bearing would stand 200 hours of operation, it was a good bearing. Today indium-treated bearings are expected to last as long as the engine.

"In one engine test[1], connecting rod bearings lined with Cd-Ag-Cu (cadmium-silver-copper) alloy, treated with approximately 0.20 percent In (indium) were run for over 5,000 miles at high speed. At the conclusion of the test these bearings showed only slight evidence of etching,

---

[1]Technical Paper No. 900, American Institute of Mining and Metallurgical Engineers, 1938.

Metallurgy

whereas the untreated bearings run in one of the connecting rods were replaced three times because of their badly corroded condition. The acid number of the oil used reached a value of 3.3.

"In another engine the fatigue life of Babbitt-lined bearings ranged up to a maximum of about 60 hours, under the severe conditions used for testing [4250 rpm, full throttle, with oil temperatures 250° to 260°F (121° to 127°C) at the bearings]. Cd-Ag-Cu bearings treated with 0.4 to 0.5 percent indium run under these test conditions, and in the presence of oil containing 0.5 percent of oleic acid, were in excellent condition after 120 hours, showing no evidence of corrosion and only a very slight indication of fatigue cracks.

"Such tests on bearings, in engines run at high speed and with corrosive oils, verify the information obtained at laboratory tests—namely, that the indium treatments as described prevent corrosion of the Cd alloy bearings without impairing the fatigue resistance or other bearing metal properties."

One bearing manufacturer tested indium-treated copper-lead bearings against the present micro-Babbitt type bearings for automobiles. Operating under a load of 6500 psi (44,800 kN/m^2) twice that of the micro-Babbitt bearings, copper-lead-indium bearings ran for 500 hours as compared with 30 hours for the micro-Babbitt bearings.

Indium-treated bearings in one instance gave 50 hours use under pressures from 10,000 to 15,000 psi (69,000 to 103,000 kN/m^2) where previous bearings were destroyed after 10 hours.

Other examples are many and from varied sources, but the results all remain the same, i.e., indium treatment materially improves all bearing surfaces.

Brazing alloys. There are two distinct classes of brazing alloys or solders. Several solders are available which have relatively low melting points, below 600° to 700°F (315° to 370°C), but there is a very great need for a brazing alloy with good flowability and strength for heat ranges between 700° and 1100°F (370° to 595°C). Since indium melts at 311°F (155°C), and since it alloys very readily with copper, silver, and other elements, several alloys, using indium, have been produced which melt at from 795° to 965°F (423° to 519°C). Because indium adds greater wettability to brazing materials and, in larger proportions, greater strength, it becomes a very desirable constituent of some of the conventional brazing materials.

VANADIUM AND VANADIUM ALLOYS

Vanadium is one of the more important elements used in alloy with steel. It occurs in the mineral vanadite and in iron ores, fire clay, and granite. The richest known vanadium deposit is in Peru. It is difficult to produce pure vanadium, and only small amounts are produced commercially. Two methods of vanadium production often used are the reduction of vanadium dichloride with hydrogen, and the electrolysis of vanadium trioxide in fused calcium vanadate.

Vanadium is a hard, silvery-white metal but, if heated to a suitable temperature, it is sufficiently tough and malleable to be hammered and rolled into rods or drawn into wire. It can be highly polished and will not tarnish when exposed to air. The physical properties of vanadium are as follows: melting point, 3110°F (1710°C); boiling point, 5432°F (3000°C); density, 5.68 g/cc, and electrical resistivity, 26 microhms.

Vanadium has a more powerful effect upon the properties of steel than any other element except carbon, comparatively small amounts of it are necessary, as illustrated in Figs. 20-4, 20-5, 20-6 and 20-7.

Fig. 20-4. Photomicrograph of 0.50 percent carbon steel, air-cooled at 1800°F (about 980°C), magnified 100 times; note coarse grained structure.

Fig. 20-5. Same as Fig. 20-4, with 0.27 percent vanadium content. Note effect of vanadium in reducing grain structure of the steel.

Metallurgy

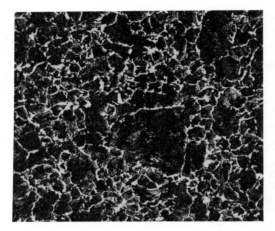

Fig. 20-6. Photomicrograph of 0.50 percent carbon steel, air-cooled at 1500°F (815°C), magnified 100 times; note coarse grain structure.

Fig. 20-7. Same as Fig. 20-5, with 0.27 percent vanadium content; note effect of vanadium in reducing the size of the grain structure of the steel.

The best results are obtained from 0.1 to 0.2 percent of the vanadium alloy, but if more than 0.3 percent is used, strength is greatly decreased. From 0.1 to 0.15 percent vanadium increases the tensile strength of low and medium-carbon steels about 50 percent, with no decrease in ductility. The tensile strength of steel, with about 0.2 percent vanadium and 0.8 percent carbon, is not changed by the vanadium, but the elastic limit and ductility are increased.

Vanadium produces a very small grain size in steel. Grain size reduction and its control by the use of vanadium are major aids in approaching perfection in steel making. The grain structure of vanadium steels can be slightly coarsened to the extent of insuring ease of machining and smoothness of the machined surface. The reheating temperature for steels of any composition is established by their vanadium content, and reheating can be accomplished at the moderate temperatures readily obtained in industrial heat-treatment furnaces. Grain structure coarseness is neither very high nor is it irregular from melt to melt. Vanadium steels return to their fine-grained condition on reheating to the temperatures customarily employed for annealing, normalizing, quenching, etc.

The fineness of structure in vanadium steels extends to the distribution of the carbide. Even in extremely low melting steels, the primary carbide is affected and more uniformly distributed in smaller particles. This is true even in steels with large amounts of free carbides. This consistent and great reduction of grain size, and subdivision of structure within the grains, are naturally reflected in many of the mechanical properties of the steel, principally in an increase in yield point and yield ratio, with a simultaneous improvement in ductility and toughness.

There is a very distinct and useful effect in the behavior of vanadium steels upon quenching. Carbon-vanadium steels, quenched from customary heat-treatment temperatures (just high enough to secure maximum surface hardness), are shallow hardening. If a carbon steel and a vanadium steel are both quenched from the same temperature, they will both have the same surface hardness, but the vanadium steel will possess lower core hardness and a much thinner, fully hardened outer layer. The grain size of the vanadium steel at 1800° F (980° C) is the same as that of the carbon steel at 1500° F (815° C), and vanadium steel coarsens slowly. The phenomena described are largely the result of fine grain size which effects more rapid transformation upon cooling and thus causes greater difficulty in the retention of hard martensite. Increase in the hardening temperature slowly increases grain size in the vanadium steel and, because of this, as well as the action of the solution of additional vanadium-rich carbides that render the matrix less rapid in its reactions, depth of hardening increases.

Observations such as these have led investigators to the conclusion that the carbides are alone responsible for the grain characteristics of vanadium steels, that is:

Vanadium-rich carbides are relatively stable and dissolve somewhat more slowly than iron carbide even at temperatures well above those usually employed in heat treatment.

At any temperature at which part of these carbides remain undissolved, they inhibit grain growth or serve on cooling as points of initial crystallization; thus they maintain fine grain size.

When temperature and time are sufficient for the carbides to become largely dissolved, the grain coarsens.

Iron and vanadium are mutually soluble in all proportions. Only up to about 1.0 percent vanadium content, however, are the carbon-free alloys hardenable. Beyond this amount gamma iron does not exist at any temperature. Hence, martensite, on which the hardness of quenched steels depends, cannot be formed. The introduction of carbon increases the range of compositions in which martensite is possible, so that with 1.0 percent carbon, for example, some degree of hardening may be obtained with slightly more than 6.0 percent vanadium in the steel.

When vanadium is added to carbon steels, it has a strong affinity for carbon, forming vanadium carbide. These carbides are very stable. To secure maximum hardening of vanadium steels, the carbides must go into solution. Hardening depends upon composition and heating conditions insofar as they influence the amount of carbon in solution, the amount of carbon as vanadium carbide remaining undissolved, and the formation of gamma iron as limited by the amount of dissolved carbon and vanadium.

Quenching temperatures that cause little or no vanadium carbides to enter into solution secure the maximum benefit from the grain-refining power of vanadium and produce exceptional combinations of useful strength and toughness. However, increased depth of hardness, on tempering at relatively high temperatures, may be obtained through the use of quenching temperatures high enough to partially dissolve the vanadium carbides present in the steel, yet insufficient in temperature or time to result in appreciable grain growth.

Carbon vanadium steel is the most commonly used steel except for tool steels. Most frequently it is used in large forgings, but the uses of small sections are increasing in number. The superior mechanical properties of carbon-vanadium steel, coupled with its fine machining qualities in the normalizing state, and its low degree of distortion, are responsible for this use.

In many instances, carbon-vanadium steel is tempered after normalizing, in which case the following heat treatment is generally used: Normalizing-heat at 1600° to 1650° F (870° to 900° C); cool in still air. Tempering-heat of 1100° to 1200° F (595° to 650° C); cool in still air or allow to remain in the furnace.

The addition of 0.15 percent vanadium to plain carbon steel of this type raises the strength and load-sustaining capacities at higher temperatures.

Carbon-vanadium steel has also been made in the 0.30 percent to 0.35 percent carbon range for use in small shafts, arms, connecting rods, and other machine parts subject to impact loading. These parts are normally quenched and tempered. The use of very mild alloy steel of this type in carburized gears is possible where (1) a moderately high core hardness is needed to support the case and prevent fatigue failure by repeated deflection; (2) uniformity of case depth and of hardening of both case and core is vitally important; and (3) moderate impact strength of the carburized tooth is required.

The manufacture and use of rolled or forged steel in large sections present considerably different problems from those encountered in manufacture of small masses. The dangers of internal ruptures, insufficient hardening, and sharp hardness gradients in the finished product necessitate uniform and not too rapid heating, time to insure total diffusion for hot working and heat treatment temperatures, uniform temperature gradients in cooling, and magnitude and time of occurrence of volume changes in the transformation range in different locations in the mass. In some instances, where smaller masses are concerned, any one of several compositions might satisfy manufacturing and design requirements.

In heavier masses, the sensitivity to fabrication processes limits to a much smaller number the selection of steels suitable for a particular set of properties. This sensitiveness to steep temperature gradients, i.e., rapid cooling, often prevents the development of a desired hardness by a quenching and tempering practice.

Some forgings of moderately large section may be quenched and tempered to provide in this hardened condition very satisfactory service. When properties that appear to demand such treatment are called for, special consideration must be given to all pertinent details, such as composition, size, manufacturing practice, and the nature of the service.

Vanadium steel forgings have supplied heavy industrial needs for many years. Carbon-vanadium steel appeared early in the century, at the beginning of commercial vanadium production. It extended rapidly to the construction of locomotives

Zirconium, Indium and Vanadium

and heavy machinery, where it still retains a preeminent position. The demand of recent years for a steel of the hardness and wear resistance of the carbon-vanadium type, with materially increased ductility and impact strength, led to the production of manganese-vanadium steel for forgings. Still more recently, a chromium-molybdenum-vanadium steel was introduced for large sections in which full penetration of high hardness and commensurately high resistance to sudden loading are required. A vanadium-molybdenum-manganese steel has been produced which has a high elastic-limit, ductility, and impact strength without liquid quenching.

Vanadium Cast Steels

The tremendous growth in the size and power of heavy machinery such as that used in the fabricating industries, mining, and transportation presents the foundryman with the job of producing not only sound castings of greater size and thinner sections but articles of greater complexity and ability to sustain both static and dynamic loads.

Vanadium is among those elements whose widened use is a result of this growing appreciation of alloy cast steels. The incorporation of vanadium into the composition of steel castings results in a vast improvement of their properties. Steel castings containing vanadium have a higher elastic ratio than castings without vanadium but of otherwise like composition and heat treatment. At the same time, they possess at least equal ductility in tension and a considerably higher impact strength and wear resistance. Their grain size is markedly smaller and more uniform, while grain growth at heat-treatment temperatures is decidedly retarded.

Other components of the composition being the same, and the details of steelmaking, pouring, etc., almost constant, a vanadium-containing steel will exhibit less marked dendritic segregation and greater freedom from Widmanstatten patterns within the grains than a vanadium-free steel before heat treatment.

Without advocating the use of non-heat-treated alloy steel castings, it is clear that, for equal hardness and strength, the vanadium steel, with its superior distribution of microconstituents, possesses greater ability to sustain stress or to deform rather than rupture under suddenly applied overloads. These characteristics are extremely important in avoiding the cracking of intricate castings during heating and cooling cycles in the early stages of manufacture.

The structure of the vanadium-containing steel exerts considerable influence upon ease of diffusion of the several constituents at heat-treatment temperatures. It also unquestionably bears a relation to the response of vanadium steels to simple heat treatment. Mechanical properties may be obtained by double normalizing and tempering that are not equaled even by the more drastic and often dangerous liquid quenching and tempering of many other alloy steels. This is of great importance in large complex castings. New uses of high-strength cast steels are available

Metallurgy

through suitable adjustment of composition, combined with even single normalizing and tempering, or through single normalizing alone.

The macrostructural advantage of vanadium steels, i.e., less marked dendritic formations, persists after heat treatment. Microstructurally, equally sharp distinctions between vanadium-containing and vanadium-free steels develop, causing or accompanied by pronounced differences in the mechanical properties.

Table 20-2 illustrates the influence of small amounts of vanadium upon the tension and impact values of some normalized alloy steels. The simultaneous increase in both yield point and resistance to impact is consistent with the magnitude of the changes that result from the incorporation of vanadium. No appreciable alteration of tensile strength, elongation, or reduction of area accompanies these increases. The ability to support high static loads and suddenly applied overloads without rupture, shown in these comparative tests, is responsible for numerous and diverse engineering applications.

Of the several vanadium cast steels now in regular production, carbon-vanadium cast steel was the earliest produced commercially. The ASTM tentative specification for physical properties requires the following minimum values for normalized and tempered castings: yield point, 55,000 psi (380,000 kN/m^2); tensile strength, 85,000 psi (585,000 kN/m^2) elongation in 2'', 22 percent; and reduction of area, 42 percent.

Representative values obtained from such normalized and tempered carbon-vanadium steel of about 0.35 percent carbon are shown in Table 20-3.

The usual heat treatment employed for the development of these properties consists of normalizing at 1600° to 1650° F (about 870° to 900° C) and tempering at 1050° to 1200° F (about 570° to 650° C). In the case where double normalizing is employed, the temperatures approximate 1775° to 1850° F (970°-1010° C) and 1575° to 1625° F (855 to 885° C).

Chromium-Vanadium Steels.

Table 20-4 shows the chromium vanadium steels which have enjoyed popularity for many years.

Low-carbon grades of chromium-vanadium steels are frequently used in the case-carburized condition. Those of intermediate carbon content are quenched and tempered to various hardnesses for axles, shafts, gears, springs, etc., while the very high-carbon steels serve for tools, ball and roller bearings, wearing plates, and other fully hardened parts.

Low-carbon chromium-vandium steels, of the types represented by SAE 6115 and 6120, are used for case-carburized parts, such as automobile and aircraft engine gears, camshafts, and piston pins, Fig. 20-8. They give a hard, tough, and strong case of high wear-resistance. The low rate of drop in the carbon content as the core is approached causes case and core to be strongly bound together so that there is little tendency to flaking, powdering, or flowing under pressure.

The usual heat-treatment procedure designed to give the very best properties is as follows: carburize at 1650° to 1700°F (900° to 930°C) and cool in the pots; oil

TABLE 20-2. TENSION AND IMPACT TESTS OF CAST ALLOY STEELS DOUBLE NORMALIZED AND TEMPERED.

TYPE	TYPICAL CHEMICAL COMPOSITION							YIELD POINT (psi)	TENSILE STRENGTH (psi)	ELONGA-TION (percent in 2")	REDUCTION OF AREA (percent)	IZOD VALUE (ft-lb)
	C	Si	Mn	Ni	Cr	Mo	V					
Mn (a)	0.35	0.40	1.40	...	...	...	...	60,850	102,650	27.5	58.8	25.8
Mn-V (a)	0.35	0.40	1.40	...	...	...	0.10	74,500	100,700	30.5	61.8	57.5
Mn (b)	0.35	0.40	1.40	...	...	...	...	67,500	108,500	25.9	54.5	19.5
Mn-V (b)	0.35	0.40	1.40	...	...	...	0.10	77,700	103,300	27.5	57.6	52.8
Mn-Mo (a)	0.35	0.40	1.50	...	...	0.15	...	69,800	114,050	21.0	41.6	18.8
Mn-Mo-V (a)	0.35	0.40	1.50	...	...	0.15	0.10	76,050	118,650	23.0	39.1	30.5
Cr (a)	0.30	0.40	0.80	...	1.00	...	...	60,400	94,500	27.5	54.7	28.3
Cr-V (a)	0.30	0.40	0.80	...	1.00	...	0.10	64,850	94,300	27.5	57.1	59.3
Cr-Mo (a)	0.30	0.30	0.50	...	1.00	0.15	...	53,800	90,900	28.0	58.4	28.5
Cr-Mo-V (a)	0.30	0.30	0.50	...	1.00	0.15	0.10	62,800	88,800	31.0	61.5	52.3
Ni-Cr (a)	0.30	0.40	0.60	1.30	0.50	...	...	61,850	92,900	27.0	54.4	48.0
Ni-Cr-V (a)	0.30	0.40	0.60	1.30	0.50	...	0.10	69,650	94,500	28.5	55.8	71.3
Ni (a)	0.28	0.35	1.00	1.50	...	...	...	53,450	93,250	27.0	56.6	35.5
Ni-V (a)	0.28	0.35	1.00	1.50	...	...	0.10	69,900	91,200	28.5	59.4	74.3

Experimental induction furnace melts. Average of several commercial open-hearth melts heat-treated in small furnaces

TABLE 20-3. MECHANICAL PROPERTIES OF CARBON-VANADIUM CAST STEEL.

YIELD POINT (psi)	TENSILE STRENGTH (psi)	ELONGATION (percent in 2")	REDUCTION AREA (percent)	IZOD VALUE (ft-lb)
59,900	93,800	23.5	46.3	29.0
55,800	91,000	25.0	44.9	30.8[1]
63,000	90,000	25.0	42.0	35.0[1]
62,300	93,400	24.5	47.8	20.8

[1]Double normalized and tempered.

Metallurgy

TABLE 20-4. PERCENTAGE COMPOSITION OF CHROMIUM-VANADIUM STEELS OF THE 6100 SERIES.

STEEL NO.	CARBON RANGE	MANGANESE RANGE	PHOS-PHORUS, max	SULFUR, max	CHROMIUM RANGE	VANADIUM min.	VANADIUM desired
6115	0.10–0.20	0.30–0.60	0.40	0.045	0.80–1.10	0.15	0.18
6120	0.15–0.25	0.30–0.60	0.40	0.045	0.80–1.10	0.15	0.18
6125	0.20–0.30	0.50–0.80	0.40	0.045	0.80–1.10	0.15	0.18
6130	0.25–0.35	0.50–0.80	0.40	0.045	0.80–1.10	0.15	0.18
6135	0.30–0.40	0.50–0.80	0.40	0.045	0.80–1.10	0.15	0.18
6140	0.35–0.45	0.50–0.80	0.40	0.045	0.80–1.10	0.15	0.18
6145	0.40–0.50	0.50–0.80	0.40	0.045	0.80–1.10	0.15	0.18
6150	0.45–0.55	0.50–0.80	0.40	0.045	0.80–1.10	0.15	0.18
6195	0.90–1.05	0.20–0.45	0.30	0.035	0.80–1.10	0.15	0.18

(Society of Automotive Engineers)

quench from 1600° to 1650°F, (870° to 900°C), water quench from 1450° to 1475°F (about 790° to 800°C), and temper at 375° to 425° (191° to 218°C).

While this treatment is the optimum one for both case and core, many purposes are served by a single oil quench, which is carried out after reheating to 1625°F (885°C). This heat treatment, followed by tempering for stress relief, is the customary one for small automotive pinions. Chromium-vanadium steels are more applicable to single quenching than many of the other alloy steels, because of their fine grain in case and core, even after the higher temperature necessary for single quenching operation. Tempering may be employed after single quenching.

Typical tension and impact properties in the core of case-hardened parts (one inch diameter) of these low-carbon chromium-vanadium steels are given in Table 20-5.

The fabrication of welded pressure vessels is another application of chromium-vanadium steel of the SAE 6115, 6120, and 6125 ranges. They are generally arc-welded with heavily flux-coated chromium-vanadium electrodes. Most of these vessels are only tempered after welding for stress-relieving purposes, but some are fully annealed.

Steels 6120 and 6125 respond to cyanide hardening and are used for gears, bolts, washers, small stampings, and forgings. The nitrides formed are extremely fine, and there is no tendency to develop the undesirable, coarse, acicular (needle-shaped) structure, thus insuring case toughness and resistance to spalling (chipping). The transition from the hard surface to the core is gradual and, even in long-time cyaniding, there is no danger of embrittlement of the core.

Chromium-vanadium steel in the SAE 6120 and 6125 carbon ranges is also used in boiler construction, superheater tubes, tubing for the chemical industry, bolts, and pressure vessels and welding rods.

Manganese-vanadium steel combines in large masses, in the normalized and tem-

Zirconium, Indium and Vanadium

pered condition, the properties of high strength or hardness with excellent ductility and impact resistance.

Some other vanadium alloy steels to meet special needs are as follows:

A chromium-molybdenum-vanadium

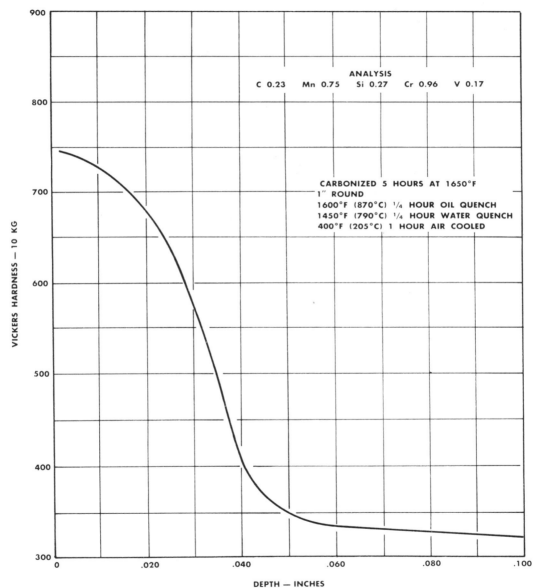

Fig. 20-8. Depth of hardness characteristics of vanadium-chromium steel, SAE 6120.

Metallurgy

TABLE 20-5. TENSION AND IMPACT TESTS OF THE CORE OF CARBURIZED CHROMIUM-VANADIUM STEELS (C-0.19, Mn-0.68, Si-0.18, Cr-0.83, V-0.15).

HEAT TREATMENT (degrees F)	YIELD POINT (psi)	TENSILE STRENGTH (psi)	ELONGATION (percent in 2")	REDUCTION OF AREA (percent)	IZOD VALUE (ft-lb)
Carburized 1675 (16 hr) Pot Q. oil; 1520 oil; 350 air	70,700	114,300	25.0	53.6	54.5
Pot cool; 1520 water; 350 air	95,450	137,950	17.0	38.0	21.3
Pot cool; 1660 oil; 1520 oil; 350 air	62,350	111,100	25.0	52.8	57.5
Pot cool; 1540 oil; 350 oil	60,450	109,150	24.0	50.8	37.0

steel was developed to meet the need for a steel capable of developing a uniform high hardness throughout moderately large sections and at the same time commensurately high ductility and impact strength.

A nickel-chromium-molybdenum-vanadium steel produces quenched and tempered forgings of moderately large section and high strength. This composition has the advantage of low-carbon content which is possible in view of the high total alloy content.

The manganese-molybdenum-vanadium steels have high yield-point in moderately large sections that have been given only a normalizing and tempering treatment.

Other major uses of vanadium are in the manufacture of spring steels, cast steels, cast irons, and the making of fine high-grade tool steels. See Table 20-6.

Vanadium Cast/Iron

Vanadium has been added to irons for decades. Its beneficial effects were first utilized shortly following the introduction of vanadium alloy steel. Renewed interest in the properties of vanadium cast irons was brought about by a demand for steel to meet severe requirements. Vanadium is used in cast iron except where resistance to structural decomposition at elevated temperatures, or uniformity of hardness and strength throughout a heavy section, is the single characteristic of importance. There is a complexity in cast-iron metal-

TABLE 20-6. VANADIUM PRODUCTION.[1]

	SHORT TONS
World production	19,949
U.S. production	4,887
consumption	5,227
USES: alloy steels—killed grades Nitrovan (78 to 80 percent Va 6 to 7 percent N 10 to 12 percent C)	

[1] 1972 *Metals Yearbook*

lurgy caused by the existence of carbon in two distinct forms, the manner in which these forms change from one to the other, and the influence of raw materials and manufacturing methods upon carbon behavior.

REVIEW

1. Describe the Van Arkel—de Boer process for refining zirconium.

2. What are the directional properties of zirconium sheet?

3. Describe the precautions necessary in the handling of zirconium powder, sponge, and turnings.

4. What properties does indium impart to alloys?

5. Why is indium used as an ingredient of solders and brazing alloys?

6. Why is indium used in making bearing metals?

7. What does vanadium do to the grain structure of steel?

8. How is depth of hardness obtained in vanadium steels?

9. What precautions should be taken to avoid insufficient hardening and sharp hardness gradients in vanadium steels?

10. What are the limitations of vanadium cast irons?

GLOSSARY

Some of the most commonly used terms employed in the working, treating, and testing of metals and alloys are defined here. As many of the terms used in the metal industry have more than one meaning or interpretation, the student may find occasion to refer to other definitions. No attempt has been made to cover all applications of metallurgical terms here. For a more complete discussion, the student is referred to the *Metals Handbook* published by the American Society for Metals.

A

Ac1: Identifies the critical temperature during heating of iron-carbon alloys where the first major change of structure takes place, i.e., pearlite to austenite.

aging: Process of holding metals or alloys at room temperature after subjecting them to shaping or heat treatment, for the purpose of increasing dimensional stability or to improve their hardness and strength through structural changes, as by precipitation.

allotropic change: Change from one crystal structure of a metal to another that has different physical charasteristics.

amorphous: To describe material that is noncrystalline and which has a random orientation of its atoms or molecules, resulting in no distinct cleavage planes such as those found in the crystal state.

alpha iron: Normal iron in a body-centered cubic structure below 1415°F (769°C). Magnetic.

Glossary

annealing: Subjecting to heat treatment. This usually involves heating, followed by relatively slow cooling of metals or alloys for the purpose of decreasing hardness and increasing the ease of machining or the cold-working characteristics. Annealing may be used to (a) remove effects of strain hardening resulting from cold work, (b) remove stresses found in castings, forgings, weldments and cold-worked metals, (c) improve machinability and cold-working characteristics, (d) improve mechanical and physical properties by changing the internal structure, such as by grain refinement, and to increase the uniformity of the structure and correct segregation, banding, and other structural characterisitcs. In general, annealing practice should be carried out so as to obtain the type of structure that is required, depending upon the end-point or use to which the metal or alloy is to be applied. See *full annealing, malleablizing, normalizing, patenting, processes annealing, spheroidizing.*

Ar1: Identifies the critical temperature during cooling of iron-carbon alloys where the change from austenite to pearlite occurs.

as-cast structures: The crystalline structure before stress relief through rolling or hammer forging.

ASTM: *Abbreviation for American Society for Testing Materials.*

austempering: Heat treatment in which the operations of hardening and tempering are combined by quenching steels, previously heated to the austenitic temperature range, in a molten salt bath maintained at the elevated temperature between the knee of the s-curve and the temperature where martensite forms.

austenite: A solid solution structure with the gamma iron acting as a solvent; a solid solution of carbon or iron carbide in iron, determined microscopically as a constituent of steel under certain conditions.

B

beta iron: Iron in body-centered cubic structure between 1414° and 1666°F (769° to 906°C). Non-magnetic.

BOF: Basic oxygen process of steel making using oxygen under pressure to oxidize excess carbon and impurities.

banded structure: A structure with a woody appearance caused by working segregated alloy such as steel; the bands are commonly formed by layers of ferrite and pearlite.

binary alloy: An alloy composed of two elements.

blister steel: Bars of carburized wrought iron with small blisters on surface. A product of an early steel making process.

blowhole: Hole found in metals resulting when gases, trapped during solidification, cause a porous condition to develop; a defect in an ingot or casting caused by a bubble of gas.

bloom: A mass of iron from furnace, before shaping.

box annealing, pack annealing: Annealing process in which the metal is placed in a suitable container or box, with or without packing material; this practice protects the metal from oxidation.

bright annealing: Process of subjecting metal to high heat, with subsequent cooling, during which operation the atmosphere is controlled in order to prevent any oxidation or discoloration.

briquette: A mass of metal powders or ore dust molded together under considerable pressure, with or without heat, into a brickshaped block; also spelled *briquet.* The process is called *briquetting.*

Metallurgy

burning: Over-oxidation of metals with loss of ductility and strength during heating of metals under oxidizing conditions. Burning results in grain growth and permanent injury to metals.

C

carbon steel: Steel whose physical properties are chiefly the result of the percentage of carbon contained in it; and iron-carbon alloy in which the carbon is the most important constituent, ranging from 0.04 to 1.40 percent. It is also referred to as *plain carbon steel* or *straight carbon steel.* Minor elements also present in carbon steel include manganese, phosphorus, sulfur, and usually silicon.

carbonization: Process of converting into a residue of carbon by the action of fire or some corrosive agent, as the driving off of volatile matter from such fuels as coal and wood by fire. This term is used incorrectly as referring to the operation of carburizing.

carburizing: The process of combining or impregnating with carbon, as in adding carbon to the surface of low-carbon steel for the purpose of case-hardening; also, as in heating steel to above its critical temperature in the presence of a carbonaceous gas.

casehardening: A heat-treatment process, applied to steel or iron-carbon alloys, by which a harder outside layer is obtained over a softer interior; depth of increased hardness depends upon length of treatment.

cast steel: Molten steel cooled and solidified in a mold.

cementite: Chemical compound of iron and carbon, containing 93.33 percent iron combined with 6.67 percent carbon by weight; also called iron carbide. Chemical formula for cementite is Fe_3C.

cleavage plane: The line of fracture related to atomic alignment of crystal particles.

cold drawing: Reducing the cross-section of a metal by pulling it through a die while its temperature is below the recrystallization temperature.

cold rolling: Reducing the cross-section of a metal by means of a rolling mill while the metal is cold or below its recrystallization temperature.

cold working: The permanent deformation or crystal distortion of a metal below its lowest temperature of recrystallization, resulting in work hardening.

combined carbon: Carbon combined with iron or other alloy elements found in steel to form chemical compounds which usually exhibit great hardness and brittleness.

crucible steels: High-grade steel produced when selected materials are melted in a closed crucible and cast into a mold.

cyaniding: Casehardening of low-carbon steel by heating in contact with a molten cyanide salt, followed by quenching.

D

decalescent point: The critical temperature at which sudden absorption of heat takes place in steel or iron-carbon alloy during the heating cycle; the absorption of energy is accompanied by a transformation of pearlite to austenite.

decarburization: Loss of surface carbon when steel is heated in an oxidizing atmosphere, resulting in a soft low-carbon skin on steel.

delta iron: Iron in body-centered structure between 2554°F and melting, which is about 2800°F (1399° to 1537°C). Magnetic.

dendrite: Crystal having tree-like shape; may be found in many cast metals.

direct process: Iron made from ore, by reduction, without purification.

Glossary

drawing: See *tempering*.

E

elongation: Permanent elastic extension which a metal undergoes during tensile testing; the amount of elongation is usually indicated by the percentage of an original gage length.

endurance limit: See *fatigue strength*.

extrusion: Forcing plastic metal through a die to produce a new form.

eutectic alloy: Alloy of a composition that solidifies at a lower temperature than the individual elements of the alloy and freezes or solidifies at a constant temperature to form a fine mixture of crystals made up of two or more phases.

eutectoid steel: Steel of a composition that will form a pearlite structure from austenite during slow cooling through its critical temperature. In plain carbon steel, its composition is approximately 0.85 percent carbon.

F

fatigue failure: Progressive cracking that takes place in metals that are subjected to repeated loads.

fatigue strength: Maximum repeated load a metal will carry without developing a fatigue failure.

finishing temperature: Temperature of metal at the moment it leaves the last pass in a hot-rolling mill, or any hot-forging or forming operation.

forging: Deforming into new shape by compressive force.

fracture: Ruptured surface ends of metals that have been broken.

fracture test: Breaking or rupturing of metals to determine their resistance to failure, to examine grain size, and to reveal any evidences of defects on the fractured surfaces.

free ferrite: Iron phase found in a steel which has less than 0.85 percent carbon that has been slow-cooled from above its critical temperature range; the iron phase is alpha iron, low in carbon content.

full annealing: Heating of steels or iron alloys to above their critical temperature range, soaking at the annealing temperature until they are transformed to a uniform austenitic structure, followed by cooling at a predetermined rate, depending upon the type of alloy and structure required; in general the cooling rate is relatively slow.

G

gamma iron: Iron in a face-centered cubic structure between 1666° and 2554°F (906° to 1401°C). Non-magnetic. Dense body structure leads to contraction from beta phase.

gray iron: A cast iron with 2 to 4 percent carbon, in which the carbon is mostly in the form of graphite.

grain: Crystal made up of atoms similarly orientated.

grain growth: Increase in size of a crystal by its stealing atoms from its neighboring crystals or grains; decrease in number of grains and increase in average size.

H

hardening: Operation of quenching steels from the austenitic temperature range so as to produce martensite or a hard structure.

heat treatment: Any operation involving the heating and cooling of metals or alloys

Metallurgy

for the purpose of obtaining structural changes.

high-carbon steel: Steel with carbon content usually below 1.3 percent carbon, but may range from 1.0 to 2.0 percent.

high-speed steel: Special alloy steel used for high-speed cutting and turning tools, as lathe bits; so named because any tools made of it are able to remove metal much faster than tools of ordinary steel.

high-strength cast iron: Cast gray iron with a tensile strength in excess of 30,000 psi (206,900 kN/m^2).

high-sulfur steel: Steel which has a sulfur content ranging from 0.12 to 0.33 percent and which then exhibits free cutting properties. Made expressly for screw machine products.

homogeneous metal: A metal or alloy, as steel, of very uniform structure, as opposed to heterogeneous or segregated metal with a nonuniform structure.

hot quenching: Cooling of heated metals or alloys in a bath of molten metal or salt, instead of using water or oil cooling medium.

hot short: Metal that is brittle and unworkable above room temperature. Sulfur in steel causes a *hot short* condition.

hot working: Plastic deformation of metals or alloys when their temperature is above the lowest temperature of recrystallization so that there will be no work-hardening or permanent distortion of crystals or grains.

Huntsman's Steel: A uniform quality steel developed for clock springs early in the 19th century.

hydrogen brittleness: A brittle and nonductile state resulting from penetration of metal by nascent hydrogen.

hypereutectic: Containing the minor constituent in an amount in excess of that contained in the eutectic mixture; a eutectic alloy that contains more than the eutectic ratio of composition, or one to the right of the eutectic point in the alloy diagram.

hypereutectoid steel: Steel containing more than 0.85 percent carbon in a plain carbon steel, or a steel containing a mixture of pearlite and free cementite.

hypoeutectic: Containing the minor element in an amount less than that in the eutectic mixture; a eutectic alloy containing less than the eutectic ratio of composition, or one to the left of the eutectic point in the alloy diagram.

hypoeutectoid steel: Steel with less than 0.85 percent carbon content in straight carbon steel, or steel containing a mixture of pearlite and free ferrite.

I

impact test: Measurement of the amount of energy required to rupture metals with sudden or shock loads.

inclusions: Dirt, oxides, sulfides, and silicates that become mechanically mixed with metals.

ingot: A metal usually cast in a metal mold that forms a material for the plastic shaping industry.

ingot iron: Commercial pure iron; low-carbon steel that has nearly the chemistry and properties of pure iron.

intercrystalline failure: Rupture that follows the grain or crystal boundaries of a metal.

isothermal transformation: Transformation that takes place at a constant temperature where the radiation of heat from the metal body is counterbalanced by an equal evolution of heat from the metal undergoing the transformation.

K

killed steel: Steel which has been sufficiently deoxidized during the melting cycle to prevent gases from evolving during the solidification period.

L

lamellar pearlite: Pearlite that has layers of cementite and ferrite in its structure.

lap: A folding over of surface layers of metal without welding; a surface defect or discontinuity formed from improper hot working of metal.

ledeburite: Eutectic of iron and carbon associated with some types of cast iron.

liquidus: Refers to *liquidus curve,* a freezing-point curve in an alloy diagram representing relationship of concentration to temperature systems comprising a solid phase and a liquid phase.

luder's lines: Lines which appear on surfaces of cold drawn metals resulting from plastic flow and stress concentrations.

M

M-point: The temperature during the cooling of austenite where martensite begins to form from austenite, i.e., in the quench-hardening operation.

macro-etch: A deep-etch technique for revealing large structures, fibers and flow lines, in steel.

macroscopic: Large enough to be observed by the naked eye; as structural details that are visible at a low magnification, usually under 30 magnifications.

macrostructure: Structure that is visible at a low magnification or with the naked eye.

malleableizing: Annealing operation used in connection with the change of white cast iron to a malleable cast iron; this process changes the combined carbon of white cast iron to a temper or graphitic carbon form.

martempering: Hardening of steels or producing of martensite by quenching steels from above their critical temperature in a molten salt bath quenching steels from above their critical temperature in a molten salt bath maintained at an elevated temperature just above or near the M-point, then slowly cooling in air to room temperature.

martensite: Structure obtained when steel is treated to achieve its maximum hardness, as by heating and quenching in the hardening operation. Martensite has a needle-like microstructure.

mechanical working: Plastic shaping of metals by any such method as rolling, forging, pressing, etc.

microstructure: Structure that is visible only at a high magnification, with the aid of a microscope after proper preparation, such as polishing and etching.

miscible: Ability of elements to dissolve or alloy with each other.

modulus of elasticity: Value obtained when the load or stress applied to a metal, expressed in pounds (psi) is divided by the value of elongation or deformation determined for a given gage length, when the load or stress applied is within the elastic limit of loading.

mottled iron: Cast iron with a structure consisting of a mixture of free cementite, free graphite, and pearlite.

N

network structure: Structure in which a network or cellular envelope of one constituent may completely surround the grains of another constituent.

Metallurgy

Neumann bands: Parallel bands or lines which appear within a crystal or grain as a result of mechanical deformation by the mechanism of twinning.

nitriding: Surface treatment applied to ferrous alloys for the purpose of carehardening, obtained by heating alloys in contact with dissociated ammonia gas, releasing nitrogen gas to steel.

normal structure: Structure found in steel after slow cooling from above its critical range of temperatures; this structure has the usual standard appearance-free ferrite and pearlite, pure pearlite, or free cementite and pearlite, depending upon the carbon content of the steel. Steel with this structure is referred to as *normal steel*.

normalizing: Treatment usually applied to steel castings and forgings consisting of heating to above the critical temperature range of the steel followed by cooling in air.

nucleus: Center or beginning of crystal formation; central portion of an atom.

O

overheating: Heating of metals or alloys to the zone of incipient melting; heating to the point of developing a very coarse grain size.

oxidation: See *burning*.

P

psi: Abbreviation for *pounds per square inch*.

patenting: Treatment usually applied to steel wire consisting of heating to well above the critical temperature range of the steel, followed by cooling or quenching the wire in a molten lead or salt bath maintained at a temperature range of 800°F to 1100°F (about 425° to 595°C). Also, air patenting is employed where the heated steel wire is air-cooled to room temperature, as in the standard normalizing operation.

pearlite: Eutectoid alloy of iron and carbon, which contains 0.85 percent carbon; structural constituent found in steel that has been cooled slowly from above its critical temperature range, consisting of layers or plates of cementite and ferrite and appearing colorful under the microscope.

pig iron: The product of the blast furnace cast into blocks convenient for handling or storage; iron alloy as recovered from the ore. A brittle material of high-carbon content (5 percent).

pipe: The cone-shaped cavity or hole found near the top of a casting or cast ingot resulting from natural shrinkage and uneven cooling during the casting and solidification period. It is usually filled with oxide and other dirt.

pit: A rough surface found on metals resulting from slight depressions on the surface.

plasticity: Ability of a metallic state to undergo permanent deformation without rupture.

powder metallurgy: Process by which shapes resembling a cast or forged metal are made from metal powders by pressure and heat.

precipitation hardening: Process of hardening by heat treatment where a hard phase is precipitated from solid solution at room temperature or at some elevated temperature.

primary crystals: The first crystals to form during solidification of a cast metal or alloy.

process annealing: Treatment applied to cold-worked steels consisting of heating to a temperature below the critical temperature range of the steel followed by slow cooling.

Q

quenching: Process of fast-cooling metals or alloys such as steel in the process of hardening, as air quenching, oil quenching, water quenching, etc.

quenching crack: Crack or failure occurring from rapid cooling of alloys such as steel in the hardening operation.

R

recalescence: The sudden liberation of heat by a metal when cooling through a certain critical temperature, as iron at 1652°F (900°C). The recalescent point in a steel occurs when it is cooled through the lower Ar_1 critical temperature and liberates a small amount of heat.

red-hardness: Hardness maintained by cutting tools when heated to about a red heat when in use, as in high-speed steels.

red-shortness: Brittleness, lack of ductility or malleability that occurs with some metals or alloys at a red heat.

reduction of area: Diminution of cross-section area, usually expressed in percentage of the original area; reduction that a metal or alloy undergoes during a tensile test; it is also called *contraction of area.*

refractory: Capable of being heated to high temperatures without fusing or softening as in fire brick.

regenerative quenching: Heat treatment involving the use of a double heating and quenching of some carburized steels for the purpose of refining the grains of both the core and case.

resilience: Elastic properties exhibited by materials such as a spring; capability of a strained body to recover its size and shape after deformation, especially by compressive stresses.

rough fracture: Irregular surface resulting from the fracture of a coarse-grained metal or alloy.

S

scale: The dark oxide coating found on metals as a result of heating in an oxidizing atmosphere such as air.

seam: A longitudinal crack or surface defect resulting from a surface cavity that becomes elongated during hot rolling or forging.

season cracking: Failure by intercrystalline fracture that some alloys such as brass undergo when in a highly stressed condition and exposed to corrosive media.

secondary hardening: An increase in hardness that occurs with some high-alloy steels when subjected to tempering after hardening.

segregation: The separation of liquid solutions into uneven concentrations of alloys during cooling.

self-hardening steels: Steels that become martensitic or fully hard by air cooling from above their critical temperature or from the austenitizing temperature.

shear steel: The reheating and welding of broken cement steel bars into a homogeneous product. An early steel making process.

slip bands: Traces of slip planes that appear in a polished surface of metal when it is stressed so as to produce plastic flow after polishing and etching. These slip bands are the result of crystal displacement or blocks of crystal fragments which

Metallurgy

have moved with respect to the crystal orientation.

soaking: Allowing steel or alloys to remain at an elevated temperature long enough to become uniformly heated and to complete the desired change in structure.

solidification range: Range of temperature that extends from the beginning of freezing to the end of freezing or crystallization.

sorbite: Structure found in steel resembling pearlite but of a finer and more granular mixture of cementite and ferrite than found in normal pearlite.

spheroidized pearlite or spheroidized cementite: Structure found in steel containing balls or spheroids of cementite in a matrix of ferrite.

spheroidizing: Annealing process which is applied to steel when the cementite or carbide is desired in a globular or spherical shape. In this treatment, the annealing cycle may follow that of full annealing, but the cooling rate employed from the annealing temperature is usually much slower in order to obtain the cementite in spheroidized form.

steel: An alloy of iron with up to 1.4 percent carbon, usually less.

steeling: Welding of a strip of steel onto an iron base. An early process not now in use.

stress raiser: Notch or discontinuity on the surface of metals that produces stress concentrations.

structural shapes: The making of S, H, and I beams, and angles by a universal mill for use in steel construction.

subatmospheric (cryogenic) treatment: Subjecting steels that are in the hardened condition to low temperatures, as low as minus 180°F (82°C), for periods up to 6 hours, for the purpose of stabilization and to aid in the transformation of austenite to martensite.

surface hardening: See *casehardening*.

T

temper carbon: The nodular graphitic carbon found in iron-base alloys of the malleable or pearlitic malleable cast type.

tempering: Application of heat to quenched, hardened steels or to steels in the martensitic condition. Temperatures employed in tempering are from slightly above room temperature to approximately 1200°F (649°C). Tempering can also be accomplished by the *austempering* or *martempering* methods. Tempering is also called *drawing*. See *austempering* and *martempering*.

twin bands: The microscopic appearance resulting from twinning.

twin crystal or twinning: A crystal distortion that takes place when metals are subjected to shock loading which causes a change in orientation resulting in differently-orientated crystal fragments within a grain or crystal. See *Neumann bands*.

W

welding: The crystallizing into union of metals or alloys by methods of fusion or pressure welding.

work-hardening: The increase in hardness and strength that takes place during any cold-working of metals.

wrought iron: A commercial form of iron that is tough, malleable, and relatively soft; less than 0.3 percent carbon.

INDEX

A

Admiralty metal, 416
air furnace, 270
AISI steel numbering system, 253–255
allotropy, 76, 146, 150, 155
alloys, 131–149
   properties, 148–149
   types, 134–143
alloy cast iron, 298–299
alloy diagram, 134
alloy powders, 332–333
alloy steels, 90, 152, 231–250
   austenitic, 236, 243–244
   cementitic, 236
   ferritic, 240–242
   martensitic, 235–236
   pearlitic, 234–235
alpha-beta brass (Muntz metal), 415–416
alpha iron, 151, 155, 159
aluminizing, 220
aluminum, 285, 363–367, 374–388
   clad, 375–376
   properties, 375
aluminum alloys, 376–383
aluminum bronze, 419–420
aluminum casting, 376–381
aluminum-copper alloy, 385, 386
aluminum-indium, 446
aluminum-silicon, 380–381
amorphous cement, 82
annealing, 61, 82, 165, 168–171, 192, 282, 289–290, 412–415, 440–444
anodic treatment, 376
antimony, 76, 140, 357, 371
antimony, lead, 140–141, 149
arc furnace, 30–32
arsenic, 357
atmosphere furnace, 199
austenite, 154, 159, 171, 172, 194

B

Babbitt metal, 430
basic oxygen process, 28–29
bauxite, 363–364
bearing metals, 427–436
   cadmium base, 430–431

   characteristics, 427–429
   composite, 434
   copper base, 431–434
   indium, 445–448
   lead base, 429–430
   lead-calcium base, 430
   tin base, 430
   zinc base, 434–435
bending, 66
beryllium, 396, 421–422, 446
beryllium-copper alloy, 421–422, 446–447
Bessemer process, 27–30
beta iron, 151, 155
bismuth, 76, 137–140, 357, 372,
blast furnace, 5–6, 13, 19–22 350, 356
blister copper, 352–353
blister steel, 10
bloom, 5–7, 46
blowhole, 40–41
body-centered cubic, 76, 151
BOF furnace, 28–29
Bonderizing, 225
brass, 408–416
brazing, 305, 322–324, 448
Brinell hardness, 113
brittleness, 97–98
bronze, 416–421

C

cadmium, 76, 137–140, 362, 428
   bearing alloy, 430–431
cadmium-bismuth alloy, 137–140
capped steel, 42
carbon, 181–186, 270–271, 285
carbon-arc welding, 307–308
carbonitriding, 219
carbon steel, 90–92, 94–95 152–153, 156, 172, 175, 181–186, 196, 231
carburizing, 9–11, 193, 205–212
cartridge brass, 411–412
case, 207, 213, 216, 223
case hardening, 213, 223
casting, 22, 45, 284, 299–301, 357–358 376–377, 382

cast iron, 45, 120, 269–301
   alloy, 298–299
   ductile, 294–296
   gray, 270–285
   malleable, 288–294
   mottled, 276
   nodular, 294–296
   white, 270, 274, 286–288
Catalan furnace, 5–6
cementation, 10
cementite, 154, 156–158, 181, 207, 236, 274, 275, 334
cementitic steel, 236
Chapmanizing, 212–214
charcoal, 2, 4
Charpy test, 109
chilled cast iron, 286–288
chromium, 190, 238–240, 278, 285, 367
   chromium-nickle-iron alloy, 242–243
   iron-chromium alloy, 240–242
   iron-chromium-carbon alloy, 242–243
chromium steel, 238–240
chromium-vanadium, 454–458
chromizing, 218, 219
cleavage plane, 76–77
coatings, 224–228
cobalt, 76, 371
coefficient of expansion, 103
cold crystallization, 84–85
cold drawing, 63–65
cold rolling, 58–60, 383
cold short steel, 27
cold work, 58–72,
   annealing, 170–171
   effects, 71–72
coke, 15–17
composite bearings, 434
compressive strength, 96, 280
constitutional diagram, 135, 233
continuous casting, 42–43
copper, 93, 269, 350–354, 367, 403–424
copper alloys, 134–136, 141, 232, 403–424
   copper-lead bearing, 431–434
   copper-manganese-nickel, 422–423

Metallurgy

copper-silver, 141–142
copper-tin, 416–418
copper-zinc, 408
Monel metal, 423–424
nickel-iron (Invar), 424
nickel-silver, 424
corrosion, 71–72, 98–100, 117–118
critical point, 150
critical temperature diagram, 161, 162, 166
crucible furnace, 35–36
cryogenic treatment, 175–176
crystalline state, 75–80, 158
polycrystalline, 80
crystallization, 77–80, 84–85, 151, 158–159

D

decarburization, 185–186, 193–194
deformation, 77, 80–85
delta iron, 151, 155
dendritic structure, 39, 40, 79
destructive testing, 105–124
die-casting, 361, 388–392
diffusion processes, 218–220
ductile iron, 294–296
ductility, 96

E

elasticity, 96
elastic limit, 80, 96, 106
electrical conductivity, 99, 100
electrical resistance, 100
electric arc furnace, 30–32
electric steel furnace, 199
electrochemical series, 98
electrolytic refining, 353, 357–358, 365
electron beam welding, 312–313
elements (metals), 99
equilibrium diagram, 153
eutectic alloy, 136–141, 146, 147, 154
eutectic line, 137
eutectic point, 137
eutectoid, 133, 147, 156, 173
extrusion, 55–56

F

face-centered cubic, 76, 151
fatigue strength, 96, 280–281
ferrite, 8, 155–156, 240–242
ferrite solubility, 155–156
ferroalloy, 372
flame hardening, 220–221

flanging, 67
flash welding, 321
flat drawing, 66–67
fluidized bed process, 25
flussofen, 7
forging, 95
foundry, 299–301
fracture test, 122–123
free machining, 169–170
freezing of alloys (cryogenic treatment), 175–176
friction welding, 314
fume, 350
furnaces, 1–8, 27–36, 197–202, 270, 348–350
atmospheres, 200–201
sintering, 339–340

G

galvanizing, 63, 225
gamma iron, 151, 155, 159
gas shielded metal-arc welding, 308–309
gas welding, 305–306
germination, 84
grain growth, 77–78, 83–84, 158–161
grain refinement, 450
grain structure, 80, 159–161, 164, 182–184, 450–451
size, 122, 159, 182–184
graphitic carbon, 122–123, 272–273, 285, 294–295
gravity concentration, 348
gray cast iron, 120, 123, 270–285
Group I pearlitic steel, 234–235
Group II martensitic steel, 235–236
Group III austenitic steel, 236
Group IV cementitic steel, 236

H

hardenability, 8, 266–268
hardening, 171–173, 181–182, 184, 188–192, 283–285
hard-facing, 227
hardness, 97–98, 127–128, 181–182, 192
heat resistance, 100, 281–282, 298–299
heat treating furnace, 197–202
heat treatment, 187–192, 197–202, 218, 282–285
aluminum alloys, 384–386
heterogeneous solution, 136

high brass, 411
high-carbon steel, 92, 123, 152
high-frequency furnace, 32
high-speed molybdenum steel, 187, 189, 247
homogeneous solution, 135
hot forging, 53–55, 167
hot pressing, 340–341
hot rolling, 46–53, 382–383
hot strip mill, 49
hot working, 46–58, 382–383
effects, 56–58
Huntsman's steel, 11

I

impact, 108–109
impurities in steel, 94
indium, 445–448
induction furnace, 32–33
induction heating, 221–223
inertia welding, 315, 316
ingot, 37–42, 45,
types, 41–42
inoculant, 301
intermetallic compound, 142
iron, 1, 81, 87, 151, 342
iron-carbon alloys, 146, 152–153, 343
iron-carbon diagram, 146–148 149–153
iron-chromium alloys, 240–242
iron-chromium-carbon, 242–243
iron-nickel, 1
iron ore, 1–2, 14–15

K

Kelly process, 27–28
killed steel, 41
kiln, 23–24

L

laser welding, 313–314
lead, 93, 354–359
lead-antimony alloy, 149
lead-base bearing metals, 429–430
lead-calcium bearing metals, 430
leaded bronze, 418–419
lead-zinc diagram, 142
light metals and alloys, 374–401
limestone, 17–18, 23, 32, 33
liquid heating bath, 201, 202, 211–212

Index

liquidus, 135, 137
low brass, 411
low-carbon steel, 92, 123, 152

M

M-point, 175, 179
macro-etching, 123–124
Magnaflux test, 125–126
magnesium, 76, 367–368, 392–396
magnesium-aluminum alloy, 394
magnetic iron, 151, 155
magnetic steels, 250
magnetic testing, 125–126
malleable cast iron, 120, 288–294
   alloy, 292
   black heart, 290–291
   pearlitic, 292–293
   white heart, 291
malleablizing, 289–290
manganese, 32, 92–93, 245, 269, 285, 367, 370, 420
martensite, 171, 172, 173, 192
martensitic steel, 172, 173, 181, 190, 235–236, 242–243
matte copper, 349, 352
mechanical properties of metal, 95–98, 102, 103
medium-carbon steel, 92, 152
melting point, 102
metal forming, 66–70
   bending, 66
   blanking, 67–68
   flanging, 67
   flat drawing, 66–67
   punching, 67–68
   roll forming, 69–71
   spinning, 68
   stamping, 66–67
   stretch forming, 68–69
   upsetting, 67
metals, 95–103
metal powders, 330–335
metal spray, 225–226
microscope, 118–122
molding powders, 335–338
molybdenum, 187, 190, 248, 278, 286
molybdenum steel, 187, 190, 246–247
Monel metal, 136, 423
Monotron test, 116
muffle furnace, 198–199
Muntz metal, 415–416

N

nickel, 1, 232, 278, 362–363
nickel bronze, 421–422
nickel-silver, 424
nickel steel, 237–238, 269
nitriding, 214–218, 285
nodular iron, 294–296
non-destructive testing, 125–129
non-ferrous metals, 347–373
non-magnetic iron, 151
normalizing, 165–168
notched bar test, 109

O

oil flotation, 348
open-hearth furnace, 28, 33–35, 89
ore dressing, 347–348
ore hearth, 356
oxides, 93, 227–228
oxidizing, 98, 306, 357
oxy-fuel processes, 305–306

P

Parkerizing, 225
pearlite, 156, 157, 159–160, 172, 175, 273–274
pearlitic gray iron, 274
pearlitic steel, 181, 234–235
pearlitic malleable cast iron, 292–295
peritectic reaction, 143–146
phase diagram, 150–151
phosphate coatings, 225
phosphor bronze, 418
phosphorus, 27, 32, 93
pig iron, 3, 21–22, 27
plasma arc welding, 310–312
plastic flow, 81
plasticity, 74, 97
plating, 224–225
polishing machine, 118–119
powder metallurgy, 329–346, 435
precipitation hardening, 387–388
press break forming, 66
primary troosite, 175
projection welding, 319–321
proportional limit, 106
punching and blanking, 67–68

Q

quenching, 173–175, 177–180, 190, 232

R

radiography, 126
recrystallization, 82–83, 159
red brass, 411
reducing, 4, 8, 306, 307
reduction process, 2, 4, 22–25
refining, 27–43
resistance, electrical 100
resistance, heat, 100–102
resistance welding, 317–321
retort, 24–25
reverberatory furnace, 348–349, 356
rimmed steel, 41–42
roasting furnace, 349, 350
Rockwell hardness, 115, 181, 192
roll forming, 69–70

S

SAE steel numbering system, 253–255
salt-spray test, 117
scleroscope, 115–116
screw machine steel, 95, 207
season cracking, 415
segregation, 41, 56–58
semikilled steel, 41
shallow hardening, 182–186
shearing, 10, 80–81
Shephard grain size test, 122–123
Siemens-Martin furnace, 33–35
silicon, 32, 93, 232, 269, 276–278, 285
silicon-aluminum alloy, 380–381
siliconizing, 219–220
silicon steel, 93, 249–250
silver-copper alloy, 141
silver-lead-indium bearing metal, 435–436
sintering, 14–15, 338–340
   atmosphere, 339–340
skin, soft, 185–186
slip interference, 82
slip planes, 76–77
smelting, 348–350
soaking pit, 45–46
soldering, 324–325
solidification, 136, 154–155
solid solution alloys, 140, 148–149
solid state, 146–148
solid state welding, 314–316
solidus, 135, 137
space lattice, 75
spalling, 185

471

Metallurgy

specialty refining, 36–37
specific gravity, 103
spheroidizing, 169–170
spinning, 68
sponge iron, 25–26
spots, soft, 186
stainless steels, 240, 244, 264–265
    austenitic, 243
    ferritic, 240–242
    identification, 264
    martensitic, 242–243
    specifications, 264
steadite, 271, 275
steel, 3, 87, 90–95
steeling, 9
Strauss solution, 117–118
strength and plasticity, 95–96
stress-relief, 82–83, 171, 282
stress-strain curves, 107, 112
stretch forming, 68–69
structural steel, 49–53, 94–95, 156–157
stuchofen, 7
stud welding, 322
submerged arc welding, 310
sulfur, 27, 93
surface hardening, 205–223
    aluminizing, 220
    calorizing, 220
    carbonitriding, 219
    carburizing, 205–212
    Chapmanizing, 212–214
    chromizing, 219
    flame hardening, 220–221
    induction heating, 221–223
    nitriding, 214–218, 215, 217

siliconizing, 219–220
surface treatments, 205–228

T

temperature measurement, 129
temper carbon, 291, 292–293
tempering, 9, 190, 193–197, 283–285
temper mill, 60
tensile strength, 96, 105–108, 280
terne plate, 62
thermal curves, 132
Tig welding, 309
time-temperature curves, 132–134, 174
tin, 368–369
tin-base bearing metal, 430
tinning, 61–62
tin plate, 62, 94
titanium, 374, 397–401
tool steel, 95, 157, 255–264, 344
toughness, 97, 195–197
transverse strength, 280
tubular products, 65–66
tungsten, 190, 330, 344, 371
tungsten carbide, 245, 344
tungsten steel, 187, 189–190, 245–246
Type I alloy, 134–136
Type II alloy, 136–141
Type III alloy, 141
Type IV alloy, 142
Type V alloy, 142–146
Types of steel, 253–268

U

ultrasonic hardness testing, 127
ultrasonic welding, 314, 316
upsetting, 67

V

vanadium, 190, 448–459
vanadium cast-iron, 458–459
vanadium-chromium, 454–458
vanadium steels, 248–249, 453–454
Vickers hardness test, 114

W

warping, 184–185
wear resistance, 281
welding, 87, 303–327
    effects of, 325–327
white cast iron, 123, 269–270, 286–288
wootz steel, 10
work hardening, 82
wrought aluminum alloy, 377–378, 381–383
wrought iron, 2, 4–5, 9–10, 22–23, 123, 296–298

Y

yield strength, 96

Z

zinc, 76, 359–362
zinc-base bearing metal, 434–435
zirconium, 437–445